Telecommunications Cables

Telecommunications Cables

DESIGN, MANUFACTURE AND INSTALLATION

Harold Hughes

JOHN WILEY & SONS
Chichester • New York • Weinheim • Brisbane • Toronto • Singapore

Baffins Lane, Chichester,
West Sussex, PO 19 1UD, England

National 01243 779777
International (+44) 1234 779777

e-mail (for orders and customer service enquiries): cs-books@wiley.co.uk

Visit our Home Page on http://www.wiley.co.uk
or
http://www.wiley.com

Other Wiley Editorial Offices

John Wiley & Sons, Inc., 605 Third Avenue,
New York, NY 10158-0012, USA

VCH Verlagsgesellschaft mbH,
Pappelallee 3, D-69469 Weinheim, Germany

Jacaranda Wiley Ltd, 33 Park Road, Milton,
Queensland 4064, Australia

John Wiley & Sons (Canada) Ltd, 22 Worcester Road,
Rexdale, Ontario M9W 1L1, Canada

John Wiley & Sons (Asia) Pte Ltd, 2 Clementi Loop #02-01,
Jin Xing Distripark, Singapore 0512

British Library Cataloguing in Publication Data

A catalogue record for this book is available from the British Library

ISBN 0 471 97410 2

Typeset in 10/12pt Sabon by Aarontype Limited, Easton, Bristol
Printed and bound in Great Britain by Bookcraft (Bath) Ltd

This book is printed on acid-free paper responsibly manufactured from sustainable forestation, for which at least two trees are planted for each one used for paper production.

To

Rosemarie

Contents

Preface

My intention in writing this book is to provide information to those newly entering the field of telecommunication cables, which will enable them to become competent cable designers with an adequate background knowledge of cable making processes and installation problems.

The level of knowledge assumed is that of mathematics up to simple calculus, a nodding acquaintance with differential equations, and of science up to basic mechanical, electrical and optical theory. A familiarity with exponentials, natural logarithms, hyperbolic functions and the concept of complex quantities using the operator j is required. A facility with simple spreadsheet working on a personal computer is assumed for ease of computation but is not essential.

I have presented usable formulae, with enough background to understand where they come from, into which the data can be entered and the required answer calculated using basic arithmetical processes. Where empirical formulae have been presented, their results are compared with analytical theory or, when this was not possible, with actual cable measurements. The theory is illustrated with examples and measurements which show the accuracy attainable in practice.

SI units are used throughout except where field units or traditional units appeared more appropriate.

Optical fibre theory is necessarily more mathematical than that of electrical transmission lines but I have tried to emphasize the physics while giving numerical methods of solving the scalar wave equation. These appear to work with adequate accuracy using spreadsheets.

The references given are by no means exhaustive, but they acknowledge the sources that I personally have found useful. Where copyright permissions have been granted I have acknowledged this in the appropriate places. Where I have been unable to contact copyright holders, or where I have failed to recognize copyrights, I sincerely apologize.

I am grateful for the support and encouragement I have received in the preparation of this book from my former colleagues and employers at BICC Cables Ltd, UK and ATC (pty) Ltd in S. Africa. In particular, I am indebted to J E Delves Broughton for most of Chapter 1.

1

Conductors

METAL PROPERTIES

Conductors are normally made of metal wire. The properties of some metals, in decreasing order of electrical conductance, are given in Table 1.1. The electrical and thermal conductivities are given relative to copper and show an excellent correlation. This is because both these properties are governed by the mobility of the electron 'cloud' that resides in the metal lattice.

Copper is the most usual conductor material. It has a high electrical conductivity and has very good ductility. It is common to draw 6.35 mm diameter copper rod down to 0.1 mm through three successive drawing operations without any intermediate annealing of the wire. This results in an area reduction, or a length increase, of 4032 : 1. This makes the drawing of copper comparatively simple and wire drawing speeds of up to 60 m/s (216 km/h) can be attained.

The melting and continuous casting of rod from electrolytic cathode copper is now common practice and has the advantage of the rod being supplied, bright pickled, in wrapped 5 tonne coils. The absence of inclusions and imperfections gained by this process results in a marked improvement in productivity, especially on wire diameters below 0.5 mm.

A typical analysis of supplied copper rod is given in Table 1.2 The oxygen content of 0.03% is deliberate as its inclusion improves conductivity and casting properties. This type is known as 'Electrolytic Tough Pitch Copper'. Silver in significant quantities has a profound effect on raising the annealing temperature, but the amount in this analysis is quite acceptable. Fire refined copper has such large traces of silver that the annealing temperature is too high for in-line resistance annealing.

The basic properties of copper wire are given in Table 1.3

WIRE DRAWING

By definition, a rod becomes a wire only after drawing through at least one die. Until the mid-1930s wire drawing was still a craft as the dies used comprised a metal plate of ductile cast iron containing a succession of holes of decreasing diameter. This was

Table 1.1
Properties of metals

Metal	Relative electrical condtvty.	Relative thermal condvty.	Coefficient linear expansion $\times 10^{-6}/°C$	Specific heat cal/g/°C	Melting point °C	Density kg/m^3	Tensile modulus GN/m^2	Bulk modulus GN/m^2	Rigidity modulus GN/m^2	Poisson's ratio	Ultimate tensile strength MN/m^2
Silver	106	108	19	0.056	960	10 500	77	105	28	0.37	
Copper	100	100	16.6	0.093	1083	8 920	110	135	44	0.34	240
Gold	72	76	14	0.031	1063	19 300	80	165	28	0.42	
Aluminium	62	56	23	0.217	658	2 700	70	75	25	0.34	160
Magnesium	39	41	26	0.247	651	1 740	41	33	17		
Zinc	29	29	30	0.092	419	7 100	80	35	36	0.23	
Nickel	25	15	13	0.109	1452	8 800	210	170	78	0.30	
Cadmium	23	24	31	0.055	321	8 650	50	42	21	0.30	
Cobalt	18	17	12	0.104	1480	8 700					
Iron	17	17	12	0.113	1530	7 880	210	160	77	0.28	
Steel	12	12	12	0.107	1400	7 700	220	160	80	0.28	1334
Platinum	16	18	9	0.032	1773	21 450	170	245	63	0.39	
Tin	15	17	21	0.054	232	7 330	190	53	19	0.33	
Lead	8	9	28	0.031	327	11 340	16	41	6	0.44	

Table 1.2
Typical analysis of copper rod

Constituent	Content
Copper	99.97%
Oxygen	0.029%
Sulphur	8 ppm
Bismuth	0.5 ppm
Arsenic	0.1 ppm
Antimony	1 ppm
Lead	1 ppm
Iron	6 ppm
Silver	8 ppm
Nickel	1 ppm
Cobalt	1 ppm
Silicon	0.2 ppm

Table 1.3
Properties of copper wire

Conductivity – annealed	100%	58×10^{-6} S · m	(can average 101%)
– hard drawn	98%		
Resistivity – annealed		0.017241	μΩ · m
Temp. coeff. of resistance		0.00393	/deg °C
Density		8.92	g/cm^3
Coeff. of linear expansion		16.6	$\times 10^{-6}$/deg °C @ 20 °C
Tensile strength – annealed		240	MN/m^2
– hard drawn		460	MN/m^2
Maximum tension without stretch (equiv. to mass/1000 m of wire – approx. UTS/3)		80	MN/m^2

called a 'Wortel Plate' and was the personal possession of the craftsman. The drawing angles in the holes were produced by a tapered punch suited to the craftsman's needs. Die wear was rapid, and repeated 'single-hole drawing' into coils using a single drawblock/coiler unit was the normal practice.

Around 1933–1934 tungsten carbide dies were developed in Germany which were vastly superior in wear characteristics. This made multi-die machines a practical proposition with consequent increases in speed and efficiency.

Drawing Machine Definitions

Drafting – Linear speed increase between successive drawblocks.

Drawblocks – Flat, or slightly tapered, driven pulleys which haul the wire through the preceding die.

Wet drawing – The cooling and lubrication necessary to permit high drawing speeds with good die life and product quality.

Slip – This is the accumulated difference between the drafting elongation of successive dies relative to the surface speed increments of successive drawblocks.

Capstan – The final drawblock, situated externally to the 'wet machine' enclosure and which is not lubricated. This controls the actual wire drawing speed.

The capstan is a device used at many stages of cable making, and is shown in Figure 1.1.

Capstan is a torque multiplier with the property

$$\frac{T_{in}}{T_0} = \varepsilon^{\mu\theta} \tag{1.1}$$

where θ is the angle of contact and μ is the coefficient of friction. For a modest coefficient of friction of 0.25 and a single turn around the capstan, the angle of contact $\theta = 2\pi$ radians and $T_{in}/T_0 = 4.8$. For two turns on the capstan the multiplying ratio of the tensions is 23.1 and for three turns it is 111.

The difficulty with complete turns on a flat capstan is that the turns move across the capstan. To overcome this, a tapered capstan is used, where the incoming turn nudges the existing turns down the taper. Another method is to use a double capstan with shallow grooves to define the turn positions as shown in Figure 1.1

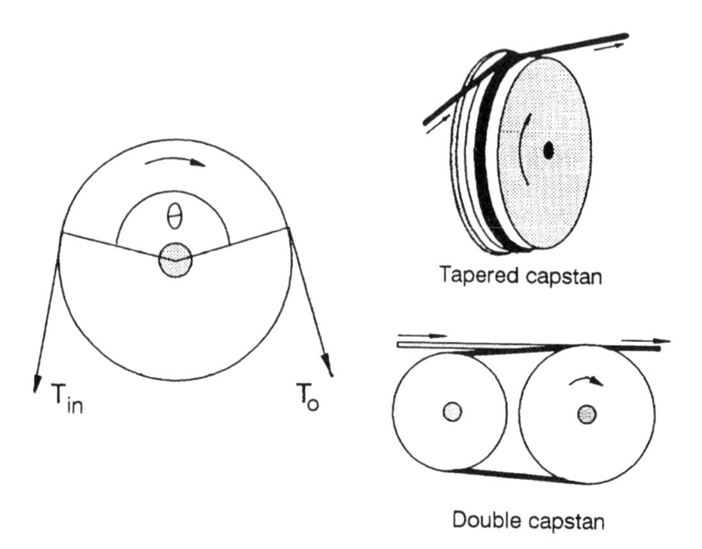

Figure 1.1
The capstan

Wire Drawing Dies

Wire drawing dies are used in sets which comply with the given input and output requirements to suit a particular machine. The diameter difference between successive dies in a set is called the 'drafting' and is generally expressed as a percentage elongation of wire length. The following draftings are commonly used

	Output diameter	Elongation	Area reduction
Rod Breakdown	2.3 mm	26%	20.63%
Intermediate wire	1.3 mm to 0.3 mm	21%	17.36%
Fine wire	0.2 mm to 0.05 mm	15%	13.04%

These elongations are related to the increase in speed of successive drawblocks in a particular machine (the machine drafting), which is generally 1% less than the die drafting. This allows for a certain amount of accumulative slip in the machine. The die drafting must never fall below the machine drafting otherwise the wire tension between the drawblock and the die might lead to breakages. The die drafting should be kept in the range 0 to +2% of the machine drafting. This creates exacting demands on the die reconditioning, since on the smaller dies a diameter tolerance of 0.001 mm may be required.

Since the mass and volume of the wire entering the die (or the machine) at diameter D, and leaving the die (or machine) at diameter d, is constant, the length of wire l, entering in a fixed time and that exiting, L, in the same time are related:

$$\frac{\pi}{4}D^2 l = \frac{\pi}{4}d^2 L \quad \text{therefore} \quad \frac{D^2}{d^2} = \frac{L}{l} = \frac{V_{out}}{V_{in}} \tag{1.2}$$

The fractional area reduction of the wire is

$$\frac{D^2 - d^2}{D^2} = \frac{D^2/d^2 - 1}{D^2/d^2} \tag{1.3}$$

For example in a 21-die machine with 21% die drafting, each die reduces the wire diameter in the ratio $\sqrt{1.21} = 1.1$ and the overall diameter reduction is $1.1^{21} = 7.4$. Thus the ratio of output to input speed is $7.4^2 = 54.8$. For an acceptable input speed of 1 m/s the output speed is thus 54.8 m/s (197 km/h).

In order to sustain these speeds a high degree of perfection is required in

Copper quality
Die condition and drafting
Lubrication and cooling
Drawblock surface condition
Machine condition

Experience has shown that a speed of about 40 m/s over a diameter range of 0.15 mm to 0.5 mm gives the optimum balance between production speed and breakages.

Die Profile

The two basic requirements of die profile, following the bell-shaped entrance, are the reduction angle, which for copper must be $18 \pm 2°$, and the length of the parallel bore which must not exceed 50% of the exit diameter. Wear in dies occurs at the point of contact in the drawing cone and results in a circular wear ring. This ring must be removed regularly by polishing while still maintaining the correct reduction angle. Consequently after two or three repolishings the parallel bore becomes so shortened that the die must be taken up to the next size.

Drawing Die Material

For rod breakdown producing wire above 1.5 mm diameter, tungsten carbide dies are used. The expected life of a set of dies is about 25 tonne of copper. At the end of this time they are resized to the next diameter and a new finishing die is introduced. Thus a finishing die has effectively 10 lifetimes (250 tonne of copper).

For wires below 1.5 mm diameter, highest quality industrial diamond dies are used and have lifetimes running into years, so long as they are carefully serviced to polish out the wear ring. Servicing intervals depend on diameter as follows

Wire diameter (mm)	0.1	0.2	0.4	0.5	0.63	0.9	1.25
Wt. copper drawn (t)	1.0	4.0	16	25	30	50	80

'Compax' dies formed by sintering man-made diamond powder with tungsten carbide, with a lifetime of about 100 times that of plain tungsten carbides are available but at a cost 60 times greater. However the extended lifetime and infrequent servicing can reduce the overall cost to half or one-third of that of tungsten carbide dies. Intermediate and fine wire 'Compax' dies are inferior to diamond dies and only become viable above 1.5 mm diameter. The die slug is encased, by 'shrink fitting', in a mild steel casing for reinforcement and to build up the bulk for heat dissipation purposes.

Wire Drawing Machine

A sketch of a three-die drawing machine with two drawblocks and a dry capstan is shown in Figure 1.2. The drafting of this machine is 20% and the drafting elongation of each die is 21%. The linear velocity of the capstan is V. The wire entering the final die is thus at $V - 21\%$ which is slower than the linear speed of the preceding drawblock, $V - 20\%$, so that there will be slip between the wire and the drawblock, unless the drawblock is tapered so that the wire can move down to a smaller diameter suited to its speed.

By mounting several drawblocks of differing diameter on each spindle (not recommended) or by folding back the wire line, compact machines of up to 21 dies can be constructed. Any slip is also present at all drawblocks and is accumulative, 1% per drawblock, so that on a 21-die machine, the slip at the first drawblock (input end) will be 21%.

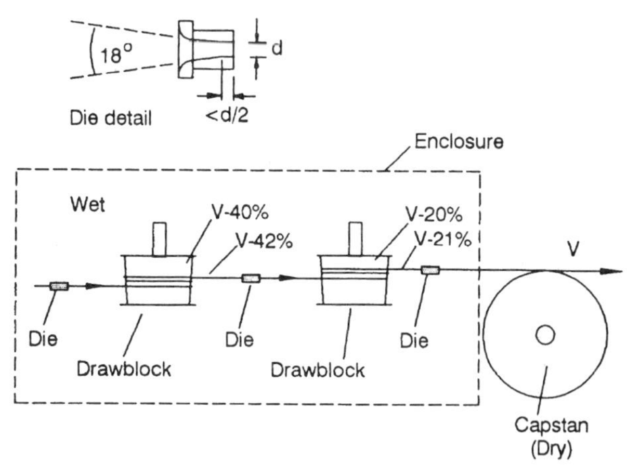

Figure 1.2
Wire drawing die and wire drawing machine principles

Drawblock Surface Material

The progressively increasing slip towards the input end of the machine causes wear on the drawblock surface and abrasion on the wire. In order to reduce the rate of wear, very hard surface materials are used, some of which are listed below

Material	Suitability	Life
Hard steel	copper only	fair
Tungsten carbide coating	copper only	very good
Ceramic-white (aluminium oxide)	copper only	very good
Ceramic-yellow (Zirconium oxide)	copper and tinned copper	very good
Hard chromium plating	copper and tinned copper	good (cheapest first cost)

The use of zirconium oxide blocks is now universal where there is a requirement for drawing both copper and tinned copper wire on the same machines, since the the same die sets and lubricants can also be used for both.

Drawing Power Requirements

As mentioned, it is quite common to draw wire continuously down to the final desired size without any intermediate annealing. With the exception of the annealed entry rod, (which work hardens within the first few dies), the tensile strength of the copper reaches a steady 460 N/mm^2 for the rest of the drawing process. The following rules then apply

(a) Power required is proportional to tensile strength (constant).

(b) Power used is proportional to area reduction in each die (uniform through the die set).

(c) Because the coefficient of friction does not vary, the power is directly proportional to wire speed.

(d) The same weight of copper is passing through each die in unit time. Therefore each die has the same rate of work, i.e. the horsepower used at each die is the same.

These rules make it simple to calculate the HP required for a given input to output diameter ratio and the speed at which the machine can be run within its power rating. Table 1.4 gives the power requirements over a range of output wire diameters, listed at intervals of 26% elongation (12.2% diameter, or 20.6% area reduction), for die drafting of 26%, 21% and 15%.

These values may be obtained from the empirical formula

$$\text{HP} = 1.29 \cdot d^{1.8} \times (\text{fractional area reduction per die}) \times (\text{no. of dies}) \times V_{m/s} \quad (1.4)$$

For example if we produce 0.511 mm wire from 2.305 mm input wire at a finish speed of 25 m/s:

Using 26% die drafting requires 13 dies	Using 21% die drafting requires 16 dies
HP/die at 0.511 mm is 0.40 at 5 m/s	HP/die at 0.511 mm is 0.33 at 5 m/s
×13 dies and 25/5 m/s gives 26.0 HP	×16 dies and 25/5 m/s gives 26.4 HP

Thus the same total area reduction requires the same total horsepower at the same speed.

Table 1.4
Required horsepower per die

Wire diam.	HP per die at 5 m/s (1000 ft/min)		
mm	26%	21%	15%
2.906	9.1	7.6	5.7
2.588	7.5	6.2	4.7
2.305	6.1	5.0	3.8
2.053	4.9	4.1	3.1
1.828	4.0	3.3	2.5
1.628	3.2	2.7	2.0
1.450	2.6	2.2	1.6
1.291	2.1	1.8	1.3
1.150	1.7	1.4	1.07
1.024	1.4	1.17	0.88
0.912	1.13	0.95	0.71
0.812	0.91	0.77	0.57
0.723	0.75	0.62	0.47
0.645	0.61	0.51	0.38
0.573	0.49	0.41	0.31
0.511	0.40	0.33	0.25
0.456	0.32	0.27	0.20
0.405	0.26	0.22	0.16

Lubrication and Cooling

The horsepower expended in drawing the wire ends up as heat which must be continually removed by spray lubrication of dies and drawblocks with an emulsion of water and oil. Water has a high specific heat which is excellent for carrying away heat but has poor wetting ability hence the inclusion of oil to reduce the surface tension. For copper wire the emulsion is alkaline with a pH of 8–9. The percentage of oil varies but in general the heavier the wire the greater the concentration as follows

Rod breakdown	7–10% oil in water
Intermediate wire	4–7 % oil in water
Fine wire	3–4 % oil in water

The emulsion collects at the bottom of the machine enclosure and is constantly recirculated and cooled by using heat exchanging coils in the holding tank to maintain a temperature of 35–45 °C. Sustained operation above 50 °C results in excessive die wear and breakdown of the emulsion. Depending on tank capacity, oil concentration and pH are checked twice daily. The pH is controlled by addditions of caustic soda.

The oil is also essential for lubrication in the die. The emulsion is carried at high speed into the die entrance angle and reacts at the interface to form an organometallic lubricating film. The fluid pressure depends on the speed and the ability of the lubricant emulsion to adhere to the wire surface as it approaches the die. The choice of wire drawing oil is based on performance against price. The performance is measured by the stability of the oil over long periods (months or even years) and the effect on die wear.

The water too has a profound effect on the stability of the oil. Hard water with >150 ppm calcium carbonate is known to have a deleterious effect. Soft water with 20–60 ppm calcium carbonate is preferred.

Over a period of time, copper dust or fines accumulate in the emulsion and tends to block the spray jets and dies. This can be rectified by filtration units (expensive and difficult to control) or by using settlement and weirs in the holding tank. Settlement is only satisfactory if the oil does not form metallic soaps around the copper particles, making them bouyant.

ANNEALING

For use in cables, the conductor is required to be annealed. A wire elongation of >20% is normally specified. Copper is annealed by heating to about 450 °C followed by quick cooling, (unlike steel which requires very slow cooling). Continuous in-line resistance annealing is now universally used for the following reasons

(a) The annealed plain- or tinned-copper wire is immediately available after wire drawing.

(b) The thermal efficiency is high at about 70% resulting in large power savings compared with batch annealing.

(c) The uniformity of elongation along the wire can be controlled within 2% of the desired figure.

(d) There is a large saving of space and no costly handling of spools into and out of the batch annealer.

(e) The process is fast and can easily match the wire drawing speed of 60 m/s.

Figure 1.3 shows a sketch of an in-line resistance annealer. Pulleys A, B and C are belt-driven from the wire drawing capstan and are arranged to have the same peripheral speed. These pulleys run on insulated shafts and are provided with slip-rings and brush gear which convey current from the annealer supply to the wire. The idler pulleys are also insulated. The wire between B and A is the preheat leg and is generally about twice as long as the annealing leg B to C, thus the current flowing in the latter is almost twice that in the preheat leg. The annealing leg is enclosed in a tube carrying low-pressure steam which excludes air and prevents oxidation of the wire. This is not required on the cooler preheat leg. The leg from C to 5 is enclosed in a water cooling tube.

Some typical results, all on the same machine, are shown in Table 1.5.

From these few results it will be seen that, at a given degree of annealing (about 30% elongation):

- For a given speed the annealing voltage is constant and independent of wire diameter.
- For a given speed the annealing current is proportional to wire cross-sectional area.
- The applied voltage is proportional to the square-root of the speed.
- The power in W · h/kg, is approx. constant.
- The theoretical power, without thermal losses, is ≈45 W · h/kg. Hence the efficiency is about 70%

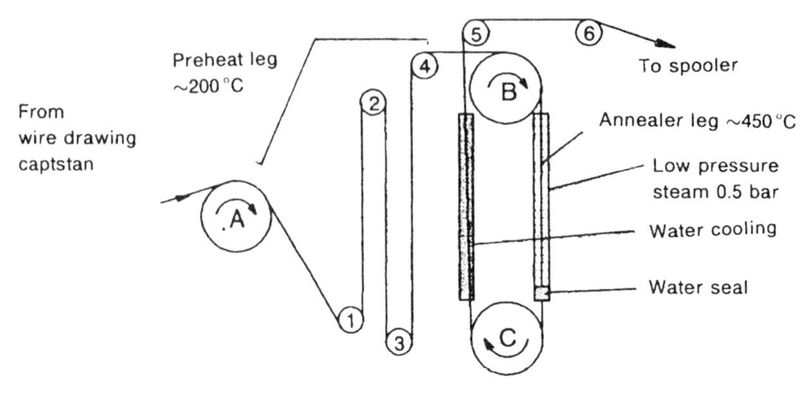

Figure 1.3
In-line resistance anealer

Table 1.5
Typical in-line resistance annealing data

Wire diam. mm	Speed m/s	Mass Cu kg/h	Anneal V	Preheat temp. A	Preheat temp. °C	Anneal temp. A	Anneal temp. °C	Total power A	Total power kW	Total power W·h/kg
0.51	25	167	50	75	199	135	445	210	10.50	62.8
0.64	25	262	50	115	210	205	467	320	16.00	61.1
0.91	20	424	44	215	188	385	430	600	26.40	62.3

Temperature Calculation

The temperature of the wire cannot be measured directly. It must be inferred from the rise in resistance relative to that calculated for the wire at 20 °C.

$$R_T = R_{20}\{1 + \alpha(T - 20)\}\ \Omega/\mathrm{m} \tag{1.5}$$

$$T = \frac{1}{\alpha}\left(\frac{R_T}{R_{20}} - 1\right) + 20\ {}^\circ\mathrm{C} \tag{1.6}$$

$$R_{20} = \frac{4\rho_E}{\pi d^2}\ \Omega/\mathrm{m} \tag{1.7}$$

where α = temperature coefficient of resistance. ρ_E is the electrical resistivity in $\Omega \cdot \mathrm{m}$.

Power Calculation

The heat required to raise 1 kg of metal with specific heat S, by ΔT °C is

$$H = \Omega T \times 1000_{\mathrm{kg}} \times S_{\mathrm{cal/g/^\circ C}}\ \mathrm{cal/kg}$$

1 calorie = 4.18 W · s hence

$$\text{Power required} = \frac{\Delta T \cdot 1000 \cdot S \cdot 4.18}{3600}\ \mathrm{W \cdot h/kg} \tag{1.8}$$

For copper with $\Delta T = 430$ °C and $S = 0.093$ this gives 46.4 W · h/kg. The power required is also given, in the absence of heat losses, by

$$\frac{V \times (V/(R_T L))}{(\pi d^2/4)\rho_m \times v \times 3600}\ \mathrm{W \cdot h/kg} \tag{1.9}$$

where ρ_m is the density in kg/m^3, d = wire diameter in m, L is the leg length in m and v is the velocity in m/s.

Equating (1.8) and (1.9), substituting for K_T from (1.5)–(1.7) and simplifying gives

$$V^2 = (4180S \cdot \rho_m \cdot \rho_E) \cdot v \cdot L \cdot \Delta T(1 + \alpha\Delta T)$$

$$V = k\sqrt{v \cdot L \cdot \Delta T(1 + \alpha\Delta T)} \tag{1.10}$$

and

$$I = \frac{\pi d^2}{4\rho_E} \cdot k\sqrt{\frac{v \cdot L \cdot \Delta T}{(1 + \alpha\Delta T)}} \tag{1.11}$$

This confirms the observations from Table 1.5. With a preheat and an annealing leg the situation is a little more complex but using the above it is easily found that, for a common voltage feed,

$$\frac{L_p}{L_a} = \frac{\Delta T_a - \Delta T_p}{\Delta T_p} \cdot \frac{1 + \alpha\Delta T_a}{1 + \alpha\Delta T_p} \tag{1.12}$$

subscript a refers to annealing leg and p to preheat leg. ΔT's are temperatures relative to 20 °C. To allow for heat losses V should be increased by 15% i.e. for copper k is increased from 0.245 to 0.282.

Similarly the curent ratio in the two legs is given by

$$\frac{I_a}{I_p} = \sqrt{\frac{L_p\Delta T_a(1 + \alpha\Delta T_p)}{L_a\Delta T_p(1 + \alpha\Delta T_a)}} = \frac{\Delta T_a}{\Delta T_p}\sqrt{\frac{\Delta T_a - \Delta T_p}{\Delta T_a}} \tag{1.13}$$

In order to resolve these ratios and to determine the preheat and anneal temperatures, there is no substitute for accurate measurement of the individual currents and leg lengths. For an anneal temperature rise of 455 °C and a preheat temperature rise of 180 °C, Equation (1.12) gives the preheat length as 2.5 times the anneal length which is the ratio commonly found in commercial annealers. A larger length ratio would increase the temperature ratio for a given anneal voltage V.

Figure 1.4 shows the relation between annealing voltage and wire elongation determined experimentally on an annealer with $L_p = 3.81$ m and $L_a = 1.52$ m, when running 0.51 mm wire at 20 m/s. The curve will apply to any wire diameter on the same machine so long as V is scaled by the square root of the speed in accordance with Equation (1.10). If the length of the annealing and preheat lengths remain in the same ratio, changing to another machine with a different leg length would require the annealing voltage, derived from this curve, to be adjusted by the square root of these lengths again in accordance with Equation (1.10).

For most purposes, it is desired to work within ±2% of a wire elongation of 25%. This is on a steep portion of the annealing curve and, at 20 m/s, corresponds to a voltage of 38.2 ± 0.9 volt. Thus accurate measurement and very tight control of the annealing voltage is required.

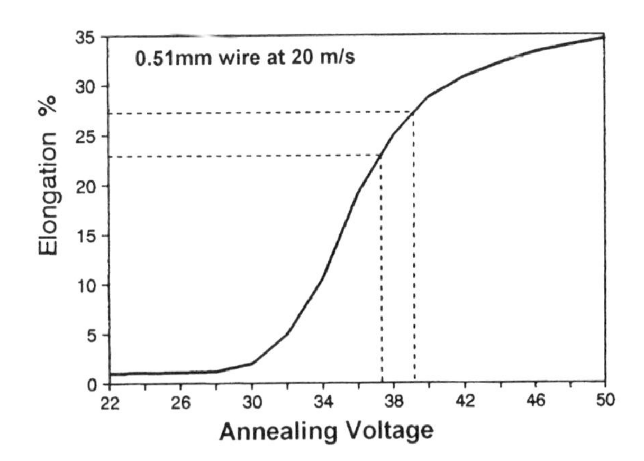

Figure 1.4
Annealing curve for copper wire

Ramp Annealing

Owing to the inertia of a wire drawing machine it needs to be ramped-up to speed, and also slowed down, fairly slowly. This can translate into quite large lengths of wire at start-up and finish. In order to maintain the desired degree of anneal in these portions of wire it is necessary to automatically ramp-control the annealing voltage according to the square root of the instantaneous machine speed in accordance with Equation (1.10).

Annealer Maintainance

In order to obtain consistent results in accordance with the design, it is essential that maintainance of contact wheel tyres (which are subject to spark erosion) and contact slip rings and brushes on pulleys A, B and C is of a high order. If there is significant resistance (e.g. $>10\,\text{m}\Omega$) in the contact to the wire, the recorded anneal voltage will be in sufficient error to affect the determination of temperature and elongation. Also it is essential that the meter readings, tachometer and the ramp annealing control remain accurate.

Limitations of Resistance Annealing

At 25 m/s the residence time of the wire in a 1.5 m anneal leg is $1.5/25 = 0.06$ s or 3 cycles of a 50 Hz anneal current. If the speed is increased to 60 m/s the residence time becomes comparable to the a.c. period and intermittent annealing will start to occur. For this reason the applied voltage for high speed annealing should be d.c.

It has also been found that at 40 m/s in an anneal leg of 0.4 m, although 0.1 mm and 0.15 mm wires were correctly annealed, 0.2 mm wire at the correct voltage and current could only be annealed below 30 m/s. Above this speed it ceased to be annealed. At 30 m/s the residence time for 0.4 m is 0.013 s. This may be a fundamental limitation depending on the wire diameter, e.g. residence time(s) $> 0.09 \times$ diameter(mm). The reason for this is not clear.

Wire Take-up Methods

The rod breakdown is normally from 6.3 mm–8.0 mm down to 2.3 mm–3.0 mm without annealing. The hard drawn output wire is collected in a stationary vertical cage or 'pak' in flat overlapping turns. The capacity of this pak is about 600 kg. As both the inner and outer ends are readily accessible, the inner end of one pak can be joined by cold-pressure welding to the start of the next pak for continuous feed to the next operation.

Intermediate wire drawing from 2–3 mm down to 0.125 mm for insulating is followed by annealing. This drawing and annealing is often carried out in tandem with a thermoplastic insulating extruder, thus eliminating the need for a take-up spool. Where it is not a tandem operation the take-up is at high speeds onto spools. The method of using a double spooler with automatic cut-over never reached a success rate of cut-over of more than 75–80% and has been superseded by single spool take-up of up to 500 kg capacity. To avoid hard wire on start-up and finish, accurate ramp annealing control is necessary. This weight of wire at 0.5 mm enables about 2 h of running time at 40 m/s. With a standardized, highly balanced, 127 mm bore reel on a cantilever shaft, it is possible to take up 0.3 mm wire at 50 m/s on a 500 kg winding with a diameter control of 0.002 mm. It is important to maintain accurate traversing of the winding at all times.

Fine-wire drawing machines with an input of 0.45 mm–1.25 mm are used to produce wire of 0.1 mm to 0.35 mm. The output wire is annealed and the take up is also on to spools of up to 75 kg capacity.

Wire Breaks

Wire breaks at any stage are a nuisance, in wire drawing they can cause excessive down time of the machines. Analysis of the breaks under a microscope can determine their cause, for example

Cup and Cone – Usually due to bad drafting in previous dies even back to rod breakdown.

Inclusion – A sudden break in the wire with no 'neck down'. Often the inclusion is left embedded in the wire ends. A magnet may be used to determine if it is a ferrous inclusion.

Tubing – A concentric hole extending several diameters into the wire. Probably due to gas bubbles from the continuous casting process.

Tension break – Revealed by 'neck down' of the wire at both sides of the break. Wire has been subject to excessive tension in drawing.

Standard Wire Sizes in Telecommunication Use

Some common conductor sizes in telecommunication cable use and their approximate equivalents are listed in Table 1.6

Table 1.6
Standard wire sizes for telecommunication cables

Diam. mm	AWG	Diam. in	lb/mile	kg/km	R_{dc} Ω/km
0.30	29	0.012	2.25	0.64	239.5
0.32	28	0.0126	2.5	0.70	210.5
0.40	26	0.0159	4.0	1.13	134.7
0.50	24	0.020	6.5	1.84	85.06
0.63	22	0.025	10	2.82	54.4
0.90	19	0.0355	20	5.65	27.0
1.20	17	0.0473	36	10.2	16.0
1.30	16	0.050	40	11.3	13.6

ELECTROTINNING

Tin coated copper wire was originally used to combat the tarnishing and corrosion of copper wire by the sulphur in its insulation of vulcanized rubber. Its use was continued under polyvinylchloride insulation because of its ease of soldering. Until the 1950s the method of coating the wire was first to draw it to the required final size and then to pass it through a bath of molten tin, through a wiping die, and on to a take-up spool. This was a slow process and it was difficult to control the tin thickness. In 1951 a continuous electrotinning process was patented by J. Delves Broughton (STC, UK). This electrotinning process was capable of producing a highly uniform tin coating on 2–3 mm copper wire which was sufficiently adherent to withstand subsequent wire drawing and annealing processes down to the smallest sizes.

Electrotinning Machine

A sketch of an electrotinning machine is shown in Figure 1.5. It consists of a wire cleaning section in which electrolysis in a caustic soda solution is used to clean and degrease the wire, followed by water rinsing and air drying. The clean wire is then electroplated using tin anodes in an acid stannous sulphate solution and finished with a final water rinse and air drying. The water rinsing is done in plastic tubes with a high speed counter-flowing water stream and drying is done in the same manner but using a high speed air jet.

The wire path from the input is via a mechanical servo-tensioning pulley, across the rubber weir, through four passes in the cleaning solution, passing over sheaves at each end. From the cleaning trough it passes through the rinsing and drying tubes and enters the plating trough where it makes twelve passes around sheaves on the left and a flat driven capstan on the right. Exiting the trough it passes through air-wipe, water-rinse and air-drying tubes before proceeding to the take-up.

The tin anodes are cast with 45° corrugations to increase the surface area and rest on the bottom of, and in contact with, the plating troughs. The positive outputs of the rectifiers are connected to the troughs which are supported on insulators. The capstan

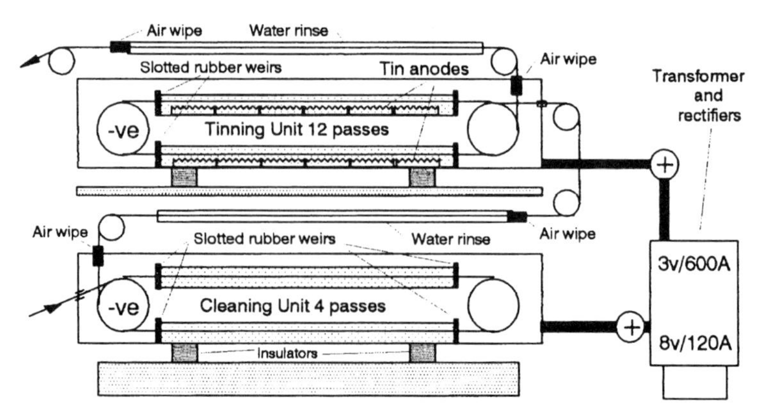

Figure 1.5
Electrotinning machine

and the left-hand sheaves in the cleaning trough are connected via slip-rings to the negative (earthed) terminal of the rectifiers. Thus the wire line becomes the cathode in both troughs.

The cleaning bath electrolyte is a solution of sodium hydroxide and sodium metasilicate in water and the plating electrolyte is a solution of stannous sulphate, sulphuric acid and phenolsulphonic acid in water. The plating solution also contains an anti-foaming agent. Both solutions are continuously recirculated through holding tanks by submersed pumps. The troughs and other parts are made of stainless steel to withstand corrosion.

Electrochemical Reactions

When strong electrolytes, such as the ones involved here, are dissolved in water almost complete dissociation into electrically charged ions occurs. On application of an electric field the positively charged ions move towards the negative electrode, the cathode, and the negatively charged ions move towards the positive electrode, the anode.

In the cleaning solution $\quad NaOH \Leftrightarrow Na^+ + OH^-$
At the cathode $\quad Na^+ + \varepsilon \rightarrow Na \quad \text{then} \quad Na + H_2O \rightarrow NaOH + H \quad (1.14)$
At the anode $\quad OH^- - \varepsilon \rightarrow OH \quad \text{then} \quad OH \rightarrow \frac{1}{2}H_2O + \frac{1}{2}O$

Thus hydrogen is evolved at the cathode at twice the volume of the oxygen evolved at the anode, per ion discharged. The hydrogen evolving at the surface of the wire lifts off foreign matter leaving the copper wire chemically clean. The liberated hydrogen and oxygen gases are in an explosive ratio and must be removed by an exhaust fan system and vented to the atmosphere.

In the plating solution $\quad SnSO_4 \Leftrightarrow Sn^{++} + SO_4^{--}$
At the cathode $\quad Sn^{++} + 2\varepsilon \rightarrow Sn$ which plates out on the wire surface $\quad (1.15)$
At the anode $\quad SO_4^{--} - 2\varepsilon \rightarrow SO_4 \quad \text{then} \quad SO_4 + Sn \rightarrow SnSO_4$

which replenishes the ions in solution.

Faraday's Law

The capture of electrons from the cathode by the positive ions and the donation of electrons from the negative ions at the anode constitutes a current flowing in the external circuit. In 1820 Faraday established that

$$M = I \cdot z \cdot t \text{ g} \tag{1.16}$$

where M is the mass of substance liberated in electrolysis, I is the current in amps, t is the time in seconds and z is the electrochemical equivalent for the substance. Faraday established that z is the mass of substance in g liberated by 96 490 coulombs of charge. Now the electronic charge is $\varepsilon = 0.16022 \times 10^{-18}$ C and $96\,490/\varepsilon = 6.0223 \times 10^{23}$ which is Avogadro's number, N. This is the number of molecules in the molecular weight of the substance (or for elements, the number of atoms in the atomic weight). Since the charge on the ion is $\varepsilon \times$ valency, the mass of the element liberated per 96 490 C exchanged is the atomic weight divided by the valency of the ion. Thus

$$z = \text{mass per coulomb} = \text{Atomic weight}/(\text{valency} \times 96\,490) \text{ g/coulomb} \tag{1.17}$$

Hence for divalent tin,

$$z = 118.69/(2 \times 96\,490) = 0.000615 \text{ g/A/s}$$

and for monovalent silver,

$$z = 107.88/(1 \times 96\,490) = 0.001118 \text{ g/A/s}$$

Plating Thickness

For the plating process used, the wire is passing through the electrolyte at v m/min and a current I A is flowing, depositing metal with a density ρ kg/m^3 on a wire diameter d m, to a thickness T m. Then

$$\text{mass deposited is} \quad M = T \times \pi d \times \rho \times v = I \cdot z \times 60/1000 \text{ kg/min}$$

$$\text{hence} \quad T = \frac{I \cdot z \cdot 60}{1000\pi d \cdot \rho \cdot v} \text{ m}$$

$$\text{which for tin becomes} \quad T = \frac{I \times 0.000615 \times 60}{1000\pi d \times 7330 \times v} = \frac{I \times 1.6024}{d \cdot v} \times 10^{-9} \text{ m} \tag{1.18}$$

Experience has shown that the plating thickness is normally 95% of this figure. This is known as the plating, or cathode, efficiency. The remaining 5% of the plating current is probably carried by hydrogen ions. Hence to use Equation (1.18) to determine T, the actual current used should be multiplied by 0.95. Thus for 2.32 mm diameter wire at a speed of 110 m/min using a plating current of 600 A the deposited tin thickness will be

$$T = \frac{600 \times 0.95 \times 1.6024}{0.00232 \times 110} \times 10^{-9} = 3.58\,\mu\text{m}$$

This thickness of plating is suitable for drawing down to 0.4 mm diameter. For wires of 0.35 mm–0.15 mm the plating thickness on the 2.32 mm wire should be doubled by halving the plating speed to 55 m/min.

Current Densities and Residence Times

In the cleaning section there are four passes of the trough with a total immersed length of wire of 15.24 m. The immersed surface area of a wire of diameter 2.32 mm is thus

$$\pi d \cdot l = \pi \times 0.00232 \times 15.24 = 0.111\,\text{m}^2 = 11.2\,\text{dm}^2$$

and for a current of 120 A the current density at the wire surface is $120/11.2 = 10.7\,\text{A/dm}^2$. The residence time at 110 m/min is $15.24/110 \times 60 = 8.3$ s. Tests have shown that with drawn wire of average cleanliness, at this current density, effective cleaning takes about 5 s.

In the plating section there are 12 passes with a total immersed length of wire of 67.07 m and an immersed area of 2.32 mm wire of

$$\pi d \cdot l = \pi \times 0.00232 \times 67.07 = 0.49\,\text{m}^2 = 49.0\,\text{dm}^2$$

and for a plating current of 600 A the current density at the wire surface is 600/49.0 $= 12.24\,\text{A/dm}^2$. The residence time at 110 m/min is $67.07/110 \times 60 = 36.6$ s. This current density gives a pale grey, matt finished plating of excellent adhesion which will easily survive the rigours of further wire drawing and annealing down to diameters of 0.1 mm.

The tin anodes of the plating section have a corrugated surface area of 400 dm^2 and so the current density at the anodes is $600/400 = 1.5\,\text{A/dm}^2$. At this current density they can be operated continuously for 24 h before requiring cleaning (by scrubbing in boiling water).

Tin Consumption

For the plating conditions given above at 600 A and 110 m/min the weight of tin deposited at a cathode efficiency of 95%, is

$$600 \times 0.95 \times 0.000615 \times 3600/1000 = 1.26\,\text{kg/h}$$

The mass of wire plated is

$$110 \times 60 \times \pi d^2/4 \times \rho = 110 \times 60 \times (4.23 \times 10^{-6})_{\text{m}^2} \times 8920_{\text{kg/m}^2} = 248.0\,\text{kg/h}$$

Thus the mass of tin is 0.5% of the mass of copper.

At the anodes which operate at 100% efficiency, the loss of tin is $1.26/0.95 = 1.33$ kg/h. The original mass of 42 tin anodes is 210 kg. When their mass drops by 95% they must be replaced since the corrugations become eroded and the anodes become too thin. This occurs in $0.95 \times 210/1.33 = 150$ h.

Solution Maintainance

As with all electroplating cleanliness of the plated surface is essential. The only time this becomes a problem in this process is when the cast rod has bad surface blemishing or tarnishing. Normally this is removed by the pickling process after casting.

Solution volumes and composition must also be maintained. There is a continuous slow loss of solution due to drag-out on the wire and evaporation. In the case of the plating solution, the difference between the anode and cathode efficiencies causes it to become rich in tin so that addition of stannous sulphate is rarely required. The acids need to be replenished at 100 h running time intervals and the volume made up with water. Also more anti-foaming agent needs to be added per shift.

In the case of the cleaning solution, the sodium hydroxide content needs to be checked daily, and the content must not fall below 75 g/l. Owing to its cleaning action the solution becomes contaminated and should be changed every four months.

Performance Limits

Tests have shown that with the solution compositions given above, the cathode current density can be increased to about 17 A/dm^2 and still maintain an excellent plating and subsequent drawing performance. This would correspond to a speed increase of $17/12.2 \times 110 = 153$ m/min (40%) with a corresponding increase in plating current to 836 A. Under these conditions however the cathode efficiency falls to 90%. This would eventually lead to problems with solution control and excessive 'treeing'.

Laboratory tests have shown that by increasing the electrolyte concentrations of stannous sulphate and other components in proportion, current densities of 20 A/dm^2 can be achieved with cathode efficiencies of 97%. However these improved cathode current densities were not carried through to drawing performance of the tinned wire to check the 'toughness' of the plating, although the plated tin appearance was excellent and adjudged capable of standard drawing performance. To maintain the anode current density at 1.5 A/dm^2 the corrugation angle would need to be increased to 60° included angle.

It is a long-standing piece of electroplating wisdom that high plating current densities produce granular and friable coatings, only marginally improved by plating solution changes. This has been illustrated in recent years by the introduction of stannous fluoroborate solutions for tin plating. Apart from the highly corrosive fumes and effluents of this electrolyte which are deleterious to personnel and building structures, the plating troughs have also to be rubber-lined to withstand the corrosion and separate connections have to be made to the anodes. At the recommended current densities of 20 A/dm^2 the tin coating was found to be coarse and granular and was quite incapable of being drawn to fine-wire diameters and still passing the persulphate test. This was so even with a reduced current density of 10 A/dm^2.

Tin Thickness Measurement

Tin thicknesses on wire are measured by deplating the tin from the wire in an acid stannous chloride electrolyte, by making the wire the anode (positive) to a tubular copper cathode. About 50 mm of wire is used and a current of 100 mA is passed through the cell. When the tin is completely deplated the current suddenly drops to about 90 mA. The time, t seconds, for this to happen is measured using a stop-watch.

$$\text{Mass of tin removed in g} = \pi d_{\text{m}} \cdot l_{\text{m}} \cdot \rho_{\text{kg/cu·m}} \cdot T_{\text{mm}} = I_{\text{A}} \cdot z_{\text{g/A/s}} \cdot t_{\text{s}}$$

whence

$$T = \frac{I_{\text{mA}} \times t_{\text{s}}}{d_{\text{mm}} \times l_{\text{mm}}} \times 0.02671\,\mu\text{m} \tag{1.19}$$

Drawing of Electrotinned Wire

Tinned copper wire of 2 to 3 mm diameter can be drawn to intermediate or fine wire sizes without intermediate annealing with the same diamond dies and lubricants as plain copper wire. Drawblocks should however have hard chrome or zirconium oxide tyres as noted previously and it may be desirable to add one extra turn for tinned wire because of the lower coefficient of friction of the tin surface.

The thickness of the tin coating decreases in direct proportion to the diameter reduction. It has been found that in order to pass the persulphate and solderability tests (see below) the final wire should have a minimum tin thickness of 0.5 μm. To standardize electrotinning conditions it is usual to plate 2.32 mm wire with 3.5 μm of tin for drawing down to 0.4 mm wire, and to increase the plating thickness to 7.0 μm (by halving the plating line speed) for drawing to fine-wire sizes of 0.3 mm to 0.15 mm.

Final annealing of tinned copper wires after drawing to size is accomplished on the same resistance annealers as for plain copper wire. Owing to the higher contact resistance of tinned copper, the arcing to the contact wheel tyres is increased and rolled nickel strip tyres are superior in this application.

The temperatures of 450–500 °C reached in the annealing leg are well above the melting point of tin (232 °C) and this causes a tin/copper alloy to form at the interface. The time interval during which this occurs is very short (0.05 s at 30 m/s) and the amount of free tin absorbed is of the order of 0.15 to 0.3 μm. It is the free tin on top of this alloy layer that gives the tin its solderability and protective properties.

Persulphate Test

This is a test for the degree of cover and protection of the copper wire by the tin layer. It consists of immersing 1 m of solvent-degreased wire formed into a helix of diameter about $70d$ to $35d$ (depending on the wire diameter, d) in a specified ammonium persulphate solution at 18 °C for 10 min. Any exposed or incompletely protected copper is dissolved by this solution which turns it a blue colour depending on the amount of tin dissolved. The amount of copper is determined by comparison with a similar solution with known amounts of copper sulphate added. The specification requires that this amount shall be less than 40 mg/dm^2 of wire surface.

Solderability Test

In this test, samples of wire are pressed into a molten bead of specified solder at a specified temparature: the average time for the solder to wet the tinned wire should be less than 2 s.

Other Platings

Other metals have been plated successfully on to copper wire and steel wire by similar continuous plating machines. Silver plating and nickel plating of copper, and copper plating and zinc plating of steel are examples that have been commercially used. In all cases the products are much superior and show a cost saving over other methods. Silver and nickel plated wires are required when insulating with polytetrafluorethylene (or some of its chemical cousins) since copper catalyses the breakdown of this material. Zinc electroplated steel wire is a great improvement over hot-dip galvanized steel wire, using much less zinc for the required protective cover, being much more concentric in application and far more adherent.

HANDLING OF CONDUCTORS

The following comments apply to conductors at all stages of cable making and use, whether bare or insulated, twisted or cabled. The design of transmission parameters is often critically dependent on the average and point values of wire diameter. A 0.5% change in wire diameter (0.003 mm in 0.63 mm) can change the transmission parameter of characteristic impedance by 1%. As a single occurence in a line, this may not be of great importance but as frequencies of transmission are raised to 100 MHz and more, the occurence of repeated faults of this size, particularly if they repeat at regular intervals of about 1–10 m, can produce resonances which can seriously affect the transmission performance. Consequently it is necessary to avoid situations that can damage the wire, such as snatches and snagging on reel flanges etc. Probably the more serious situation involves wire tension especially since the wire is generally annealed and hence soft.

Wire Tension Effects

Figure 1.6 shows a typical stress/strain curve for a ductile wire such as annealed copper. At low stress, up to point A, the curve is linear and the wire behaves elastically, returning to its original state when the stress is removed. If the stress is greater than A, say P, the wire starts to behave plastically, so that on removing the stress the wire will be found to have a permanent length increase, dS. The greater the stress, the greater the plastic deformation. Ultimately the strain increases to the point where the wire breaks, at a stress called the ultimate tensile stress, UTS. The wire elongation at break is the measure of the degree of annealing of the wire. For fully annealed copper wire the UTS is 240 N/mm^2 and the point A, the limit of elastic behaviour where $dS < 0.1\%$, is known as the 'proofstrain'. This is generally about one third of the UTS or 80 N/mm^2.

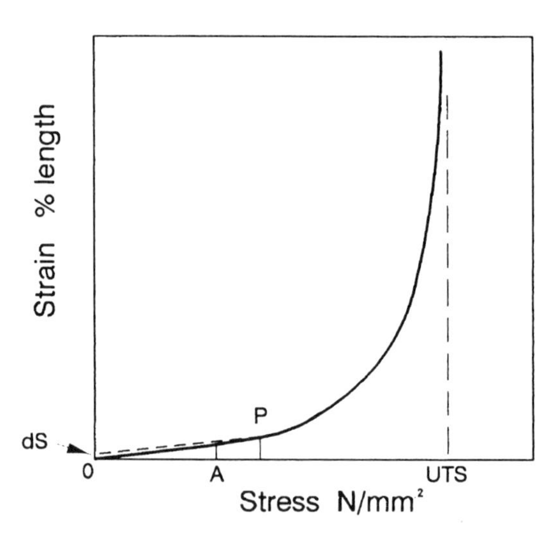

Figure 1.6
Stress and strain relation for ductile wire

During processing and handling of annealed copper wire therefore, the stress should never be allowed to exceed 80 N/mm^2. In order to allow for snatches, the average stress should in fact be less than this, at about 20 N/mm^2 For 0.5 mm diameter wire this requires an average tension of less than 3.9 N (or 0.4 kgf). In the case of perfectly twisted twin the tension will be split evenly between the two wires. If however the twisting is uneven the straighter wire will carry the majority of the tension. The same applies to unevenly stranded small pairage cables.

Tension Control

For the lightest of tensions, wire can be taken from the reel by 'flying off'. This is illustrated in Figure 1.7 and is probably the best method of handling fine wire. But the wire is given one twist for every turn removed, which can sometimes be undesirable.

If the wire is taken from a rotating reel, some light tension on the reel is required to prevent overrun. The simplest method is the band brake, also shown in Figure 1.7. Note the direction of rotation relative to the positioning of the spring (or mass). Band brakes must be kept in good condition otherwise variations in coefficient of friction can cause snatching. A smoother control of tension is provided by the use of a 'torque motor'. Essentially this is an a.c. induction motor trying to rotate the reel in the opposite direction. The voltage applied is just sufficient to provide a light torque against the reel rotation. This method of tension control is much used in tensioning twins on stationary stands, feeding into a drum twisting strander. All the torque motors can be set to the same voltage by one control.

For more sophisticated tension control, a servo-controlled d.c. motor drive can be used. The principle of servo control is the measurement of the speed or tension relative to a set value. The difference between the actual value and the set value is amplified and used to control the motor speed, as illustrated in Figure 1.7. This can also be done in a

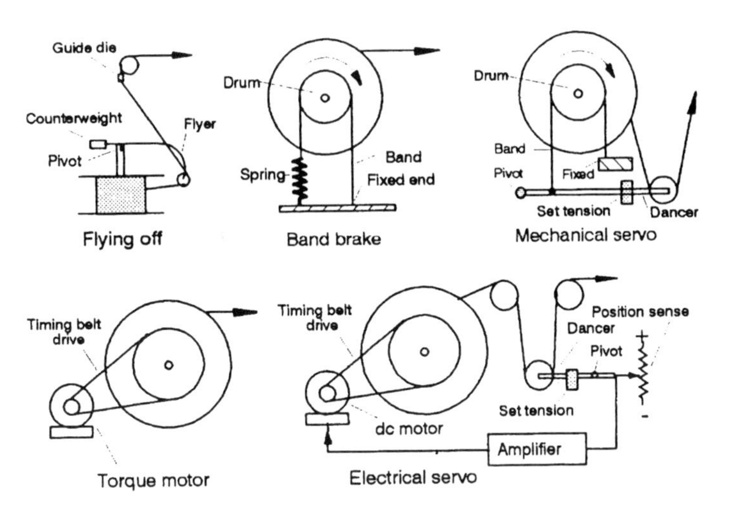

Figure 1.7
Tension control methods

purely mechanical arrangement, also shown, but the use of mechanical band brakes requires careful attention to cleanliness of the band and drum. Particularly when using servo control, it is essential that there is no backlash anywhere in the drive, otherwise there will be 'chatter', causing severe snatch on the conductor. When running, the dancer arm on the servo control should move only gently, floating about the set equilibrium position. If it oscillates wildly there is some snatch occuring in the feed from the reel, or further down the line. If this condition persists after checking, then there is probably too much amplification of the difference signal or too little damping in the servo loop. An advantage of electrical servo controls is that the difference signal can be modified if required, by differentiation or integration, to tailor the response to the control the rate-of-change or the running average of the controlled variable.

2

Insulation

INSULATING PRINCIPLES

Originally the connection between distant telegraphy or telephony stations was by means of two pole-mounted bare conductors. Typically these would be 2 mm diameter copper (or steel) wires spaced 150 mm apart and supported on porcelain insulators. At voice frequencies, the pair of conductors had an impedance of 600 Ω, a capacitance of 5.5 nF/km, an inductance of 2.0 mH/km and an attenuation of 0.08 dB/km. Since the instruments in use could tolerate a total loss of about 20 dB, route lengths of 250 km were possible without amplification (even more with telegraphy). If several parallel connections were required, magnetic and electric coupling between the pairs caused crosstalk between the circuits and wire transpositions were used to minimize this effect. However the open wire lines were susceptible to damage, theft, climatic conditions (wind or rain and dirt on insulators causing losses), wire-tapping and interference from external electric and magnetic fields (e.g. electric traction and lightning strikes). Also the proliferation of circuits, especially near exchanges, caused mechanical congestion and visual nuisance.

Thus it became necessary to develop cables containing many pairs of wires which could be handled conveniently (i.e. weight, diameter and flexibility should be manageable) and which could be buried in the ground or installed in buried ducts. Since groundwater would cause deterioration of the electrical properties or long-term corrosion, the cables (and joints) had to be protected by impermeable sheaths.

Requirements

The first requirement is to provide electrical isolation of the wires from each other by insulating the individual wires. The insulation should be flexible, resist abrasion in manufacture, have a high insulation resistance and preferably a high voltage withstand against induced surges. Since the spacing between the wires is now much smaller, the insulation should have low electrical permittivity (to minimize capacitance increase) and low dielectric losses. Also it is necessary to be able to identify the different conductors and pairs. This is done in a variety of ways using colour and/or printed marking of the insulations. Since a low mass, small diameter cable is required, the conductors need to be of smaller diameter but not so small that resistive losses in the transmission of the signal become prohibitive. Near an exchange where large numbers of pairs need to be brought together, the conductors are made smaller to keep the cable diameter reasonable. Further away, as the pairs branch out, the conductor diameters are increased to minimize resistive losses. The calculation of the pair parameters of resistance, inductance and capacitance and the use of these to calculate the transmission parameters of impedance and attenuation will be dealt with in Chapters 6 and 7.

Since there are now many pairs very close together in the cable, the effects of crosstalk are more serious and steps must be taken to combat it. This is done in pair and quad cables by twisting the insulated conductors together with a number of different twist lengths and positioning the pairs in the cable so that similar twist lengths are not placed together. This minimizes the crosstalk as will be explained in Chapter 8. For important security pairs (e.g. carrying broadcast signals) the pairs are frequently screened with metallic foils.

For even better crosstalk reduction and for better transmission properties, especially as carrier frequency systems were developed using frequencies far beyond the voice range up to 4, 12 and 60 MHz, coaxial pairs (an inner conductor surrounded by a tubular outer conductor) were developed and used in cables. For precision in achieving the required transmission parameters these were of semi-rigid construction and the low permittivity insulation of the inner conductor was mostly of air, with dielectric supports of high dimensional precision.

Insulating Materials

Table 2.1 gives a guide to typical properties of the more common materials used to insulate wires. For definitive values the manufacturers specifications should be used. Insulating paper should be of long-fibred wood pulp composition. During manufacture it should have about 6% moisture content to improve elongation but requires vacuum drying before sheathing to reduce the permittivity and loss angle and to increase its resistivity.

PVC is a composition of polyvinyl chloride with plasticiser and filler and for insulating purposes it is formulated into a 'hard grade'. This is less flexible but has higher abrasion and crush resistance. These compositions can be ignited but are self-extinguishing owing to the evolution of the halogen (chlorine).

Polyethylene (or polythene) properties vary with density and its method of polymerization, particularly its Melt Flow Index (MFI) which governs its ease of extrusion. Polymerization methods can result in a narrow range of molecular weights which tends to give a lower MFI for a given density, or a broad spread of molecular weights which increaes the MFI for a given density. The recent development of linear low density polyethylene (LLDP) seems to have captured most of the desirable properties in one product.

PTFE is polytetrafluoroethylene. On heating to about 330 °C it does not melt but gels. This means that it cannot be handled by a normal screw extruder, but it can be extruded by a ram extruder. At higher temperatures it degrades by charring. FEP is a

Table 2.1
Properties of insulating materials

Material ≫		Paper		PVC	Polyethylene density		Poly-propylene	Poly-styrene	Poly-amide (Nylon)	PTFE	FEP	Poly-ester (PET)
Property		Dry	(moist)	plasticized	Low	High						
Permittivity		2	(6)	3–5	2.2	2.28	2.2–2.4	2.4–2.7	3–4	2.1	2.1	3.1
Loss angle	rad	0.015	(0.3)	0.05	<0.0005		0.0005–0.002	0.0002	0.02–0.06	0.0002	0.0006	0.003
Dielectric strength	kV/mm	3	–	15	>30		>30	20	20	>30	>30	13
Vol. resistivity	TΩ · m	–	–	>0.001	>100		>100	>0.05	>0.1	>1000	>1000	>100
Softening point	°C	–	–	80	85	127	150	82–100	180–220	(330)	290	260
Thermal conductivity		1.2	–	3–4	8	12	3–4	2.4–3.3	5–6	6	6	5
Thermal expansion	μm/m °C	–	–	70	170	110	110	70	70	100	90	70
Specific heat	cal/°C/g	0.3	–	0.3–0.5	0.53	0.55	0.46	0.33	0.4–0.3	0.25	0.28	0.3
Density	g/cm^3	–	–	1.18–1.4	0.915	0.95	0.90	1.05	1.13–1.01	2.1–2.3	2.15	1.35
Tensile strength	MN/m^2	28	–	0.25	0.1	0.3	0.36	0.4	0.6–0.4	0.2	0.22	0.75
Elongation at break	%	1	(6)	250	400	150	250	1–2.5	300	250	300	200
Tensile modulus	MNm^2	–	–	25	1.5	4.5	12	30	20–14	4.1	3.6	25
Water absorption in 24 h	%	6		0.5	<0.01	<0.01	0.02	0.05	2–0.15	0	<0.2	0.08

fluoridated ethylene propylene which has nearly as good high temperature properties but does melt and can be handled by screw extrusion. Neither of these polymers burns owing to their high fluorine content. Since copper is a catalyst in the degradation of these polymers they should be applied over nickel- or silver-plated copper wire.

Polyester (PET) is polyethyleneteraphthalate and is sold under the trade names of 'Melinex' and 'Mylar'. It is widely used in the cable industry in the form of tapes for lapping and also, when laminated with a very thin metallic foil (usually aluminium), for screening.

String and Tape Insulating

The earliest insulations for telephone cables were paper wrappings. Initially these were also impregnated with paraffin wax, but it was later found adequate to dry the paper well before sheathing, with the added advantage of obtaining a lower permittivity. This reduced the capacity and attenuation of the pairs. By increasing the width of the paper tape a controlled degree of paper creasing was introduced which also lowered the effective permittivity of the insulation. Thus for an 0.5 mm conductor unit twin cable at a capactitance of 52 nF/km, the insulation would be a paper tape 5 × 0.064 mm at 30% overlap and 48 laps per metre, crushed by the die to a diameter of 1.1 mm. This achieved a permittivity of 1.8 when dried. For even lower permittivity, a helix of paper string was introduced under the paper tape thus producing a tube of paper supported by the string.

A sketch of this method of insulating is shown in Figure 2.1. The string and paper tape are applied in opposite directions. Essential items like string- and paper-tensioning arrangements are omitted from the sketch. For a typical carrier quad design (28 nF/km) with 100 laps per metre of 0.5 mm diameter string and 50 laps per metre of 30 × 0.18 mm tape on 1.27 mm wire, this achieved an effective permittivity of 1.45. The

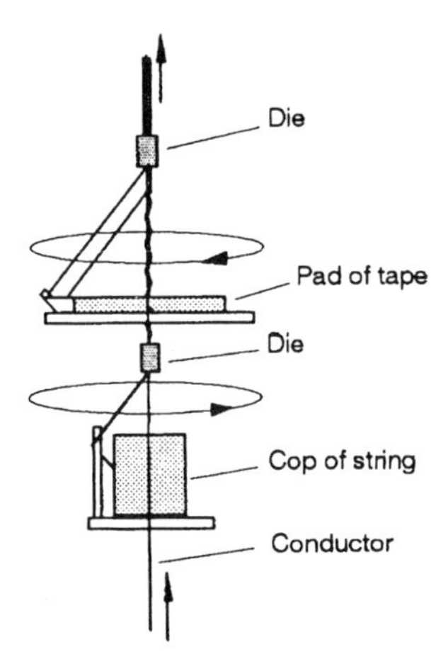

Figure 2.1
String and tape insulating

string head turns twice as fast as the tape head. If the string head is limited to 3000 r.p.m. the linear speed of the wire is 30 m/min, which is quite slow. Consequently the insulating section of the factory had a large number of insulating machines with the operators supervising up to 10 machines each. It was essential to have a detector which stopped the machine automatically in the event of a string or paper break so that the operator could effect a repair. For good quality of transmission, uniformity of the insulation was essential and this principally meant a very uniform string diameter. This was always a difficulty with paper string, made as it was, from twisted paper tape. However, paper insulated carrier quads were made in the mid 1950s which worked well up to 252 kHz (60 channels).

To achieve even higher uniformity, 'Styroflex' insulation was developed in Germany. This used an extruded polystyrene string of very uniform diameter and a polystyrene tape of uniform thickness. Since this material is very springy, the string wrapping head and die were heated to near the softening point of the polystyrene (80–100 °C). This produced a stable helix of the polystyrene string around the wire which resisted deformation in subsequent processes and enabled the construction of very uniform 120-channel quads operating up to 512 kHz. The effective permittivity achieved was as low as 1.2. An additional advantage of this insulation was its relative immunity from water absorbtion (in contrast to paper). The process however was very slow. These machines were confiscated from Berlin after the war by the Russians and transferred to Moscable in Moscow where they were still working well in 1967. At that time the principal telephone circuits between Moscow and Vladivostok were carried on 4-quad styroflex cables (under a seam-welded aluminium sheath developed in Russia).

Thermoplastic Insulation

After the Second World War, thermoplastic insulations were introduced, firstly of PVC for switchboard wiring and indoor cables, replacing lapped cotton and silk coverings and rubber insulations. In the 1950s polythene insulation was introduced for distribution cables and later, for junction and carrier cables and for coaxial leads. When semi-rigid carrier coaxials were introduced about 1952, various air-spaced polythene constructions were devised.

PRINCIPLES OF EXTRUSION

Thermoplastic materials are generally supplied in the form of cylindrical or ellipsoidal pellets with dimensions of a few millimetres. These must be melted, homogenized, compressed (releasing air bubbles), fed under pressure around the wire passing through a sizing die and cooled rapidly to set the insulation. This is done in an extruder, which is sketched in Figure 2.2.

The extruder comprises a barrel containing a rotating screw, with a feed hopper at one end and an exit port at the other, feeding a crosshead fitted with a point and die. The barrel is fitted with electric induction heaters in three to four zones, and also cooling fins and fans. Each zone, and the crosshead, is independently temperature controlled by means of embedded thermocouples and electronic controllers operating

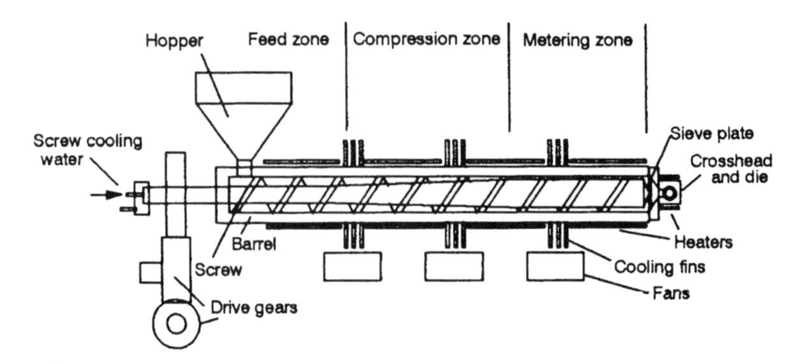

Figure 2.2
Thermoplastic insulation extruder

the heaters or fans as required. The zone and head temperatures are set to a gradually increasing temperature profile to suit the material being extruded so that it emerges from the die in a plastic state between the softening and melting temperatures.

An electric motor drives the screw via a gearbox and the motor amps must be monitored to prevent overload. The screw flight depth is larger in the first third of the screw length, decreases in the middle third to produce compression of about 2 to 4, and is again constant in the last third of its length. These sections are known as the feed, compression and metering zones respectively. The screw collects the pellets from the hopper and feeds them forward, melting and stirring them to homogenize the extrudate, any trapped air escaping backwards. Screw cooling, by means of water flowing inside it, helps in the stirring and homogenizing. Compression builds up the pressure and the metering section stabilizes the output flow. The ratio of the screw length L, to the barrel diameter D, is the L/D ratio of the screw, which differs for different materials. For PVC the L/D ratio is about 12 with a compression ratio of about 2.5 and for polythene it should be about 25 to 35 with a compression ratio of about 3.5.

At the end of the barrel is a perforated metal disc called the sieve plate which supports two or three metal gauze discs called the sieve pack. The purpose of these is to filter out any fines or mineral contamination from the melt which might cause weak spots in the insulation. The extrudate then enters the crosshead, sketched in Figure 2.3.

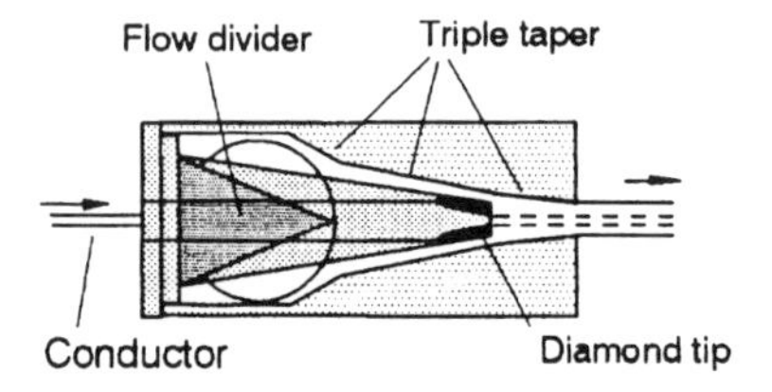

Figure 2.3
Crosshead die arrangement

The purpose of the crosshead is to turn and divide the flow of plastic material around the point through which the wire is passing. At the same time the area of the flow is reducing in order to accelerate the plastic up to the speed of the wire, through a triple-angle tapered die. The included angles of the die are about 10°, 15° and 30°. For some purposes there is a parallel section, or land, at the exit of the die but for insulation this tends to cause 'die drool' which is undesirable. For sheath extrusion the point can be adjusted to be just behind the die face, flush or protruding, depending on whether a pressure or tubing application of the plastic is required. For wire insulation the point should have a diamond tip to reduce wear.

Typical Extrusion Line

A sketch of a typical insulating extrusion line, arranged in tandem with a wire drawing machine, is shown in Figure 2.4.

Tandem Operation of Wire Drawing

The advantage of tandem operation of wire drawing and extrusion insulating is that, at 1000 m/min line speed, the mass of 0.5 mm wire insulated is about 0.2 kg/min or 12 kg/h. A 500 kg reel of 0.5 mm wire as the input would last about 42 hours, but the wire would have to be 'flown off' at 1000 m/min with serious risk of wire damage or breakage. Using a 1 tonne pak of 2.5 mm wire, the wire exits the pak at a lazy 40 m/min and the pak lasts for 80 hours before a new one needs to be joined on. Alternatively, depending on other limitations of the line, the speed can be increased to 2000 m/min for 42 hours and still be well within the limitations of the wire drawing machine. The details of the wire drawing operation up to in-line annealing are described in Chapter 1.

Preheater

The wire exits the annealer at about 30 to 40 °C. By heating the wire to about 80–90 °C before it enters the extruder crosshead, the plastic near the wire cools more slowly than the surface in the cooling trough. Thus the surface layers of plastic shrink the

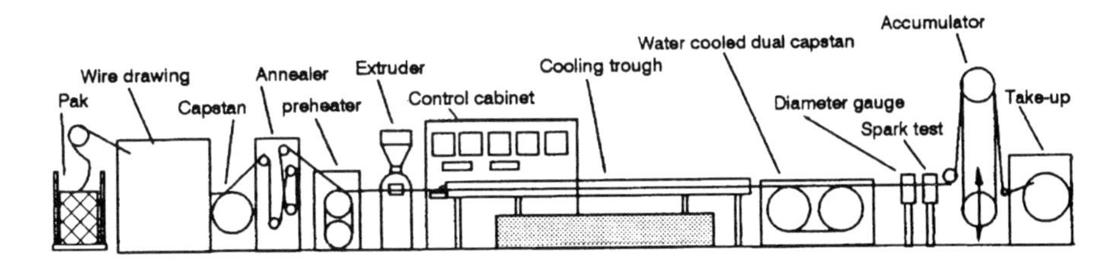

Figure 2.4
Thermoplastic insulation extrusion line with tandem wire drawing

insulation on to the wire to give a good grip which resists any tendency of the insulation to 'shrink back' along the wire when cut.

The preheater consists of two pulleys around which the wire travels for one or two turns. The upper pulley is earthed and a current is fed into the lower pulley which travels through the wire thus heating it. With care, neither of the pulleys need to be driven but it is safer to have a little driving torque applied to avoid wire stretch.

Cooling Trough

The faster the line speed and the thicker the insulation the longer it takes to cool and set the plastic. At 2000 m/min even a radial thickness of 0.5 mm of plastic requires about 10 m of trough containing water at 20 °C to set. At high speeds the insulated wire tends to tunnel through the water surrounded by hot air or vapour without efficient transfer of heat. For high-speed lines therefore it is preferable to use spray cooling of the insulation with the insulated wire passing through annular spray heads spaced down the first half of the trough. The sprays are pointed in the direction of travel to reduce any tension build-up which could cause wire stretch in the smaller wires. Even so a cooling length of 15 m is just about adequate. The water is pumped and recirculated from a holding tank beneath the trough and may need to be cooled. (This is especially so for high ambient temperatures and when cooling relatively large plastic sheaths).

The heat transferred from the preheater and the extruder to the cooling trough by an 0.5 mm wire insulated to 1.0 mm diameter with polythene at a line speed of 2000 m/min is about 6000 cal/min. This requires contact with 2 l/min of water to cool it, if the water is not to rise more than 3 °C/min.

Capstan

The line is provided with a dual capstan which is also spray cooled with water to augment the trough cooling. This capstan is the speed controlling device for the line and must be synchronized with the wire drawing capstan by means of a dancer-arm and servo control to maintain a low tension.

Diameter Control

After the capstan is situated an optical diameter gauge. This is frequently a scanning laser beam with means of measuring the occultation time due to the shadow of the insulated wire. Consequently the insulation diameter is measured extremely accurately without any contact. This information is displayed on the control cabinet of the extruder for manual control, or may be used in a servo-control loop. The line speed is generally maintained constant and the extruder screw speed is varied to change the diameter of the insulation. It is therefore important to know that the output of plastic from the extruder is reasonably linear with screw speed. Generally this is so up to a limiting speed, which depends on the diameter and *L/D* ratio of the screw, where the output tends to level off.

HV Spark Tester

Following the diameter gauge is a high voltage spark tester. This uses a ball-chain curtain through which the insulated wire passes. The ball-chains are maintained at a high voltage, generally 10 kV d.c., and any flaw in the insulation which causes a discharge is detected and counted. Such flaws might be mineral or metal dust inclusions, or splits or scrapes of the insulation that reveal bare wire. The source of these flaws must be identified and eliminated if the insulation process is to be viable.

Accumulator and Take-up

The purpose of the accumulator is to hold a sufficient length of insulated wire to enable the spool in the take-up to be automatically changed when it is full. It consists of two blocks of up to 15 free-running sheaves around which the insulated wire is looped, taking a length of up to 30 times the separation of the blocks. The bottom block of sheaves can move in the vertical direction. During normal running, the separation is small, say 1 m. During change-over of the take-up spools, the blocks separate as the cut wire end slows down. If the separation can rise to say 3 m (or an increase of 60 m of wire in the accumulator) there will be 1.8 s of running at 2000 m/min for the new spool to 'grab' the cut end and continue the take-up operation, without the line speed changing. The position of the bottom block determines the take-up spool speed, so that it can empty the accumulator before the next change-over is required. Modern high-speed take-ups using 500 mm spools can reliably effect the change-over well within 1 s. The average rotational speed of such a spool at 2000 m/min is 2300 r.p.m. and it could contain about 55 km of 0.5/1 mm insulated wire weighing 12 kg.

Cellular Insulations

For some purposes the permittivity of solid polythene even at 2.2 is too high. By mixing the polythene with gas bubbles, or cells, the permittivity can be lowered since the permittivity of gasses is approximately unity. Thus polythene containing 50% uniformly dispersed gas bubbles would have an effective permittivity of 1.6. The gas cells are required to be non-interconnecting.

This was attempted in the early days of extrusion by incorporating hydrazine, $(NH_2)_2$, which would decompose in the heat of the extruder (preferably in the high pressure metering zone) to produce hydrogen and nitrogen. These gases would dissolve in the molten plastic until the pressure was reduced at the point of the crosshead, when they would come out of solution to form cells of gas dispersed in the extrudate. Hydrazine turned out to be too vigorous and other formulations were developed. Some of these had the disadvantage of leaving residues which were almost as hard as diamond, causing the insulation to act like a saw, producing heavy abrasion of guide dies in twinning and stranding. The residues also degraded the loss angle of the dielectric. However excellent formulations with minimal side-effects are now available which can achieve over 60% blow (permittivity = 1.48).

Another way of producing cellular insulation is the gas injection method. In this method dry compressed air or nitrogen is injected into the barrel of the extruder near the end of the compression zone (a short decompression section may be built into the screw). The pressure must be sufficient for the gas to be dissolved and dispersed in the plastic. This method should be capable of greater control of the degree of blowing.

A third method of producing cellular polythene insulation was discovered inadvertently when trying to produce thin insulations of polythene by the enamelling process used for fine wire. The wire was passed through a solution of polythene in xylene or toluene followed by a sizing die and then heated to remove the solvent. However the solvent also vaporized within the coating, causing cells in the insulation, i.e. cellular polythene. Although the process was simpler than extrusion it was very slow and rather hazardous owing to the highly inflammatory solvents.

Foam Skin Insulation

Cellular polythene has a lower voltage withstand than solid polythene owing to the electric field variations at the cell walls. To overcome this, a thin skin of solid polythene can be extruded over the cellular insulation using an extra extruder in a tandem die arrangement. It is possible to keep the radial thickness of solid polythene below 20% of the total radial thickness of insulation to give minimal penalty to the permittivity and to improve the voltage withstand significantly.

Capacitance Monitor

The use of cellular or foam skin insulation introduces an extra design variable, the blowing, which must be carefully controlled. This is done by including a capacitance monitor in the extrusion line, immersed in the water at the end of the trough. In effect this uses the water in the cooling trough as a coaxial electrode around the insulation. The capacitance between this and the wire depends on the permittivity of the insulation (and its diameter). If the diameter is well controlled, the capacitance is a measure of the degree of blowing of the extrudate. As the diameter is also a function of the blowing (the diameter swells as it exits the die), balancing the two effects can be a delicate task. For chemical blowing, the degree of blow is primarily dependent on the temperature in the compression zone (governing the degree of dissociation of the blowing agent) and the diameter depends on the plastic flow rate, die size and how quickly the plastic insulation is solidified by cooling.

Semi-air Spaced Insulations

These are principally used for coaxial constructions. Even cellular polythene insulation has a limit in reducing the permittivity below 1.5. For even lower permittivity, constructions like 'thread and tube', spaced discs of polythene or 'balon' insulations are used. These are illustrated in Figure 2.5.

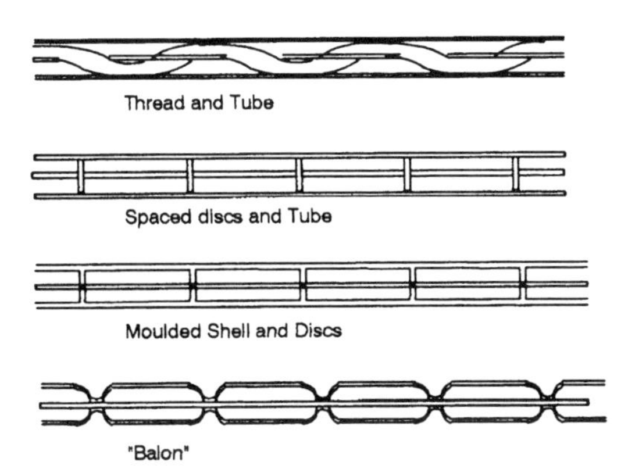

Figure 2.5
Semi-air-spaced insulations

The thread and tube insulation is essentially the same as the styroflex construction, but using polythene. The disc and tube type is constructed using slotted discs placed on to the wire, or by moulding the discs directly on to the wire using moving moulds on a caterpillar chain, followed by tape lapping. Each half of the moulded-shell type is hot-pressed into polythene tape and the two halves are hot-sealed around the wire. The 'balon' construction is made by extruding a polythene tube, with a small positive air pressure inside, around the wire, and pinching it on to the wire with a moving belt mould.

Coloured Insulation

For identification purposes it is required to produce coloured insulations. This is done by mixing the pellets of plastic with pellets containing a fairly high concentration of coloured pigments, known as colour masterbatch pellets. These pellets must use the same thermoplasic material as the base insulation. The pigments may be mineral or organic dyes. Since the radial thickness of insulation may be as small as 0.15 mm the particle size of mineral pigments must be very small e.g. $<1\ \mu m$, to avoid problems with HV breakdown. Hence organic dyes are preferable to mineral pigments but the opacity of dyes is less and higher concentrations may be required. Another drawback of mineral pigments is that some of them interfere seriously with the blowing of cellular polythene blowing agents.

The masterbatch pellets are mixed with the natural pellets at a ratio of about 1% before loading into the hopper. Alternatively a separate hopper may be used for the masterbatch with automatic feed in the correct proportion into the main hopper. Adequate mixing and dispersion by the extruder screw is necessary for uniform colouring.

Coloured PVC inks are available which adhere well to PVC insulation so that colour ring-marking in a separate process is possible, to give a greater variety of identifications. Inks do not generally adhere satisfactorily to polythene, but larger diameter cores can be printed on, using a heated type-wheel and a low-melting point polythene coated on a Mylar ribbon.

UV Protection

Ultraviolet radiation can degrade polymers such as pvc and polythene quite rapidly, causing them to lose flexibility and to become hard and brittle. This is due to free radicle formation which affects the polymerization bonds and also causes cross-linkages between the long-chain molecules. Thus for polymers exposed to UV radiation suitable protection methods are required. Organic materials which preferentially take up the free radicles can be incorporated in the polymer, but these are expensive. Normally only sheathing materials are exposed to significant levels of UV and these are generally protected by incorporating finely divided carbon black into the sheathing polymer. Since the particle size of the carbon is required to be less than 0.5 μm and well dispersed, this is a rather dirty business, and most cable manufacturers buy the sheathing polymers ready mixed with carbon black to the required standards.

HIGH VOLTAGE WITHSTAND

It is the voltage gradient with distance, i.e. the electric field, which determines the voltage withstand of the insulation. For a flat electrode configuration the field is simply the voltage on the electrodes divided by the distance between them, but for cylindrical conductors the field is strongest at the surface of the conductor and is given by

for coaxials

$$E = \frac{2V}{d \ln(D/d)} \text{ V/mm} \tag{2.1}$$

for parallel wires

$$E = \frac{2V}{d \cosh^{-1}(D/d)} = \frac{2V}{d \ln[D/d\{1 + \sqrt{1 - (d/D)^2}\}]} \text{ V/mm} \tag{2.2}$$

where D is the diameter of the outer electrode or the centre separation of the wires in mm, d is the diameter of the inner electrode or the wires in mm and V is the voltage applied.

When this field exceeds a critical field for the insulation, electrons are torn from the molecules of the insulation and accelerated towards the positive electrode. In their travel they tear off more electrons eventually causing an avalanche of electrons and producing a spark. This avalanche of electrons degrades the insulation along the path of the discharge leaving permanent damage which reduces the insulation resistance of the dielectric.

The critical field for breakdown to occur in gas is given by an empirical expression [1]

$$E_c = A\rho\left(1 + \frac{B}{\sqrt{\rho \cdot r}}\right) \tag{2.3}$$

where A and B are constants, ρ is the density of the gas and r is the radius of the cylindrical electrode. For air at 20 °C, $A\rho$ is stated to be approximately equal to 30 kV/cm and $B/\sqrt{\rho}$ to be approximately 0.3. Thus for $d = 0.127$ cm, $r = 0.0635$ cm and

$E_c = 66$ kV d.c./cm or 6.6 kV d.c./mm. Some measurements were made to determine the voltage breakdown at 50 Hz a.c., between two 1.27 mm diameter wires in air and are shown in Figure 2.6.

The measured values fall closely on the curve

$$V_b = 2\cosh^{-1}(D/d)\,\text{kV r.m.s.} = 2\sqrt{2}\cosh^{-1}(D/d)\,\text{kV d.c.} \tag{2.4}$$

If this is substituted for V in Equation (2.2) we get

$$E_c = \frac{2 \times 2\sqrt{2}}{d} = \frac{5.66}{d}\,\text{kV d.c./mm} = 4.45\,\text{kV d.c./mm} \tag{2.5}$$

Considering the difficulty of the measurement, the probability of transient spikes on the 50 Hz supply and the possible effects of humidity, this half-order of magnitude agreement with the value calculated from Equation (2.3) is satisfactory. Under the same conditions the breakdown between two 0.9 mm wires would be as shown in Figure 2.6.

Applying Equations (2.2) and (2.5) to a 0.9 mm paper string and tape insulated conductor (diameter over insulation 1.5 mm), the breakdown voltage per conductor, ignoring the paper insulation, is

$$V = \frac{E_c}{2}\,d\cosh^{-1}(D/d) = \frac{4.45}{2} \times 0.9 \times 1.099 = 2.20\,\text{kV d.c.}$$

Actual measurements on cables gave a mean value of 2.2 kV d.c. and a minimum value of 2.1 kV d.c. This is more in line with Equation (2.5) than Equation (2.3). Breakdown in a paper insulated cable tends to punch tiny holes in the paper but does not degrade its insulation resistance.

The breakdown field quoted in the literature for polythene is >30 kV/mm. (3/16″ sample for short time). If there is a void of lower permittivity on the radius considered, the voltage profile will have a step in it since the field is inversely proportional to the

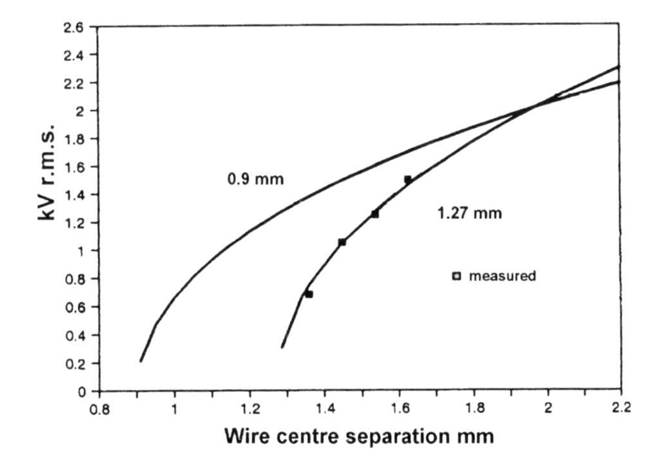

Figure 2.6
Voltage breakdown between wires in air

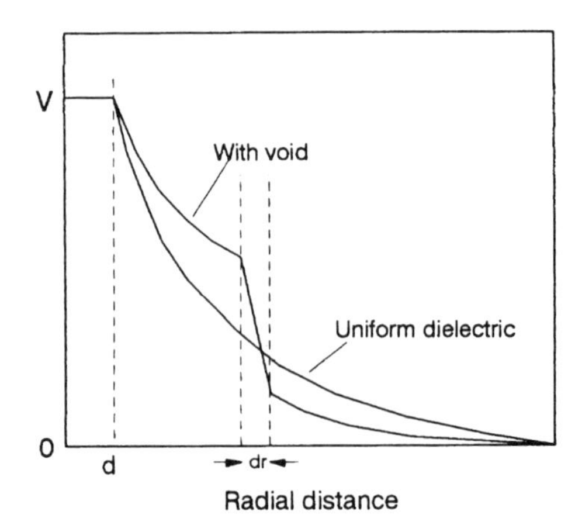

Figure 2.7
Radial voltage profile for insulation with void

permittivity. This is illustrated in Figure 2.7. Thus the field in the void is greater than the field in the dielectric. If this field exceeds the ionization potential of the gas in the void, discharges occur in the void which degrade the insulation around it. This (conductive) degradation grows into the insulation in the form of 'trees' until the field in the insulation also exceeds the breakdown level. Thus the breakdown in the void starts first, at a lower applied voltage than would be required to achieve avalanche conditions in the dielectric. This fact can be used for non-destructive testing for voids, using a high-impedance ionization detector.

Hence the voltage breakdown of insulation is not instantaneous but depends on the build-up of avalanche conditions in the dielectric. The presence of voids decreases this delay. Typical voltage versus minimum time-to-breakdown curves for polythene insulation are shown in Figure 2.8.

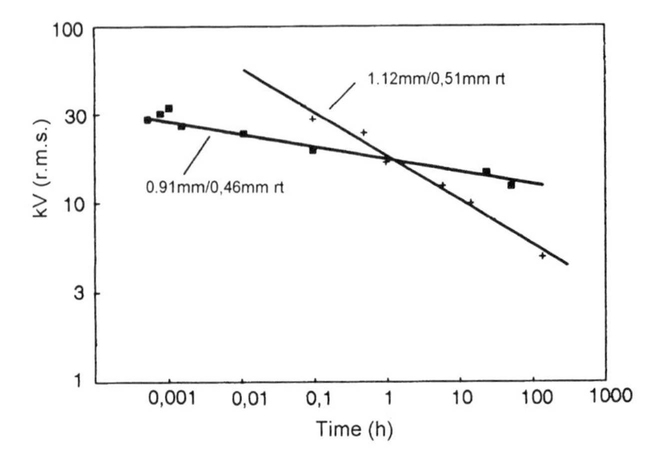

Figure 2.8
Voltage versus Time-to-breakdown in polythene-insulated cables (Reproduced by permission of BICC Cables)

These curves become straight lines on log–log axes over many decades of the time axis. The applied voltage was a.c. at 50 Hz and it was applied between adjacent wires in quads, i.e. a to c, or b to d, in a multipair telephone cable. The 0.9 mm cable was tested in short lengths of about 1.25 m, while the 1.12 mm cables were tested in long lengths. The larger slope of the curve for the longer cables is possibly due to the increased probability of voids being present. The *mean* time to breakdown in both cases also gives a good straight line at about twice the voltage for a given time.

The minimum breakdown voltage for a 5 minute test on the 0.9 mm samples is 21 kV r.m.s. or 30 kV d.c. Applying Equation (2.2) gives a critical field of 49 kV d.c./mm for the polythene insulation which is consistent with the value quoted in the literature.

INSULATION RESISTANCE

The same field configuration around the conductors that governs the breakdown voltage also governs current leakage through the insulation. This effect establishes the insulation resistance between the conductors. Interpreting the terms in Equations (2.1) and (2.2) as the effective separation of the electrodes, then for 1000 m length

$$R = \frac{\rho l}{A} = R_i = \frac{\rho_i d \cdot \ln(D/d)}{2d \times 10^3} = \rho_i \frac{\ln(D/d)}{2 \times 10^3}\ \Omega \cdot \mathrm{km} \tag{2.6}$$

$$R = \frac{\rho l}{A} = R_i = \frac{\rho_i d \cdot \cosh^{-1}(D/d)}{2d \times 10^3} = \rho_i \frac{\cosh^{-1}(D/d)}{2 \times 10^3}\ \Omega \cdot \mathrm{km} \tag{2.7}$$

for coaxial and parallel wires respectively. ρ_i is the bulk resistivity of the insulating material and R_i is the insulation resistance. The bulk resistivity in Table 2.1 is given in $\mathrm{T}\Omega \cdot \mathrm{m}$ which is $10^{12}\,\Omega \cdot \mathrm{m}$ Thus for parallel 0.9 mm conductors insulated to 1.8 mm diameter with polyethylene,

$$\cosh^{-1}(D/d) = \ln(D/d\{1 + \sqrt{1 - (d/D)^2}\}) = 1.32 \quad \text{and} \quad \rho_i \geq 100 \times 10^{12}$$

hence

$$R_i \geq 10^{14} \times \frac{1.32}{2} \times 10^{-3} \times 10^{-6} = 66\,000\ \mathrm{M}\Omega \cdot \mathrm{km}$$

This is compatible with specificatons which call for a minimum insulation resistance at 500 V, after steady electrification, to be greater than 40 000. Steady electrification is achieved when the capacitance between the conductors is fully charged to the test voltage. As the test voltage is applied through a high resistance for safety reasons, this takes a few seconds, depending on the capacitance.

For PVC insulation, the bulk resistivity varies according to the compounding, from $10\,\mathrm{T}\Omega \cdot \mathrm{m}$ to $0.001\,\mathrm{T}\Omega \cdot \mathrm{m}$. For hard grade insulation a figure of $0.1\,\mathrm{T}\Omega \cdot \mathrm{m}$ would give an insulation resistance of about $66\,\mathrm{M}\Omega \cdot \mathrm{km}$ which is compatible with specifications calling for $20\,\mathrm{M}\Omega \cdot \mathrm{km}$.

Paper insulations when properly dried can achieve insulation resistances of $15\,000\,\mathrm{M}\Omega \cdot \mathrm{km}$ at a temperature of 20 °C, but they vary with temperature of measurement by ±30% per ±5 °C.

Insulation Defects

Most telecommunication designs call for the ratio D/d to be between 1.6 and 2.3 depending on the capacitance required. This ratio needs to be achieved accurately to within about 2%. Not only does this require care in die maintainence but also in extruder running conditions. Also for small wire diameters, e.g. 0.32 mm to 0.5 mm, the radial thickness of the insulation is only a few tenths of a millimetre. Consequently it is easily damaged during manufacture, either by scrapes or splits. The former are caused by running over sharp edges and the latter by squeezing the insulation through too tight dies or its being forced against smooth surfaces, for example by centrifugal force during high-speed twinning. All dies in plastic cable manufacture should be regarded as guides only, as opposed to paper insulation where dies are frequently used for sizing insulation or cables.

Such splits and scrapes cause HV failure on testing due to the exposure of bare conductor, and are frequently referred to as pinhole faults. Actual pinhole faults are rare and can usually be traced to solid inclusions contaminating the plastic, either from metallic or mineral dust or crystalline mineral contamination from colour masterbatches (it pays to examine extruder sieve packs microscopically).

3

Twisting and the Helix

INTRODUCTION

By definition, a cable is a collection of long filaments laid together so that the assembly can be handled conveniently. In this context, filaments may be wires, insulated wires, pairs, coaxial tubes, optical fibres etc. The cable should have sufficient strength and flexibility for its purpose. By far the most common way to achieve this is to twist the filaments together to form a collection of helices. This not only forms a compact cable in cross-section but also gives flexibility, so that when the cable is bent the portion on the outside of the bend draws the necessary extra length of filaments from the inside of the bend. This implies that the cable should not be so compacted that the filaments cannot move relative to each other. On the other hand, too loose a cable will easily deform or flatten when bent or compressed. To achieve good circularity and flexibility requires a short pitch of the cable helices. For a good mechanical design of cable therefore it is necassary to understand the geometrical properties of a helix and the ways in which filaments can be twisted together. The mechanical design and the electrical or optical requirements will also react together and may produce limitations, the one on the other.

Take-up Factor

Figure 3.1 shows a section of a layer of cabled filaments with a mean diameter D. The same filament reappears at the same circumferential position at regular intervals along the cable length. This distance is the pitch of the helix, P, and is known in cable parlance as the *lay length*. If a filament is cut at point B in the diagram and unwound back to point A, the cut end will unwind by one circumference of the cable πD, and the axial length unwound is one lay length P. Thus by Pythagoras, the ratio of the filament length to the axial length is

$$T = \frac{L}{L_0} = \sqrt{1 + \left(\frac{\pi D}{P}\right)^2} \tag{3.1}$$

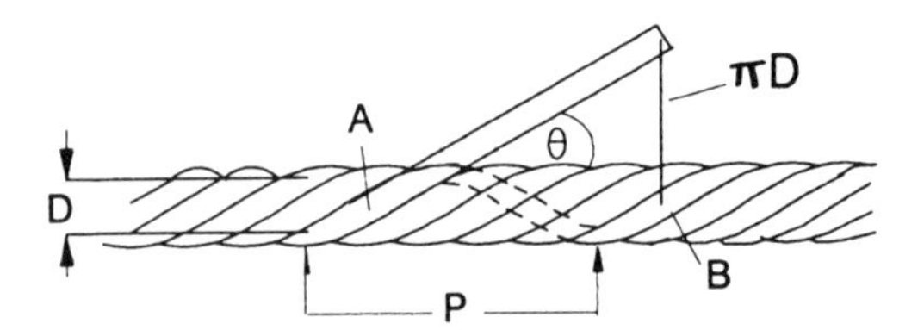

Figure 3.1
Cable take-up factor

T is called the take-up factor of the cable layer. Thus in a cable where the lay length is 10 times the mean diameter, the take-up factor will be 1.05 or 5%, hence each filament would need to be 1050 m to make a 1000 m length of cable. For a lay length 15 times the mean diameter, the take-up factor is 1.02 or 2%.

Curvature of Filament in a Helix

When a filament of diameter d is formed into a helix, each small element of the filament adopts a curvature as shown in Figure 3.2. This imposes a bending stress on the filament, the outside of the curvature being in tension and the inside in compression.

For a mean helix diameter D, the radius of curvature is

$$r = \frac{D}{2}\left\{1 + \left(\frac{P}{\pi D}\right)^2\right\} \tag{3.2}$$

The resultant strain on the filament is the relative extension of the outer surface of the filament compared to its axis

$$s = \frac{r + \frac{d}{2}}{r} \tag{3.3}$$

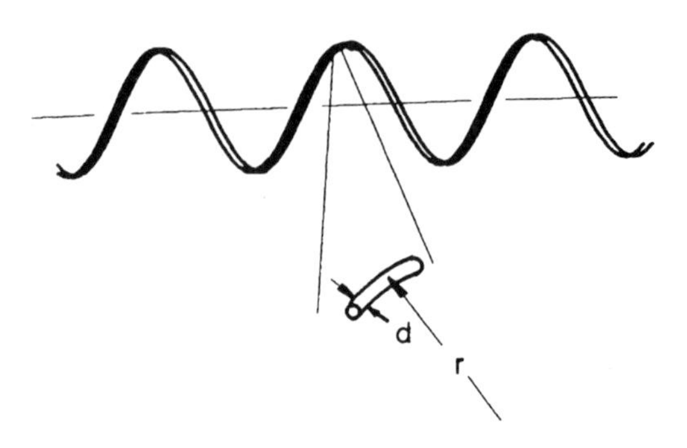

Figure 3.2
Curvature of filament in helix

For example, a layer of 12 optical fibres, each of diameter 0.125 mm and buffered to 1 mm diameter, would have a mean diameter $D = 3.9$ mm. The lay length of the layer is $P = 50$ mm. Thus $T = 1.03$ and the radius of curvature of the fibres is $r = 34.4$ mm. The strain on the outside of the fibres will be

$$s = \frac{34.4 + 0.125/2}{34.4} = 1.0018 \text{ or } 0.18\%$$

For a metal wire this strain would have very little consequence, but for an optical fibre it has significance in relation to its static fatigue lifetime. For a lifetime of 30 years, the fibres in this situation should have been proofstrain tested to $3 \times 0.18\% = 0.54\%$.

Torsion in a Filament

When a filament is formed into a helix it also twists about its own axis. If this twist is not relieved during the manufacturing process it results in torsional strain in the filament. For a rigid filament like a coaxial tube or steel wire this torsional strain can cause undesirable buckling. For an optical fibre this is an extra strain which can shorten its lifetime.

To avoid such twisting, the filament pay-off reel should be rotated about the axis of the filament during the formation of the helix. The required amount of detorsion is

$$S = 100 \cos\theta = \frac{100}{T}\% \tag{3.4}$$

where θ is the angle of the filament to the helix axis and T is the take-up factor from Equation (3.1). $S = 100\%$ corresponds to one complete back twist (360°) per lay length.

The optical fibre in the previous example, in the absence of detorsion, would have a torsional strain of

$$S = \frac{100}{1.03} = 97\%$$

If the strander forming the cable applied 100% detorsion to each fibre spool, it would overcompensate and leave a net torsional strain of 3% in each fibre. This would be unacceptable for fibre lifetime. The strander should thus have the facility of adjusting the detorsion with some precision (within 0.1%) to 97%.

Effects of Unrelieved Strains

The effects of unrelieved bend and torsional strains can be most dramatically seen in a stranded steel wire cable. Unless the ends of the cable are strongly bound, the individual wires spring out violently from each other. The use of 100% detorsion is usually practised in manufacture, even so the the net torsional strain can be significant.

The bend strain can be removed by preforming the individual wires during stranding by bending the individual wires with 'bend rollers' before bringing them together at the stranding point. In this way a 'dead' strand can be formed which does not unwind when cut. Also when armouring cables with steel wires, it is essential to use detorsioning pay-offs; even so the bend strain can cause 'bird-caging' of the wires when the cable is bent, unless the lay length is properly adjusted.

If a pair or quad with a lay length of P_1 is stranded into a cable with lay length P_2 without detorsion, an extra 360° twist is given to the pair/quad for each stranding lay. This modifies the the pair/quad lay to

$$\frac{1}{P} = \frac{1}{P_1} \pm \frac{1}{P_2} \tag{3.5}$$

This must be allowed for when designing lay schemes for the reduction of crosstalk, and also when designing for resistance, attenuation etc. per unit cable length. The positive sign in (3.5) is used if the twists are in the same direction (same 'hand'), the negative sign for opposite hand.

Stretching the Helix (Coil Spring)

A common example of the helix is the coil spring. The behaviour of the coil spring when stretched is typical of all helices and applies to filaments stranded into cables. A force is required to stretch the helix a given axial distance and when it stretches the mean diameter changes depending on whether the helix coils or uncoils under tension. To stretch the axial length L_0 of the helix by an amount δL requires a force W N and

$$\frac{\delta L}{L_0} = \frac{8WD^2}{\pi d^4} \left\{ \frac{2\sin^2\alpha}{E} + \frac{\cos^2\alpha}{N} \right\} \tag{3.6}$$

where α is the angle shown in Figure 3.3, E is Young's tensile modulus and N is the rigidity modulus (see Table 1.1).

The angle through which the end of the spring moves in the plane normal to the axis is

$$\beta = \frac{16WDL \sin\alpha\cos\alpha}{\pi d^4} \left(\frac{1}{N} - \frac{2}{E} \right) \tag{3.7}$$

Since in most cases $E > 2N$, the spring tends to coil up when stretched and D gets smaller. The reduction δL, in diameter, is related to the stretch by

$$\frac{\delta L}{L_0} = \left(\frac{\pi}{P} \right)^2 \cdot \delta D \cdot \left(D - \frac{\delta D}{2} \right) \tag{3.8}$$

An example of the application of Equation (3.8) is in the determination of how far a loose-tubed fibre cable can be stretched without straining the fibres. The cross-section of an 8-fibre loose-tubed cable is shown in Figure 3.4. Initially the 0.125 mm diameter fibres are assumed to lie along the axes of the helical tubes which have a lay length of 100 mm

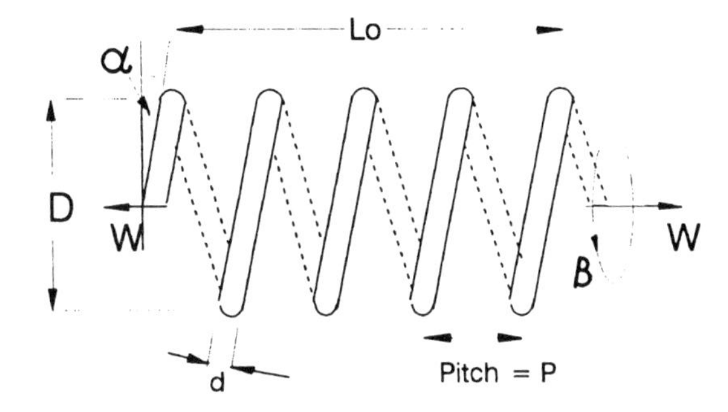

Figure 3.3
The coil spring

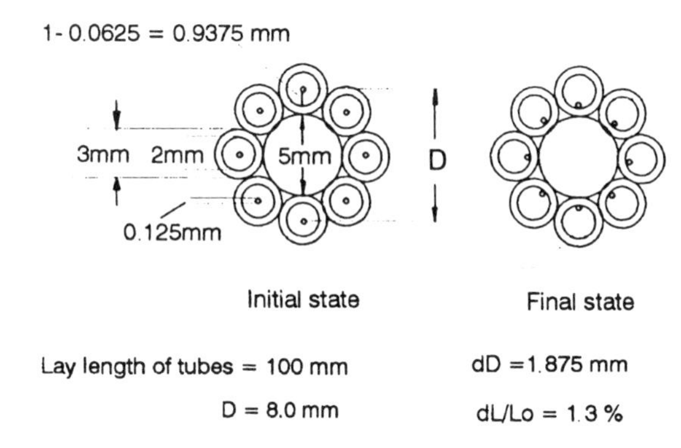

Figure 3.4
Stretching of loose-tubed optical cable

around a centre member of 5 mm diameter. Each tube has an outside diameter of 3 mm and an internal diameter of 2 mm. After stretching, all the fibre helix diameters have decreased until the fibres just touch the inside of the tubes, as shown. Application of Equation (3.8) shows that $\delta L/L_0 = 0.013$. Application of Equations (3.2) and (3.3) to the initial and final states shows that the fibre bend strain decreases from 0.1% to 0.08%.

Tape Lapping

An obvious application of the helix is the tape lapping of cables. Such lappings can be of paper or plastic tapes or woven fabric tapes. The purpose of lapping is primarily to compact the cable but they can also serve as heat barriers to protect the cable from the heat of extruded plastic sheaths. Open helical lappings of coloured or printed tapes may also be used to separate and identify layers or bundles within the cable. Another application is in the steel tape armouring of cables for protection or for electromagnetic screening. The aim in all such lappings is to achieve as smooth an application of the tapes as possible with the required gap or overlap of the tape edges. For this, the tape width, lapping angle and lay length (defined in Figure 3.5) must satisfy the following equations

$$\tan\theta = \frac{\pi D}{P} \qquad \sin\theta = \frac{w_0}{P} \qquad \cos\theta - \frac{w_0}{\pi D} \tag{3.9}$$

$$\text{Tape width} = w = w_0 \pm (\% \text{ overlap or } \% \text{ gap}) \tag{3.10}$$

w_0 is the width of tape which would achieve perfect edge-to-edge lapping of the tape.

In the case of overlapped tapes, the overlapped portion must stretch to obtain a smooth application. Sufficient tension must be applied, and the tape material must have sufficient tensile elongation to cause this to happen, otherwise tape wrinkling will occur.

$$\text{for the tape angle set to } \theta = 50^\circ \ldots P = \frac{\pi D}{1.2} \ldots \text{from Equation (3.9)}$$

$$\begin{aligned} \text{then for overlaps of} \ldots \; & 0\% \ldots w = \pi D/1.56 \\ & 10\% \ldots w = \pi D/1.4 \\ & 30\% \ldots w = \pi D/1.2 \\ & 50\% \ldots w = \pi D \end{aligned} \tag{3.11}$$

As an example of tape lapping, consider the application of a paper tape with 10% overlap to a 30 mm diameter cable. Set the lapping angle to 50° and apply Equations (3.11). P becomes 78.5 mm and $w = 67.3$ mm ($w_0 = 61$ mm). A frequent limitation in lapping is the required rotational speed of the lapping head. In this case for a desired line speed of 30 m/min, the head must rotate at $30/P = 30/0.0785 = 382$ r.p.m. If this is beyond the capabilities of the head, the use of two pads on opposite sides of the cable, each applying a tape of width $w/2$ at the same angle but with a lay length of $2P$, would halve the rotational speed to 191 r.p.m.

To maintain flexibility of the cable, steel tapes must be applied with a gap of about 20%. In order to obtain full cover, two layers of tape are applied from each lapping head with application points on opposite sides of the cable. This gives a balanced application of the tape tensions. For a 60 mm diameter cable and a 20% gap, $\theta = 55°$

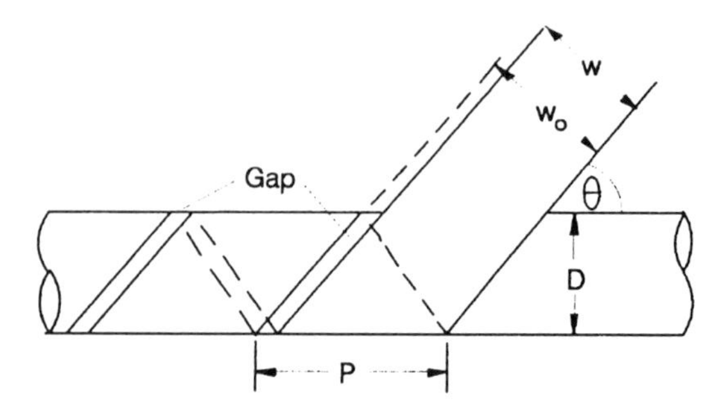

Figure 3.5
Tape lapping

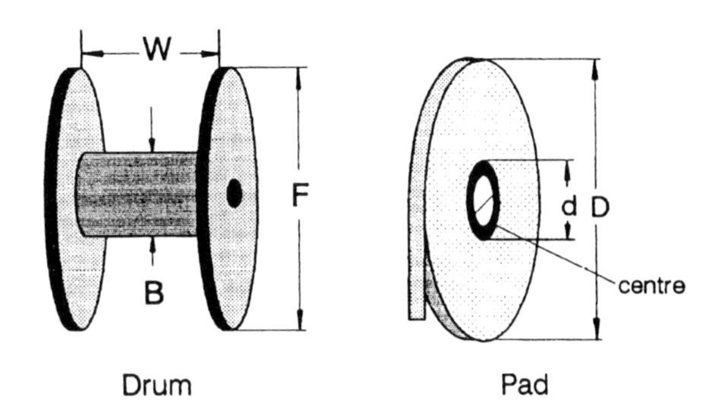

Figure 3.6
Drum and pad capacities

and Equation (3.9) gives $P = 132\,\text{mm}$. The tape width $w = (108\,\text{mm} - 20\%)/2 = 43.2\,\text{mm}$. For a tape head speed limited to 30 r.p.m. the line speed cannot be greater than $2P/30 = 8.8\,\text{m/min}$.

Drum and Pad Capacities

A winding on a drum is an example of multiple helices. The length of cable that can be accomodated on a drum is calculated from the (mean turn length) × (number of turns/layer) × (number of layers). For a perfect winding of a cable with diameter D, this leads to

$$L_{\max} = \frac{\pi W}{2} \cdot \frac{F^2 - B^2}{D^2} \tag{3.12}$$

where the dimensions are shown in Figure 3.6. For a less than perfect winding, and for a necessary clearance between the outside of the winding and the flange diameter, suitable adjustments must be made to the effective dimensions used in Equation (3.12). As a minimum, one diameter should be subtracted from W and $2D$ from F. The winding on a pad is not a helix but a spiral. The length of tape on a pad can be calculated by

$$L = \frac{\pi}{2t}(D^2 - d^2)$$

PRINCIPLES OF TWISTING

Basic 'fixed motion' twisting is shown in Figure 3.7. The supply spools are mounted in a rotating carriage. The 'filaments' are drawn off through a lay plate which defines their position and prevents crossovers. The filaments are brought together by a bell-mouthed die. The line speed is fixed by a capstan and the twisted filaments are evenly

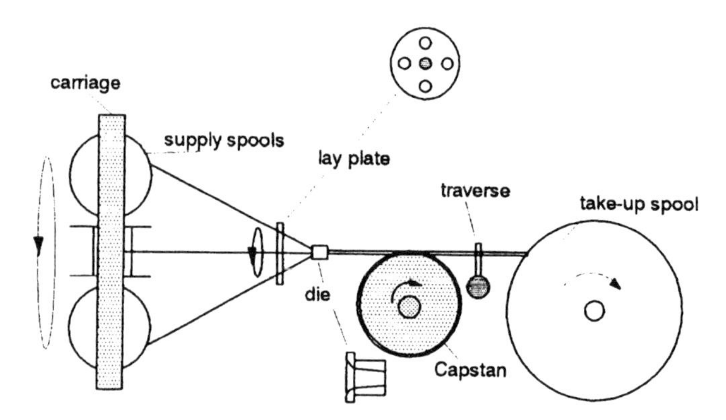

Figure 3.7
Basic fixed motion twisting

laid by a traversing fork on to the take-up spool. The disadvantages of fixed motion twisting are that the filaments are also twisted on their axes, producing torsion, and the speed is limited by the inertia of the carriage and supply spools. Consequently for short lays the rate of production is slow. By mounting the supply spools in forks on the carriage, as shown in Figure 3.8, and driving them by sun-and-planet gearing, the supply spools can be kept in a fixed orientation. This eliminates the twist in the filaments and is called detorsioning.

By changing the carriage into a cage and mounting the supply spools in line as shown in Figure 3.9(a), the inertia can be reduced and rotational and line speeds increased. Also the capstan can be eliminated if linearly varying lays can be tolerated. Alternatively the take-up spool rotational speed can be decreased automatically by using a winding diameter sensor. Another way of reducing the inertia of the rotating parts is to keep the supply spools fixed and to rotate the take-up spool on two axes as shown in Figure 3.9(b). This has the further advantage of enabling sophisticated tension control devices to be applied to each of the supply spools. Both (a) and (b) still

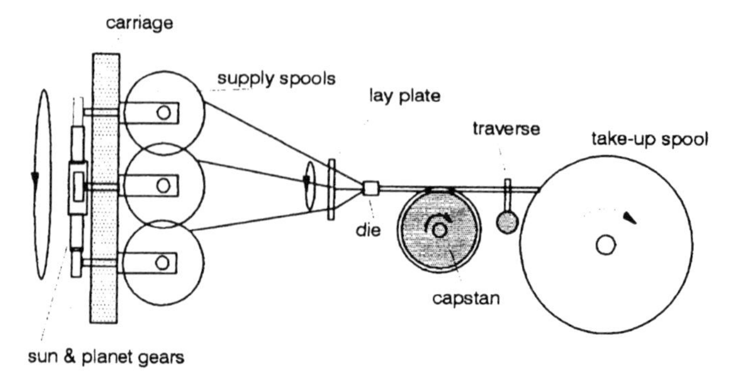

Figure 3.8
Twisting and detorsioning

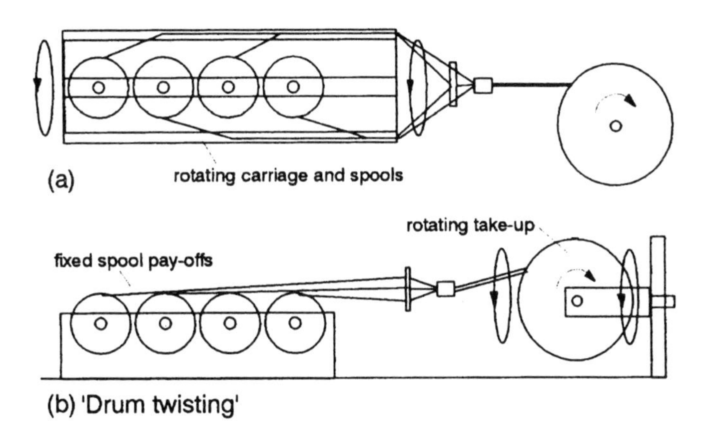

Figure 3.9
Reduced inertia fixed motion twisting

give torsion in the filaments, but this is relatively unimportant for copper conductors or cables. Method (b) became widely used for cabling 'units' of pairs, and for stranding these units into cables. As the take-up spool now became a sizeable drum, the method became known as 'drum-twisting'.

The ultimate in reducing the inertia of rotating parts is illustrated in Figure 3.10. In these arrangements both the supply spools and the take-up spools stay stationary. The input filaments are led into a die, around a 'bow', (which may or may not have guide pulleys mounted in it), and are led out onto the take-up spool. If the exit angle of the twisted filaments is at 90° to the bow axis as in (a), there will be one twist per bow rotation. This is a single twist device. In this case the take-up spool axis is parallel to the bow axis and is reciprocated to give even laying on the take-up spool. If the exit

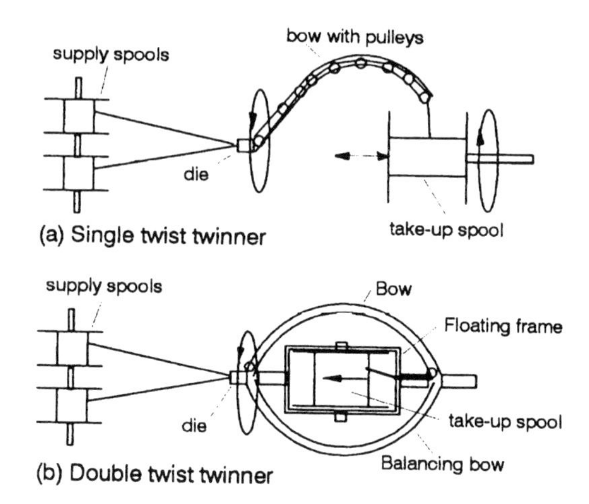

Figure 3.10
Rotating bow twisting

angle is at 180° to the bow axis as in (b), there will be two twists per bow rotation (thus doubling the speed of production). This is a double twist device. In this case the bow motion encloses the take-up which floats, 'gimbal' fashion under gravity, in a frame which also carries the take-up drive motor and a traversing fork. The frame may also carry a capstan.

In theory the tension in the filaments, the bow speed, size and shape could be arranged so that the filaments follow a catenary-shaped bow. In practice this is only partially achieved. Bow rotational speeds for twinning are frequently as high as 1500 r.p.m. (giving 3000 twists per minute)

In order for the twisting of two filaments into a pair to give equal twist to each leg (a perfectly balanced twin), the input tensions of the filaments should be closely controlled to be equal. DC servo control with dancer-arms is the usual method used, this also facilitates the ramping-up of the speed at start-up.

Another common use of the double twist method is in the manufacture of stranded wire conductors. In these the lay is relatively short (about 15 × the wire diameter), so obtaining good flexibility and production speed is very important. For this wire bunching the input die should be a crictically sized diamond die.

In the ultimate application, the Western Electric Co. also applied this bow twisting to unit making and unit cabling, utilising take-up drums up to 2 m in diameter.

Oscillating Lays

In the 1950s the Japanese introduced the concept of *SZ* twinning and cabling. In this method of providing flexibility and crosstalk reduction, the filaments are first twisted with a right-hand lay for several revolutions, followed by an equal number of twists in the left-hand direction. Various mechanical arrangements can be used to do this. In general the method relies on the oscillating lays being 'trapped' in some way, e.g. by binding with a lapping tape, before the line tension pulls the cable straight. This method was largely only practised in Japan, although the UK developed a fairly successful method using a series of oscillating lay plates, of rotational angle $n \times \theta$, where n was the lay plate number and θ was the angle of rotation of the first lay plate, usually about 300°. A string- or tape-binding was used after the oscillating head. This was used to construct up to 30-pair cables in tandem with a longitudinal paper lapping head and a sheathing extruder.

4

Cable Geometry

The designing of a cable involves specifying the wire and insulation diameters necessary to achieve the required transmission parameters. The details of how this is done will be covered in Chapters 6 and 7, but basically, the larger the wire diameter and the greater the ratio of the insulation diameter to the wire diameter, the lower will be the attenuation of the signal and the higher will be the characteristic impedance. Having set these two diameters, the physical dimensions of the cable follow from the number of physical circuits included and the manner of laying them up together to form the cable.

PAIRS, TRIPLES AND QUADS

The elementary unit of a telecommunication cable is generally two insulated wires twisted uniformly together to form a balanced pair. By twisting four insulated wires together, sometimes using a centre string or filament to improve the mechanical uniformity, a quad is formed. By using the opposite wires of the cross-section as the two elements forming a pair, the overall diameter, now containing two pairs, is smaller than two individual pairs twisted together and so is the capacitance and attenuation of the pairs for the same diameter ratio.

If the wire diameter is d, and the insulation diameter is D, a pair of two of these will have a diameter, i.e. will pass through a die of diameter, of $2D$, and a quad of four of these insulated wires will have a diameter of $(1 + \sqrt{2}) = 2.414D$ which is 60% that of two individual pairs laid together. However because the diameter of a cable made of pairs or quads depends on how they nest together, the diameter saving in the cable is not so large and is roughly 30%. But the capacitance will be about 18% less and the attenuation 9% less for the same ratio of D/d. Alternatively for the same capacitance and attenuation, the ratio D/d can be 20% less for quad formation, giving a diameter reduction of nearly 43% compared with pair construction.

For this reason quads were much favoured in the early days of telephone cables but as multichannel (frequency multiplexed) working was introduced and frequencies of transmission were increased through 108 kHz and 252 kHz up to 512 kHz it became increasingly difficult to control the crosstalk between the pairs within the quad. Also as

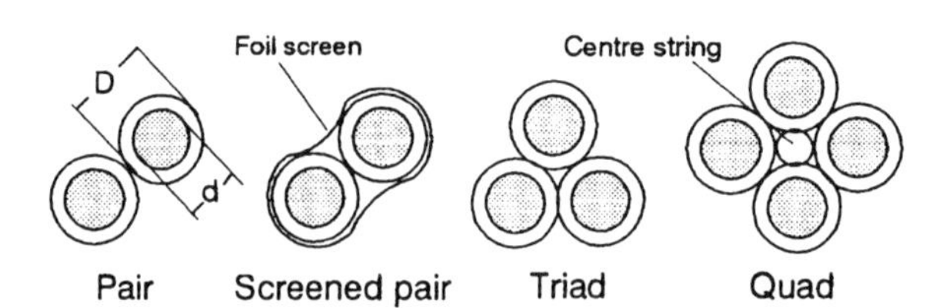

Figure 4.1
Cabling elements

the number of channels per pair increased, fewer pairs were required and diameter considerations became less important. Thus quad cables are not used to such a large extent today.

Triples, or triads, are not used for telecommunication *per se,* but they are sometimes included in combined telecommunication and instrumentation cables for three-wire instrumentation purposes. Triples require a die diameter of 2.155*D*.

All three elements can be screened with aluminium foil/polyester tape laminate to produce individually screened pairs, triples or quads with improved crosstalk between the individual elements. This may be done by a separate lapping process or the screen can be applied at the same time as twisting the element which is more economical of processing time. This increases the diameter by a minimal increment of 0.05 mm. Sometimes a stranded drain wire is included to facilitate connection to the screen. The diameter of the element in relation to the insulation diameter is given below

Element	Pair	Triple	Quad	Quad centre string
Diameter	2*D*	2.155*D*	2.414*D*	0.414*D*

Pairs are generally made on high-speed double twist bow twinners as illustrated in Figure 3.10. Triples generally have heavier conductors and are made on rotating carriage machines. Quads intended for trunk and junction service need precision twisting and are made on detorsioning twisters, Figure 3.8, but for distribution quads, double twist non-detorsioning bow machines are adequate and faster.

Semi-rigid Coaxials

These were used extensively for multichannel telephony up to 12 MHz or higher, until they were superseded by optical fibres. Their manufacture is now obsolete, however many routes are still operating over these semi-rigid coaxials in many countries. The coaxial element consisted of an inner conductor with a low-permittivity semi-air spaced insulation such as illustrated in Figure 2.5, with a copper tape outer conductor formed around it, followed by lappings of two thin steel tapes and an overall lapping of paper tape. The copper tape was slit to a precise width to match the desired diameter after forming.

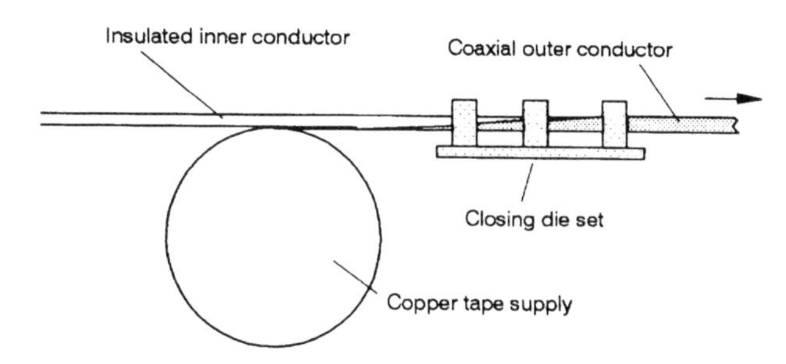

Figure 4.2
Coaxial conductor formation

The forming from flat tape to cylindrical conductor was done by a series of dies of diminishing diameter as shown in Figure 4.2. If the tape was overlapped or gapped, the diameter would be too dependent on the insulation diameter, so a butting-edge formation was generally aimed for. This was often assisted by corrugating the tape edges to increase the effective thickness of the tape edge so preventing overlapping. When stranding these semi-rigid coaxials into cables it was essential to use detorsioning stranders.

CONCENTRIC STRANDING OF CABLES AND DIAMETERS

Stranding of Cables

Cables are constructed by stranding the cabling elements together in concentric layers around a centre group of 1, 2, 3 or 4 similar elements. Each layer can accomodate 6 elements more (strictly 2π elements more) than the layer beneath it, except in the case of a single element centre where geometry dictates that the first layer contains 6 elements (i.e. an increment of 5). These 'perfect' lay-ups are shown in Table 4.1 where

Table 4.1
Perfect concentric layer construction

Centre	Bay 1	Bay 2	Bay 3	Bay 4	Bay 5	Bay 6	Bay 7	Bay 8
1	6	12	18	24	30	36	42	48
	7	*19*	*37*	*61*	*91*	*127*	*169*	*217*
2	8	14	20	26	32	38	44	50
	10	*24*	*44*	*70*	*102*	*140*	*184*	*234*
3	9	15	21	27	33	39	45	51
	12	*27*	*48*	*75*	*108*	*147*	*192*	*243*
4	10	16	22	28	34	40	46	52
	14	*30*	*52*	*80*	*114*	*154*	*200*	*252*

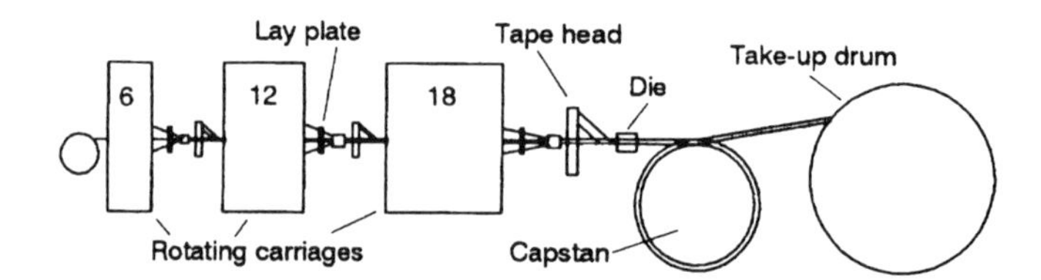

Figure 4.3
3-bay strander

the total number of elements is shown in italics. Because the increment is actually 2π, every fourth or fifth layer, depending on the lay length, can accept an extra element in addition to the integer increment of 6.

The reels of elements (pairs, quads etc) are mounted in rotating carriages, or 'bays', of the strander. The centre group can often be run into the strander without a twist. For the most flexible use of the machine therefore it would be designed according to a 4-element centre but the increment of six leads to a hexagonal rotational symmetry of the carriages. Machines with 2, 3, 5 and 8 bays are usual. For even larger cables it is possible to make them in two operations, combining the bays for the second operation to accomodate the required number of elements. The stranding economy is another advantage of quad elements. It is not common to use detorsioning carriages for large pair/quad cables, but it is considered desirable for up to 12-quad carrier cables, and it is essential for coaxials and optical fibres.

Between each bay of the strander, there is a lay plate, a closing die and a tape lapping head. This last is required for applying, in an open helix, a numbered or coloured tape for identifying the layer. The final tape head(s) apply a close-helix of overlapping tape, to protect the cable during sheathing and to control the cable diameter. The lay plates maintain the position of the elements in the layer and prevent crossovers. The dies are a close fit on the cable layers, again to avoid misplaced elements, but should not be tight, to avoid damaging the insulation.

The cable is drawn through the machine by a capstan whose diameter should be at least 15 times the cable diameter. The supply spools are prevented from overrun by simple band brakes. If each element has a back tension of 1 kgf, the capstan is required to pull at 50 kgf for a 50-element cable. A sketch of a 3-bay strander is shown in Figure 4.3.

Stranding Lay Length

The lay length of a layer is given by the ratio of linear machine speed in m/min to the rotational speed of the carriage in r.p.m. For good cable circularity and good flexibility, the lay length P should be short (about 6 to 10 times the layer diameter). Owing to the limitations on carriage rotational speed, short lays reduce the speed of manufacture. The lay take-up factor T (see Equation (3.1)) is also increased.

As the lay length P is reduced, the lay angle θ increases and the cross-section of the cabling element becomes elliptical. Owing to using a cabling increment of 6 rather than

2π, each element of diameter d_e has a circumferential space of $2\pi/6 = 1.047d_e$ into which the major axis of its cross-section must fit. The major axis length is given by

$$\frac{d_e}{\cos\theta} = d_e\sqrt{\frac{P^2 + (\pi D)^2}{P}} = d_e\sqrt{1 + \left(\frac{\pi D}{P}\right)^2} = d_e \cdot T \qquad (4.1)$$

This should be less than $1.05d_e$ to avoid riding elements. This means that P/D must be greater than 10.12 if the layer has its full complement of elements. If one element is removed, each of the remaining ones has a space $2\pi/5 = 1.257d_e$ and the P/D ratio can be reduced to 4.12 before elements start to ride.

If the cable is to be flexible enough to bend around a mandrel of diameter nD, the length of the outer circumference of the bent cable should be $(n+2)/n$ times longer than the inner circumference. This extra length for the elements comes from relative movement between the inner and outer portions of the stranding helix. There should be at least one lay length in the circumference and there should not be too much friction between the elements. Thus the cable should not be too tightly compacted by tension or lapping tapes. For $n = 12$ the relative movement required is

$$14/12 = 1.167 \equiv 16.7\% \quad \text{and} \quad P \text{ should be} \leq n\pi D = 37.7D \qquad (4.2)$$

This is a very large relative movement and to reduce it to a more reasonable 5% requires at least four lays in the circumference of the bend and so $P \leq n\pi D/4 = 9.42D$. This means that the layers cannot carry their full complement of elements for maximum cable flexibility. If they do, P/D must be greater than 10.12 to avoid riding elements and then n must be greater than 12.9 to avoid buckling of elements or flattening of the cable.

In order that the cable should be 'torsionally dead' the lay directions are of opposite hand for successive layers. This stops the elements of adjacent layers 'nesting' together. It also reduces crosstalk between adjacent layers.

Concentric Cable Diameters

Cable diameters are measured with a diameter tape whose graduations are at π mm intervals. This means that it is really the cable circumference that is measured and the reading is the effective diameter of a circular cable.

The cable core diameter of a concentric cable is

$$\begin{aligned} D_c &= \text{centre diameter} + \text{number of layers} \times \text{twice the radial thickness of the layer} \\ &= D_{cc} + 2N \cdot d_e \end{aligned} \qquad (4.3)$$

The centre diameters D_{cc} are given below as a factor times the element diameter d_e.

Number of elements in centre	1	2	3	4
Diameter factor	1	1.75	1.155	2.414

UNIT CABLE CONSTRUCTION AND DIAMETER

Unit Cable Construction

For very large cables concentric cabling becomes multi-operational owing to limitations on the size of the strander. However paper and polythene insulated quad cables were once made up to 1040 pair (520 quads) by this method. The large pairage distribution cables required near to exchanges are made by using unit construction. In these cables, units of 100 pairs are made with a long, unidirectional, stranding lay of about 1 m. This increases the speed of manufacture and makes a soft, easily deformable, unit. These units are then assembled into cables of up to 3500 pairs or more using concentric lay-ups of the units. By using a lay-up lay length of about 1 m in the same direction as the stranding lays, these are shortened to 500 mm in the final cable but the units remain relatively soft and are deformed into sector-shaped cross-sections enabling a compact, but still flexible, cable to be achieved.

A further sophistication of this method of cabling was to make the units themselves from subunits of 25, 20 or 10 pairs. Particularly in polythene insulated cables, this made individual pairs easy to identify. By using only ten colours, the 25 pairs in a subunit can be uniquely identified. By using the same colours in tape bindings the whole cable can also be uniquely identified. These advantages of identification have led to quite small cables also being made using 10 pair units.

To control crosstalk between pairs, each pair in a subunit has a different lay length (see Chapter 8). The crosstalk between subunits, and between units, is controlled by the relative rotations between them as a result of the fixed motion stranding and laying-up operations.

Advantages in manufacture also flow from this form of construction. Using drum twisting stranders (see Figure 3.9) both 10 pair subunits and up to 100 pair units can be formed in one operation, followed by laying up to the required cable size on a larger drum-twister. By using drum-twisters, not only can the rotational speeds be increased compared with rotating carriage stranders, but also tension control methods can be made more sophisticated since they no longer have to be accomodated in a rotating carriage.

A further advantage of unit construction is exploited in PCM (pulse code modulation) cables working up to 1 MHz (2 Mbit/s). It is relatively easy to introduce

Table 4.2
Colour combinations for identification

Identification colours		
White		Blue
Red		Orange
Black	×	Green
Yellow		Brown
Violet		Grey

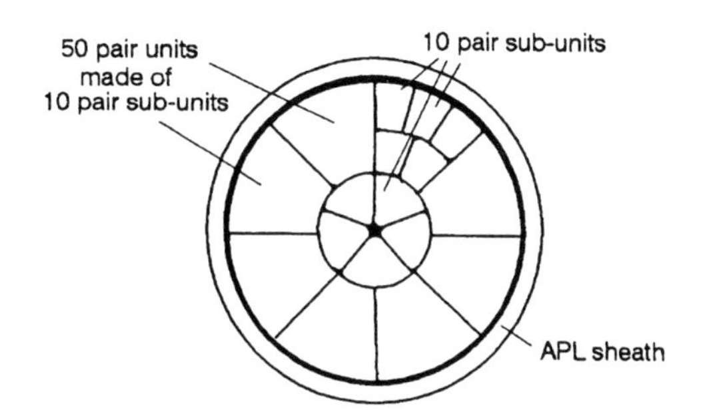

Figure 4.4
Cross-section of 450 pair unit cable

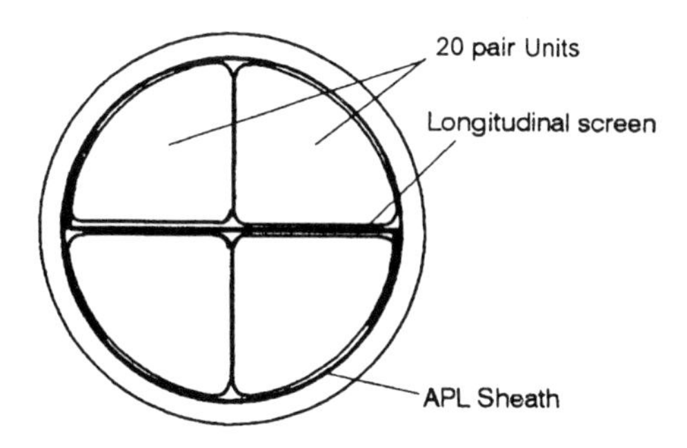

Figure 4.5
Cross-section of longitudinally screened PCM cable

a longitudinal aluminium/polythene laminate screen during cable laying-up, to separate the go-and-return units so improving the crosstalk between them. Figure 4.4 and Figure 4.5 show sketches of the cross-section of unit cables and of a longitudinallyscreened PCM cable. Figure 4.6 shows a sketch of a subunit/unit stranding drum-twister. By using a supply stand accomodating 200 reels, it is possible to load the second 100 reels while the first 100 are being stranded. This again reduces the time of manufacture.

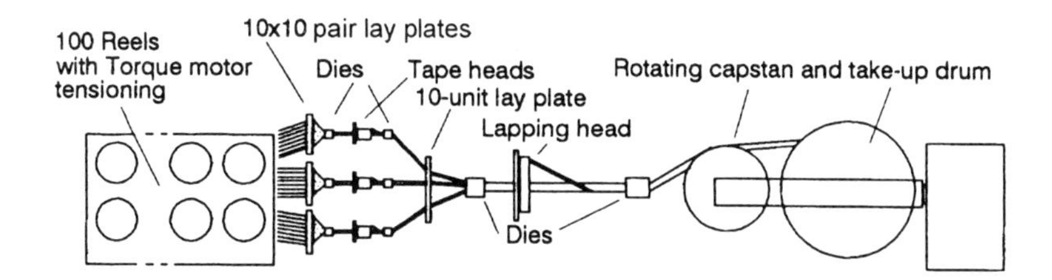

Figure 4.6
Drum-twisting strander for 10 × 10 pair subunits

Unit Cable Diameters

The subunits are stranded with nominal concentric formations. Owing to the long lay length they are soft and easily deformed. However their equivalent circular diameters are given by Equation (4.3). When they are formed into units, and the units formed into cables, they are compacted together to a degree depending on back tensions and overall lapping tape tensions. For practical values of these factors and for conductor diameters up to 0.63 mm, the cable core diameter, D_c, is given by

$$D_c = 1.87 \times D \times \sqrt{N} \pm 2\% \text{ mm} \tag{4.4}$$

where D is the insulation diameter of the wires in mm and N is the number of pairs in the cable.

This is equivalent to a space per pair of $2.69D^2$ mm^2/pair, compared with concentric formation where $D_c \approx 2.06D\sqrt{N} \equiv 3.3D^2$ mm^2/pair. Thus a unit pair construction is 80% of the area and 90% of the diameter of the equivalent concentric cable. For larger conductor sizes and higher ratios of D/d such good compacting is not as easily achieved and the cable diameter factor may rise by 6–7% (i.e. from 1.87 to 2.0).

SHEATH AND INTERSTICE DIAMETERS

Sheath Diameter

Most cables are now sheathed with a thermoplastic sheath, most usually PVC or polyethylene but sometimes using more sophisticated low-smoke zero-halogen emission compounds where fire is a serious hazard. The sheath is required to protect the cable core from damage in installation and to prevent the ingress of moisture which can degrade the transmission properties.

The cable sheath should not significantly limit the bending radius of the cable and should not itself flatten or wrinkle when bent to a diameter of 12 times its outside diameter. Experience has shown that these requirements can be met for polythene sheaths (or aluminium/polythene laminate sheaths) when the minimum thickness of the sheath is a function of the diameter over the sheath. This results in the diameter over the sheath D_0 and the radial thickness t being given by

$$\begin{aligned} D_0 &= 2.7 + 1.04D_c \quad \text{mm} \\ t &= 1.35 + 0.02D_c \text{ mm} \end{aligned} \tag{4.5}$$

where D_c is the cable core diameter given above plus any allowances needed for cable lappings.

Interstice Diameters

In cases where there are only a few elements in an outer cable layer causing an uneven appearance, filling strings or 'worms' may be laid in the outer interstices to improve the circularity. The size of these interstices (i.e. the diameter of the circular element which will fit into them) is also needed when designing composite cables of different element diameters. They are

Number of elements in layer:	2	3	4	5	6	7
Diameter factor of layer:	1.75	2.16	2.41	2.70	3.00	3.30
Outer interstice diameter:	0.67	0.48	0.41	0.38	0.35	0.34

WIRE AND TAPE ARMOURING

Wire Armouring

Wire armouring of cables may be required to provide additional tensile strength to the cable, to protect it from severe abrasion in installation and use, or to increase the weight of the cable as in subaqueous or submarine environments where currents may disturb the cable. Such wires are generally of mild steel or galvanized steel. For subaqueous or submarine use it is generally required that a 'bedding' of compounded hessian be applied to the cable sheath under the armour layer and that the armour wires be flooded with a bituminous compound followed by a futher layer of hessian tapes. The armoured cable is finally oversheathed with PVC or polythene. This provides a degree of corrosion protection to the wires.

The number of armour wires of diameter d_A which will fit over an inner diameter of D_A (i.e. with a pitch circle diameter of $D_A + d_A$) is given by

$$N \leq \frac{\pi(1 + D_A/d_A)}{\sqrt{1 + (\pi D_A/P)^2}} = \frac{\pi(1 + D_A/d_A)}{T} \tag{4.6}$$

where P is the lay length of the armouring wires and T is the take-up factor. For a stable and flexible layer of armour wires $P = 10D_A$, which results in

$$N \leq 3\left(1 + \frac{D_A}{d_A}\right) \tag{4.7}$$

For severe circumstances it may be desirable to use two layers of wire armour. These are applied with opposite hand lays to reduce torsion in the cable. The diameter over the armour and protection is derived by calculating the thickness of the protective layers and the single or double layer of armouring wires and adding twice this radial thickness to the known diameter over the cable sheath.

Steel Tape Armour

For protection against crushing and against penetration of sharp objects which would damage the sheath and allow ingress of water, steel tape armouring is used. This may be of mild steel tapes or galvanized steel tapes. Again these are applied over compounded hessian (which helps prevent damage to the sheath by the steel tapes themselves), and are similarly protected from corrosion by bitumenous compounds and a thermoplastic oversheath. Such steel tape armours are also used for magnetic protection of the cable in high electromagnetic fields (see Chapter 9) and for lightning protection (see Chapter 11). The tapes must be applied with about 20% gap to allow for cable bending. To cover a gap in the first tape, a second steel tape is required, usually applied from the same lapping head. Again in severe cases of induction, four layers of steel tape may be required. The diameter over the armouring and oversheath is calculated by adding twice the total radial thickness of the armouring to the cable sheath diameter.

5

Sheathing

Historically, sheaths for outdoor cables were made of extruded lead. Lead had the advantages of ductility making it easy to bend to shape, ease of jointing by plumber's techniques, low rate of corrosion under normal circumstances, conductivity giving electrostatic screening to the cable and impermeability to water vapour. Its disadvantages were its weight and cost. Also under adverse groundwater conditions it could corrode fairly rapidly and exhibit corrosion fractures. Indoor cables were sheathed in rubber, often with braided impregnated fabric for extra abrasion resistance. These were relatively slow to manufacture and costly.

Thermoplastic polymers, such as polyvinylchloride and polyethylene, became available in about 1938. Polythene was restricted on strategic grounds to insulation for radar cables where its excellent dielectric loss at high frequencies made it uniquely suitable. Indoor cables and subscriber drop-wire soon made use of PVC for insulation and sheathing on account of its superior abrasion resistance, ease and speed of extrusion and relative freedom from fire risks. PVC is used in a compound form with plasticiser to improve flexibility and cold-shatter and chalk fillers to reduce its cost. Much development in compounding was carried out to produce grades specific to use, hard grades for insulation and softer grades for sheathing.

When polythene became freely available in about 1948, it was rapidly adopted for insulation and sheathing of small pairage subscriber cables (up to 100 pair). Its advantages over lead were its easy extrudability, its lightness in weight, its chemical inertness (no corrosion), its low coefficient of friction and its rapidly decreasing price compared with lead. Unfortunately it provided no screening, so if required this had to be provided by lapped metal foils beneath the sheath. Also both polythene and PVC have small but significant permeabilities to water vapour over long periods, although polythene is about fifteen times better than PVC in this respect. So while polythene is regarded as an excellent vapour barrier for short-term packaging of food, fertilizer, cement etc., it leaves something to be desired for cable sheaths with a lifetime expectancy of 40 years or more. It was also found that polythene was permeable to gases, in particular to air (when used for pressurising cables to detect physical leaks) at a rate 60 times that of water vapour. Thus a small polythene sheathed cable (15 mm diameter), pressurized to 10 p.s.i., would lose almost all excess pressure over a few days even in the absence of leaks.

Table 5.1
Permeability coefficients for plastics

Permeabilities to vapours and gases ($\mu g \cdot cm/cm^2/h/mmHg$)			
Polythene/water vapour:	MFI	0.06–0.12	0.0013
@ 15 °C (+50 to 70% per 10 °C rise)	MFI	0.3–0.5	0.0019
	MFI	7	0.0021
Polythene/nitrogen	density	0.92	9.7×10^{-5}
	density	0.95	1.5×10^{-5}
Polythene/oxygen	density	0.92	32×10^{-5}
	density	0.95	5.1×10^{-5}
PVC/water vapour	30% plasticizer		0.03

The typical values of the permeation coefficients quoted in the literature (see Table 5.1) are generally determined on thin films of material and often show wide variations between determinations. The measured value, in standardized units, varies with the thickness of the sample tested, the morphology of the specimen (percentage crystallinity), the thermal history of the polymer and the temperature of measurement.

WATER PERMEABILITIES AND EFFECTS

Measurement of Polythene Sheath Permeability to Water Vapour

In a classic investigation of polythene sheath permeability to water vapour Harrison [2] determined the mean permeability of natural and carbon black loaded polythene in the form of actual cable sheaths of about 3 mm radial thickness. This was carried out by passing dry nitrogen (at less than 1 p.p.m. water content) through about 15 m of empty cable sheath immersed in a water bath at constant temperature. The emerging gas was dried in weighed dessicator tubes for a measured time. The increase in mass of the dessicator tubes gave the mass of water permeating the sheath in that time. In later experiments the dessicator tubes were replaced by an electrolytic gravimetric detector for increased sensitivity.

The mean value of P for both natural and black (2% carbon black) polythene sheaths was found to be 0.00132 $\mu g \cdot cm/cm^2/h/mmHg$ with a standard deviation of 15% at 15 ± 2 °C, substantially in agreement with the value given in the table for polythene of Melt Flow Index 0.1, although the sheaths were extruded using MFI 0.3 polythene. The values in the table apply to thin films of the order of 0.1 mm thickness. The processing of thin films often includes considerable stretching of the plastic material and quick quenching, a vastly different situation from sheath extrusion.

The permeation coefficient is calculated from

$$P = \frac{Q \cdot T}{\pi D \cdot t \cdot L(p_0 - p_i)} \tag{5.1}$$

where Q is the mass of water permeating in time t through a sheath of length L, diameter D and thickness T, for a vapour pressure difference from outside to inside of $(p_0 - p_i)$ mmHg, using consistent units.

Using Harrison's value for P, a polythene sheath of 23 mm mean diameter and 1.73 mm thickness, immersed in water at 15 °C, will be permeated by water at a rate

$$Q = \frac{P \cdot (p_0 - p_i) \cdot \pi D}{T} = \frac{0.00132 \times (12.67 - 0) \times \pi \times 2.3}{0.173}$$

$$= 0.699\,\mu\text{g/h/cm} = 11.74\,\text{g/week/km}$$

Effects of Water Permeation on all Polythene Cable

A polythene insulated and sheathed cable (without any paper lappings) leaving the factory has an internal relative humidity of about 40–50%. When installed in contact with groundwater at a constant temperature, say 15 °C, there will be a relative humidity (RH) difference from outside to inside which will drive water vapour through the sheath into the cable. The RH in the cable will rise to 100% on a time scale dictated by the moisture absorbing properties of the materials inside the cable and the permeability of the sheath. In the case of the all-polythene cable there is very little absorption of water by the polythene insulation ($<0.1\%$) or the Melinex core wrapping ($<0.6\%$). Thus the time to reach 100% RH is short (about 1 year).

If now the temperature falls from 15 °C to 0 °C, the saturation vapour pressure falls from 12.67 mmHg to 4.58 mmHg and water will condense out in the form of dew, coating all the inside surfaces. To calculate how much water condenses out, assume that the vapour acts as an ideal gas and that it behaves independently of other gases present (Dalton's Law of partial pressures). From Boyle's law and Charles's law the ideal gas equation is

$$pV = nRT \tag{5.2}$$

where p is the partial pressure in Pa(N/m^2), V is the volume in m^3, T is the temperature in Kelvins, n is the number of moles of gas and R the gas constant of 8.314 (N/m^2)(m^3)/mol/K = 8.314 J/mol/K.

For example take a 200 pair, 0.5 mm polythene insulated and sheathed cable with a melinex core wrap. For a capacitance of 50 nF/km the insulation diameter will be 0.94 mm and the diameter over the Melinex-wrapped 200 pairs will be 25.4 mm. The sheath outside diameter will be 29 mm and the mean diameter 27 mm. The sheath thickness will be 1.8 mm and the free air space in the cable will be 0.21 m^3/km. Thus for 100% RH in the cable at 288 K the saturation vapour pressure is 12.67 mmHg which is equivalent to a pressure of

$$p = \left(\frac{12.67}{760}\right) \text{atm} \times 1.01 \times 10^5 = 1684\,\text{N/m}^2\ (= \text{Pa})$$

$$n = \frac{pV}{RT} = \frac{1684 \times 0.21}{8.314 \times 288} = 0.148\,\text{mol} \equiv 0.148 \times 18 = 2.66\,\text{g/km of } H_2O$$

Table 5.2
Saturation vapour pressure of water

Temp. °C	SVP of water mmHg
0	4.58
5	6.51
10	8.94
15	12.67
20	17.5
40	55.1
60	149
80	355
100	760

where 18 is the molecular weight of water. If the temperature falls to 273 K (=0 °C), the vapour pressure falls to 4.58 mmHg = 609 Pa, hence

$$n = \frac{pV}{RT} = \frac{609 \times 0.21}{8.314 \times 273} = 0.056\,\text{mol} \equiv 0.051 \times 18 = 1.04\,\text{g/km of } H_2O$$

The mass of water condensing out is thus the difference 1.65 g/km, which occupies a volume of $1.65 \times 10^{-6}\,\text{m}^3/\text{km}$. Although this is a miniscule volume of water, calculating the total internal surface areas of the sheath, core wrap and insulation gives an area of 1445 m^2/km. Thus if the dew is deposited uniformly over this area it will form a layer of 1.14 nm, or about four molecules thickness. In practice it may not be so uniform. Also several mechanisms have been proposed which can result in 'moisture pumping' under conditions of temperature cycling, leading to a steady accumulation of liquid water in the cable. This water will tend to migrate to the lower portions of an installed cable. If it bridges insulation pinholes it will degrade the insulation resistance and may cause wire cut-off due to electrolysis (subscriber cables are polarized by the central battery). Such troubles have been reported from field investigations in the USA.

If sufficient water accumulates in the cable, the capacitance, loss angle and attenuation of the pairs increases, although on subscriber cables of moderate length this may not be a serious factor, a similar situation on a longer distance higher frequency cable could result in loss of service. It is extremely difficult to remove water from polythene cables once it accumulates. Saturation vapour pressures of water versus temperature are given for reference in Table 5.2.

Effect of Water Permeation on Paper Insulated Cable

Water permeating a polythene sheathed, paper insulated cable is absorbed by the paper in proportion to the relative humidity building up in the air spaces of the cable. This absorbed water degrades the insulation resistance of the cable as shown in Figure 5.1 which relates the percentage weight of water in the paper to both the insulation

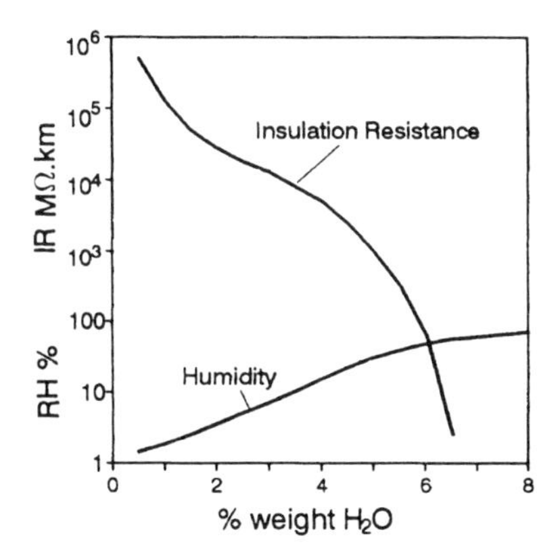

Figure 5.1
Variation of insulation resistance and internal relative humidity in paper insulated cable. (Reproduced from [2] by permission of the IEE)

resistance and to the RH of the air in the cable. These curves were obtained by measurements on an actual cable. When the insulation resistance falls to about $1\,M\Omega\cdot km$, over a time determined by the moisture capacity of the paper and the permeation rate of the sheath, the cable is considered unfit for service. This corresponds to about 6.5% by weight of water in the paper and a relative humidity of 40%.

Since this is the usual condition for the paper during manufacture, the cable must be thoroughly dried in a heated vacuum oven, before sheathing. To avoid making the paper too brittle, the drying should not be taken below about 1% wt. of water corresponding to about 2 to 3% RH and an insulation resistance exceeding $40\,000\,M\Omega\cdot km$. Figure 5.2 shows an independent determination of the water absorption of paper.

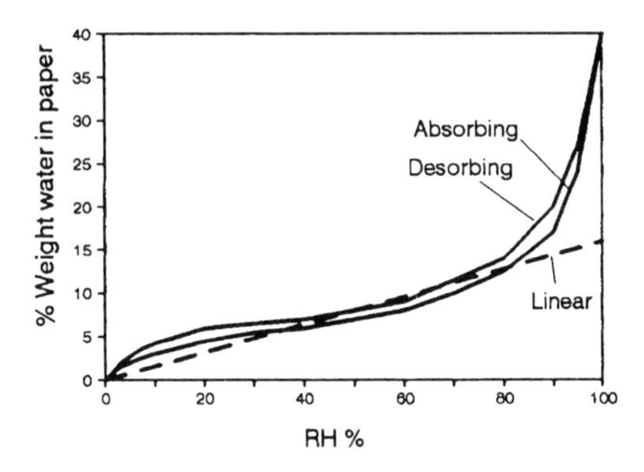

Figure 5.2
Water absorption of paper versus ambient humidity. (Reproduced from [2] by permission of the IEE)

Sheath Designs to Reduce Water Permeation

Several sheath designs which combat water permeation are available. They all rely on a layer of metal tape beneath the polythene outer sheath. In order to retain the cost advantage over lead, these tapes need to be thin. Thus they need mechanical support against crushing and also need some flexibility. This is usually accomplished by corrugating the tape circumferentially or helically.

Early European designs used multiple helical wraps of metal foils flooded with soft bituminous compounds directly beneath the polythene sheath. Later designs such as the Hackethal sheath used a seam-welded metal tape (0.5 mm steel, copper or aluminium) which was longitudinally formed oversize and then swaged to size after welding by means of helical corrugations. Line speeds were slow at about 6 m/min. The cable was then oversheathed in polythene.

The Western Electric Co devised the ALPETH sheath, in which a corrugated aluminium tape (0.2 mm thickness) is formed longitudinally around the cable between a thin inner sheath and a thicker outer sheath. The tape edges are overlapped and a flooding compound may be applied. This was actually intended to supply the electrostatic screening that a polythene sheath lacks, but it also provides a small measure of water permeation protection.

For greater protection this can be combined with a thin corrugated steel tape (0.15 mm) also longitudinally applied but with an overlapped soldered seam. In this case the aluminium tape being provided for electrical conductivity has an edge gap opposite the soldered seam. The steel tape is flooded with a thermoplastic compound before sheathing. This is the STALPETH sheath. This sheath is impermeable to water so long as the soldered seam is substantially perfect. These sheaths are sketched in Figure 5.3. Line speeds are about 40 m/min.

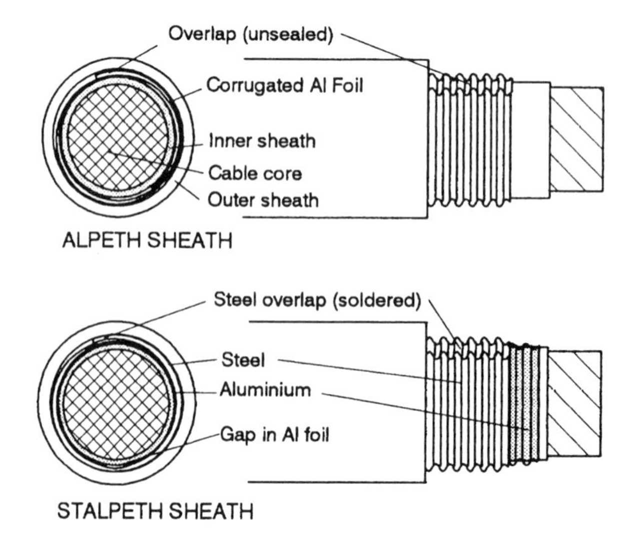

Figure 5.3
ALPETH and STALPETH sheaths

THE APL SHEATH AND WATER PERMEABILITY PERFORMANCE

The APL Sheath

None of these alternatives found favour in the UK, partly because of the elaborate equipment required and partly because they were essentially separate processes that were too slow to combine with the polythene sheath extrusion. Dr Glover of BPO and Dr Hooker of TCL together devised the aluminium/polythene laminate sheath in 1958. In this method a polythene coated aluminium foil is applied longitudinally with an overlap, over the dried cable core with the polythene surface on the outside. (Early designs used a helical application of the tape, but the longitudinal application is to be preferred since it can be carried out in line with the extrusion). When the hot polythene sheath is extruded over this with a slight vacuum inside, the hot polythene welds on to the polythene of the foil. On cooling, this produces a sheath with the thin aluminium foil firmly bonded to its inside (Figure 5.4).

The thick polythene provides mechanical support and strength to the foil and the foil forms a barrier to the radial permeation of water vapour. It does not prevent water vapour entering the sheath material, which rapidly becomes saturated in a wet environment, but it severly restricts the area from which re-evaporation into the cable can occur. Permeation is therefore limited to the 'windows' where the foil overlaps itself. Aluminium foil thicknesses are 0.075 mm, 0.15 mm or 0.3 mm with an overlap of at least 6 mm and line speeds of 60 m/min or more are used.

Measurements of Permeation Rates of APL Sheaths

Harrison measured many APL sheaths for water permeation and concluded that the improvement factor over plain sheaths for helically applied Al/Pe foil was in the range of 25 to 30 and for the longitudinally applied foil it was in the range of 60 to 120. This accords with the observation that with helical foil the overlap was not usually as well bonded as in the longitudinal case and also the length of the overlap per km of cable was about three times longer. For poorly bonded overlaps, or delaminated tape caused by severe handling, or for corrugated longitudinal foils, the improvement factor was reduced to the same range as that for helical foil.

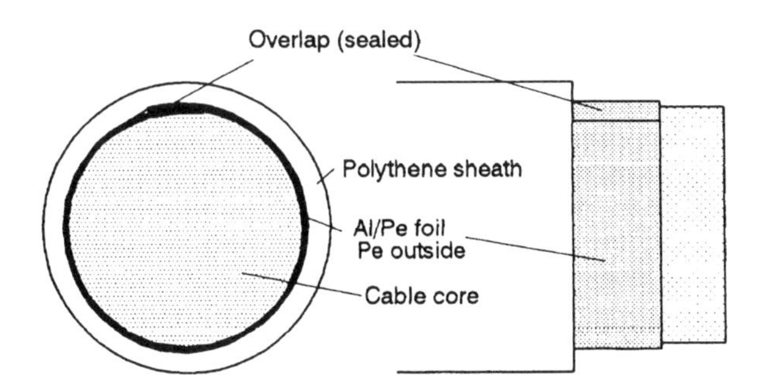

Figure 5.4
The APL sheath

Prediction of Cable Lifetimes

If the uptake of water vapour by materials inside the cable can be assumed to have a linear relation with the inside relative humidity, then the relative humidity V at time t, when outside the cable the relative humidity is E, can be calculated in the same way as charging a capacitor through a resistance

$$V = E(1 - \varepsilon^{-(t/RC)}) \tag{5.2}$$

Where the resistance $R = 1/Q$, (Q is the rate of water vapour permeation in g/year/km of cable) and C is the capacity of the materials inside the cable to absorb water in g/km. Alternatively, the time required to reach a given RH from an initial dry state, is calculated from

$$t = CR \cdot \ln\left(\frac{E}{E - V}\right) \tag{5.3}$$

For illustration the 200 pair/0.5 mm cable above with various constructions is considered, namely

Cable A Polythene insulated, Melinex wrapped, plain polythene sheath

Cable B Polythene insulated, Melinex wrapped, paper wrapped, plain polythene sheath

Cable C Polythene insulated, Melinex wrapped, paper wrapped, APL sheath

Cable D Paper insulated, paper wrapped, plain polythene sheath

Cable E Paper insulated, paper wrapped, APL sheath

The masses of the various internal components and their water capacity at 100% RH and 15 °C are given in Table 5.3

Combining these as appropriate for the cables above gives

Cable A C = 202.5 g/km of water

Cables B and C C = 1678.5 g/km of water

Cables D and E C = 40.75 kg/km of water

Table 5.3
Water capacities of cable components

Component	Mass kg/km	Water capacity % wt	C g/km
Polythene insulation	193.0	0.001	193
Paper insulation	260.0	15	39.000
2 paper wraps	9.84	15	1.476
2 melinex wraps	11.15	0.006	6.69
Free airspace	217 l/km	0.0126 g/litre	2.74

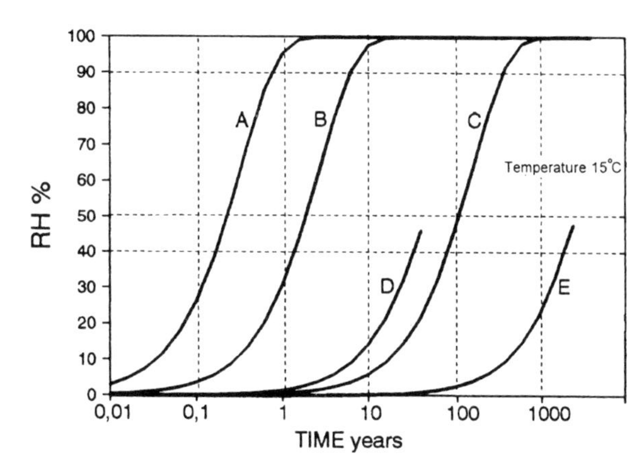

Figure 5.5
Increase in internal relative humidity in cables with time. (Reproduced from [2] by permission of the IEE)

Using Harrison's value for P, for this size of cable the value of $Q = 12.0\,\text{g/km/week} = 624\,\text{g/km/year}$ at 15 °C, so the value of R in Equation (5.2) is $1/Q = 0.00160$.

For cables C and E, the improvement factor for the longitudinal APL is taken as 60 (bottom of the range) Applying Equation (5.2) for the five cables gives the results shown in Figure 5.5 at a temperature of 15 °C.

Discussion of End-of-life Points

The effective end-of-life points for the cables immersed in water depends on the construction.

Cable A

This all-polythene cable has no significant water absorbing components. Such cables will have an internal relative humidity of about 50% when leaving the factory. If the cable temperature falls to zero immediately on installation, water will condense out and deterioration of the cable will start immediately, followed by vapour pumping as the temperature cycles. Even if the temperature stays at 15 °C the permeation of water vapour will raise the internal humidity to 100% in less than a year. If the temperature of the installation is about 8 °C the internal air capacity will be 1.72 g of water vapour compared with the 1.33 g already present, but the permeation coefficient will be lower at about 8.6 g/km/week. Even so in wet conditions the cable lifetime will be limited to a few years. This will hold for any diameter of cable of this construction

Cable B

Merely by adding two lappings of paper tape to the cable the rate of rise of the internal RH is reduced by a factor of ten. Again it will have an initial RH of 50% but the time required to rise to 99% will be about 18 years at 15 °C. At a lower temperature of 8 °C this will be extended to 25 years. However the calculations assume a linear rise of weight of water in the paper with RH. Figure 5.2 shows that this is a reasonable assumption for RH between 10 and 90% but between 90% and 99% the paper absorbs as much water again as between 0 and 90%. Thus even if the cable temperature drops to 0 °C when the RH has reached 99% by the calculation, the paper will absorb any water that condenses out. Thus it may be assumed that it will not start to deteriorate for about 25–30 years. Since both the permeation rate and the amount of paper are proportional to the cable circumference, this lifetime will not vary with cable size.

Cable C

By adding an APL sheath with an improvement factor of 60 to the construction, the lifetime, on the same basis as for cable B, will be extended beyond 1000 years! Since the amount of paper is proportional to the circumference, while the permeation is proportional to the length of the APL overlap, the lifetime will be better for larger cables but smaller for small cables. But even for a 10 mm diameter cable the lifetime will be several hundred years on this basis. Even if the APL is very imperfect and the improvement factor is only 20. the lifetime will be of the order of hundreds of years.

Cable D

The very large water capacity due to the paper insulation causes the rate of rise of RH to be slow even without an APL sheath. Since the cable is dried to about 2% RH before sheathing, and its insulation resistance falls to an unusable figure for 40% RH, its lifetime will be about 30 years at 15 °C in wet conditions. At 8 °C the lifetime will probably be about 42 years. However significant reduction of the insulation resistance down to 2000 MΩ · km should be expected when the RH rises to 20%, that is in about 15 to 20 years. The performance will not depend on the size of the cable. In a similar construction of cable the insulation resistance, corrected for temperature, has been monitored to 14 years, showing a drop from 80 000 to 3000 MΩ · km over this period as expected.

Cable E

With the addition of the APL sheath, the end of life on the same basis as Cable D will be extended to 2000 years, with deterioration of IR not being significant until 800 years, nor even noticeable for 200 years! Again the performance of this construction of cable will be better for larger cables and worse for smaller cables but not significantly in practice, even with a very imperfect APL sheath.

Effect of Overlap Imperfections

The foil laminate is bonded to the inside of the sheath preventing any radial permeation in this region, but at the overlap permeation can occur through the adhesive bond as shown in Figure 5.6.

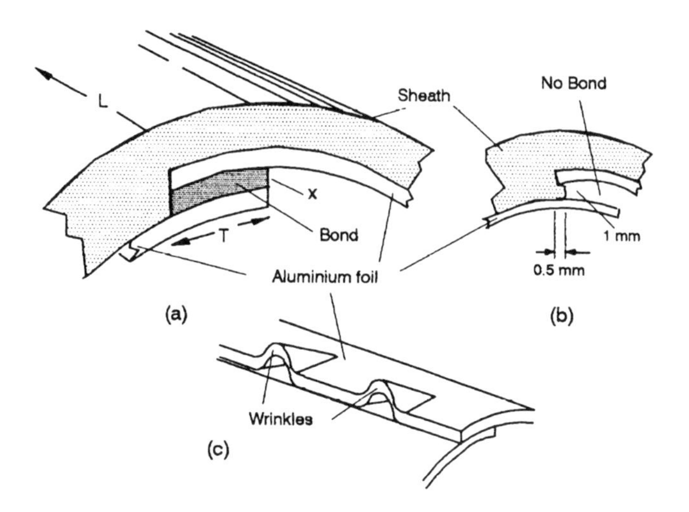

Figure 5.6
Tape overlap and defects on an APL sheath

The rate of permeation can be calculated using Equation (5.1) thus

$$Q = \frac{P \cdot (p_i - P_0) \cdot L \cdot x}{T} \ \mu g/h/cm \qquad (5.4)$$

Where x is the thickness of the bond, T is now its circumferential length in the same units as x, and L is the cable length in cm. For

$$P = 0.00132\,\mu g \cdot cm/cm^2/h/mmHg \quad \text{and} \quad (p_i - p_0) = 12.67\,mmHg$$

$$Q = 0.281 \cdot \frac{L \cdot x}{T} \ g/week/km \qquad (5.5)$$

The aluminium foil is 0.15mm thick with a polythene layer of nominal thickness 40 μm and should be applied to have a minimum overlap T of 6 mm. If the overlap and bonding are perfect down the whole cable length then $Q = 0.0019$ g/week/km. This would give an improvement factor $F = 6400$ compared with the plain sheath value of $Q = 12$ g/week/km.

For an imperfect overlap, where the sheath intrudes 0.5 mm into the overlap with a thickness $x = 1$ mm preventing the overlapping aluminium foil from bonding, as in Figure 5.6(b), the effective value of $x = \pi d/2 = \pi/2$ mm; then $Q = 0.441$ g/week/km and $F = 27$ which still gives a very adequate cable lifetime to cables C or E in Figure 5.5.

If in addition to this the edge of the foil is severely wrinkled as in Figure 5.6(c), with wrinkles having a height and width of about 3 mm, for an average of 1/3 of the total length of the cable, then the improvement factor reduces to $F = 20$ which is still an acceptable value. Note that wrinkles also reduce the sheath minimum thickness.

Thus the quality of the overlap governs the performance of the APL sheath over a wide range of possibilities. The same considerations apply to other types of overlapped foil water barriers if there is a flooding compound used.

Measurement of Water Barrier Performance

The method used to measure the permeation rates, Q, of barriered sheaths is essentially that used in Harrison's later experiments with a coulometric hygrometer. The formula given to calculate the permeation rate is

$$Q = \frac{(r_2 - r_1) \times 7.45 \times 10^{-2}}{l} \text{ g/week/100 m} \tag{5.6}$$

where l is in metres and r_1 and r_2 are the incoming and outgoing water content of the gas flow in p.p.m. when stability is reached.

Thus to measure a value of $Q = 1.2/120 = 0.001$ g/week/100 m such as would be expected on the 200 pair/0.5 mm cable with a good APL sheath,

$$(r_2 - r_1) = 0.001/0.0745 = 0.013 \text{ p.p.m.}$$

This difference has to be measured in the presence of about 0.5 p.p.m. water vapour in the incoming gas stream, per metre of sample length. This corresponds to a sensitivity of less than 3%. If sufficient precautions in setting up the sample are not taken, or if the cable sample is not long enough, errors will easily mask this difference. For reasonable accuracy the sample should be about 10 m long.

CYLINDRICAL FORMATION OF TAPES

Formation of Tapes into Cylindrical Barriers

Tape thicknesses for APL are usually 0.15 mm or 0.3 mm thickness with 0.05 mm of polythene or EEA copolymer bonded to one surface. The tapes are slit to the desired width from wider rolls. This should be done in such a way as not to stretch or damage the tape edges. Many methods of forming the flat tape into a cylindrical tape have been devised. They all have to overcome the fundamental problem that the path taken by the tape edges is longer than that of the centreline, as sketched in Figure 5.7. This means that the tape edge stretches during the formation. If the elastic limit of the tape material is exceeded, wrinkles form, as shown in Figure 5.6(c). The shorter the distance over which the tape is formed in relation to the final diameter D , the greater is the edge strain S.

If the tape is constrained to follow circular arcs of diminishing radius, R, as in Figure 5.7, the following analysis, due to A. McNamee [3], holds

$$\text{AB}^2 = dz^2 + dy^2 + dx^2 \tag{5.7}$$

$$S = \frac{\text{AB} - dz}{dz} = \frac{\text{SB}}{dz} - 1 \quad \text{or} \quad \frac{\text{AB}}{dz} = 1 + S \tag{5.8}$$

$$y = R - R\cos\left(\frac{W}{2R}\right) \quad \text{and} \quad x = R\sin\left(\frac{W}{2R}\right) \tag{5.9}$$

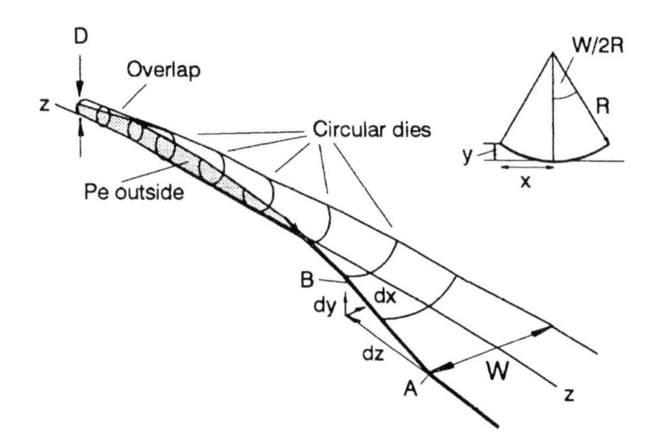

Figure 5.7
Longitudinal formation of tape into cylinder

Let $W/2R = \alpha$ and differentiating (5.9) with respect to R

$$dy = (1 - \cos\alpha - \alpha \sin\alpha) \cdot dR \quad \text{and} \quad dx = (\sin\alpha - \alpha\cos\alpha) \cdot dR \tag{5.10}$$

Squaring both of these differentials and adding gives

$$dx^2 + dy^2 = [(\alpha - \sin\alpha)^2 + (1 - \cos\alpha)^2] \cdot (dR)^2 \tag{5.11}$$

From Equations (5.7), (5.8) and (5.11)

$$\frac{\text{AB}^2}{(dz)^2} = 1 + [(\alpha - \sin\alpha)^2 + (1 - \cos\alpha)^2]\left(\frac{dR}{dz}\right)^2 = (1 + S)^2 \tag{5.12}$$

Whence

$$\frac{dz}{dR} = \sqrt{\frac{(\alpha - \sin\alpha)^2 + (1 - \cos\alpha)^2}{(1 + S)^2 - 1}} \tag{5.13}$$

To calculate the die diameter, $2R$, at any position z, Equation (5.13) must be integrated between the limits of $R = D/2$ and $R = D_n/2$, where D_n is the largest die diameter to be used. This can only be done numerically and can be done easily on a pc spreadsheet. For a 30 mm diameter cable core, this results in a graph such as that in Figure 5.8 where the strain has been limited to 0.005 or 0.5%. Notice the significant departure from linear spacing which is equivalent to a simple cone formation. The maximum strain recommended for the APL tape is 0.5%. The largest diameter to be used is recommended to be $D_n = 4W/3$. In this case $D_n = 133.7$ mm, which limits the length of formation to about 1.6 m as shown in the figure, and about 10 or 11 dies should be used. Since the length of formation is proportional to the final diameter D, for a 75 mm diameter cable core this length becomes 3.8 m. Alternatively the strain may be allowed to rise to 1% with the risk of producing a few wrinkles, when the length may be reduced to about 2.7 m.

In the early days of APL applicators, cone shaped forming horns of stainless steel sheet were used. Not only were these susceptible to damage, causing high spots which became the origin of edge wrinkles, but because they followed a linear dz/dR form;

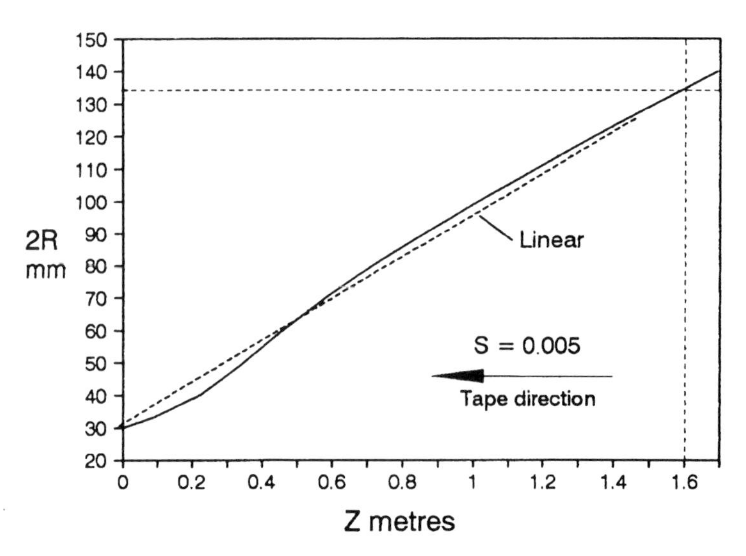

Figure 5.8
Forming die size versus position

wrinkles could be seen to arise for no apparent reason. Also, owing to physical limitations, the forming length was usually much too short for the larger diameter cables.

The slitting of the tape to the desired width can also result in edge stretch which is not noticeable until the tape is unwound. To overcome this it is possible to stretch the tape judiciously and uniformly across its whole width before commencing the forming operation. A certain amount of back tension is desirable when forming to allow the tape to conform to the circular arc dies, but too much tension can result in tape breaks.

This discussion of edge wrinkling clearly shows the advantage of transverse corrugations as used in the ALPETH and STALPETH tapes. The corrugations allow more elastic strain at the tape edges. However there must not be too much tension or the corrugations start to pull out and the desired flexibility of the sheath is diminished. For this reason Western Electric devised belt formers in which the moving belts supported and formed the corrugated tape with negligible plastic stretch. They also devised a roller system of forming the corrugated tapes through a series of cardioid arcs as opposed to the circular arcs analysed above.

Sealing the Overlap

The overlap may be allowed to seal under the combined action of vacuum and heat in the sheathing extruder head. This only works well for a wrinkle-free application of the tape. Where there may be wrinkles, caused for example by too short a formation length or varying cable core diameters, some extra means of sealing the overlap may be used. These methods vary from using a molded neoprene, resilient-cone, die following a heated region of the overlap, to a hot metal 'shoe' pressing on the overlap before entering the sheather, or best of all a hot-and-cold air-cushioned die. One difficulty of presealing the overlap is that on entering the sheather head with its vacuum, the overlap may become unsealed again as air exits the cable core.

ENVIRONMENTAL EFFECTS

Environmental Stress Cracking

Early polythenes used for cable sheathing had a melt flow index of 7. The melt flow index is a measure of the spread of molecular weight of the polymerized chains. The presence of lower molecular weight fractions acted as a lubricant to the semi-molten plastic making extrusion easier. The distribution of molecular weight in a given polymer depends on the method of polymerization used in its manufacture. In these early sheaths a number of cases of cracked sheaths occured after installation. These were eventually found to be associated with portions of the sheath which were under stress, e.g. near to sharp bends where the cable had been formed to make joints. Also there had to be a chemical agent present such as soap, detergents, alcohols or oils often present as groundwater contamination or deliberately used to test joints for leaks. This effect is called environmental stress cracking and standard test methods were devised to check the polymer's resistance to it, i.e. its environmental stress crack resistance (ESCR). It was hypothesized that the stress cracking agent was causing the breakdown of polymer bonds especially of the low molecular weight chains, thus weakening and embrittling the polymer. It was also found that incorporating 10% of butyl rubber in sheath material of MFI 2, which reduced the stress level, improved the ESCR.

With the increasing availability of narrow molecular weight distributions in low MFI polythenes, e.g. MFI of 0.5 and 0.3, the ESCR was found to increase greatly even using vicious reagents like the detergent Lissapol, and modern polymers of MFI 0.3 can withstand severe localized stress in concentrated Lissapol N for over 1440 hours.

UV Degradation

Natural polythene sheaths when exposed to ultraviolet radiation from the sun degrade by cross-linking and breaking of the molecular chains. They thus become brittle and can crack if stressed. For cables which are exposed to sunlight it is essential to protect the polythene by incorporating finely divided and well-dispersed carbon black in the polymer before sheathing. This can be carried out using masterbatches at the time of extrusion but usually the polymer is precompounded before extrusion. Buried or ducted cables can be sheathed with natural polythene so long as care is taken to protect them during storage before installation. PVC is also subject to UV degradation, and the temptation to use cream PVC sheathed cables, for cosmetic reasons, in situations exposed to sunlight should be avoided. PVC for outdoor use should also be protected by the incorporation of carbon black and then painted if required.

Flammability and Fire Risks

For indoor installation, PVC sheaths are preferred since they do not readily support combustion. However they do still present a fire hazard since in a serious fire they char and degrade producing dense smoke containing chlorine and hydrochloric acid gases which burn the eyes and lungs of fire fighters or trapped personnel. Much development work has been done to produce low-smoke and zero-halogen polymers for cables used where fire is a serious risk.

Jointing of Plastic Sheaths

The jointing of plastic sheaths, particularly of polythene sheaths, also has a long history of development. In the case of lead and lead alloy sheaths the cable industry merely took over a long established art of plumbing. No such technique was available for polythene sheaths. However, over a period of about ten years, from 1949 to 1960, successful techniques were developed which will be described in Chapter 11.

6

Primary Parameters of Lines

The transmission behaviour of lines is governed by their primary parameters of Resistance, Inductance, Capacitance and Conductance per unit length. They are usually expressed in Ohms/km, milli-Henries/km, nano-Farads/km and milli-Siemens/km. In this chapter it will be shown how these may be calculated accurately from the physical dimensions of the line. In Chapter 7 the primary parameters will be used to calculate the secondary, or transmission, parameters of the line.

Free Space Parameters

It will be necessary to use the parameters which govern the propagation of electromagnetic waves through free space, namely

The magnetic permeability of free space $\mu_0 = 4\pi \times 10^{-7}$ Henries/metre

The electric permittivity of free space $\varepsilon_0 = \frac{1}{36\pi} \times 10^{-9}$ Farads/metre

Consequently the velocity of waves through free space is

$$c = \frac{1}{\sqrt{\mu_0 \varepsilon_0}} = 3 \times 10^8 \text{ metres/s}$$

RESISTANCE AND REACTANCE OF CONDUCTORS

DC Resistance of Solid Cylindrical Conductors

This is calculated for a conductor of diameter d mm by

$$R_0 = \frac{4\rho}{\pi d^2} \ \Omega/\text{km} \tag{6.1}$$

for the resistivity ρ in $\Omega/\text{mm}^2/\text{km}$. For fully annealed copper the agreed standard resistivity is $17.241\,\Omega/\text{mm}^2/\text{km}$ (although most producers will achieve 1% lower than this) and for soft aluminium $28.21\,\Omega/\text{mm}^2/\text{km}$. Note that in these units, the resitivities are 1000 times the values quoted in $\mu\Omega \cdot \text{m}$.

Skin Effect

When alternating current flows in a conductor its magnetic field induces eddy currents in the conductor which oppose the flow of the primary current. This causes the net current to concentrate more and more to the surface of the conductor as the frequency increases. The effective resistance therefore increases and a reactive component called the internal reactance of the conductor appears (due to phase change in the net current). At very high frequency (e.g. above 10 MHz) the current is confined almost entirely to a thin skin near the surface (with a rapid exponential fall-off towards the interior) of the conductor. This is called the 'skin effect'.

In analysing these effects a dimensionless parameter u is used to normalize the frequency dependence of conductor diameter, resistivity and magnetic permeability.

$$u = \frac{d}{2}\sqrt{\frac{\mu\mu_0\omega}{\rho}} \tag{6.2}$$

μ is the relative permeability, $\omega = 2\pi \times$ frequency in Hz, d is in metres and ρ in $\Omega \cdot$ metres. For non-magnetic conductors $\mu = 1$ and for d in mm, f in kHz and ρ in $\Omega/\text{mm}^2/\text{km}$

$$u = 1.405d\sqrt{\frac{f}{\rho}} \tag{6.3}$$

Resistance and Reactance of Solid Cylindrical Conductors

The a.c. resistance and reactance of solid cylindrical conductors can be calculated exactly by means of Bessel functions. However these are not convenient to use, even using a PC. Asymptotic approximations have been published by Foch [4] and by Prache [5] for low and high values of u.

For $u \leq 2$

$$\frac{R}{R_0} = 1 + \frac{u^4}{3 \times 64} \pm 0.5\% \tag{6.4}$$

$$\frac{X}{R_0} = \frac{\omega L_0}{R_0} = \frac{u^2}{8}\left(1 - \frac{u^4}{6 \times 64}\right) \pm 0.1\% \tag{6.5}$$

For $u \geq 6$

$$\frac{R}{R_0} = \frac{\sqrt{2}u}{4} + \frac{1}{4} + \frac{3\sqrt{2}}{32u} \pm 0.05\% \tag{6.6}$$

$$\frac{X}{R_0} = \frac{\omega L_0}{R_0} = \frac{\sqrt{2}u}{4} - \frac{3\sqrt{2}}{32u} \pm 0.1\% \tag{6.7}$$

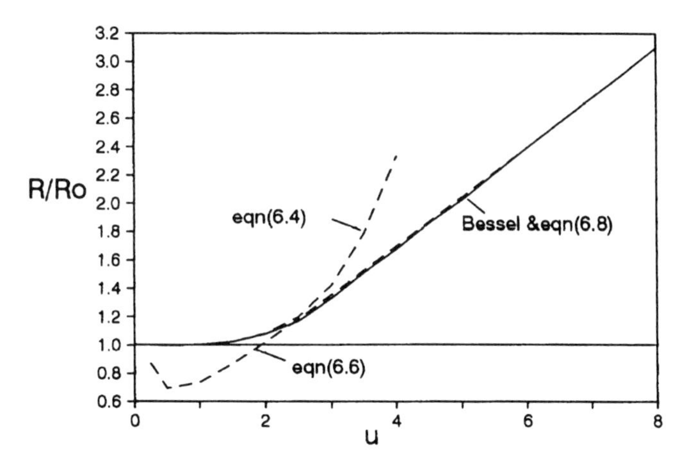

Figure 6.1
a.c. resistance of isolated conductor

L_0 = internal inductance of conductor and R_0 = d.c. resistance of conductor $\omega = 2\pi f$. These expressions are plotted in Figure 6.1 together with the Bessel curves.

However even greater convenience is given by an empirical formula due to Levasseur [6] which is accurate to better than 1% for all values of u

$$\frac{R}{R_0} = \tfrac{1}{4}(1 + \sqrt[6]{3^6 + 8u^6}) \pm 1\% \tag{6.8}$$

This is also plotted in Figure 6.1. Levasseur did not publish an expression for the reactance in his paper, but using the similarities between the resistance and reactance formulae of (6.4) and (6.5) the author found that the internal inductance could be approximated within 7% for all values of $u > 1$, by the expression

$$\frac{X}{R_0} = \frac{\omega L_0}{R_0} = \frac{0.538 \cdot u^2}{(1 + \sqrt[6]{3^6 + 8u^6})} \pm 7\% \tag{6.9}$$

Since the internal inductance of conductors in cables is usually associated with an external inductance about ten times larger, this is a satisfactory accuracy of approximation. Figure 6.2 shows the Bessel curve for reactance and the approximations. Using Equation (6.3) for a copper conductor of 1 mm diameter, $u = 2$ corresponds to a frequency of 35 kHz and, since $f \propto u^2$, $u = 6$ corresponds to a frequency of 314 kHz.

Very High Frequency Approximations

When $u \gg 1$, Equations (6.6) and (6.7) becomes

$$\frac{R}{R_0} = \frac{X}{R_0} = \frac{\sqrt{2}u}{4} = \frac{\pi d^2 \rho}{4Pdt\rho} = \frac{d}{4t} \tag{6.10}$$

where t is the thickness of the metal tube of diameter d whose d.c. resistance equals the a.c. resistance at frequency f.

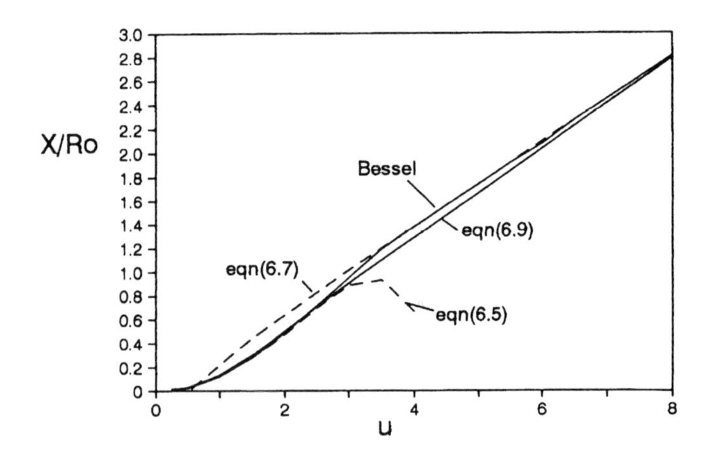

Figure 6.2
Reactance of isolated conductor

Thus

$$t = \frac{d}{\sqrt{2}u} = \sqrt{\frac{\rho}{\pi\mu f}} \tag{6.11}$$

and

$$R = X = \frac{1}{d}\sqrt{\frac{\mu\rho f}{\pi}} \tag{6.12}$$

For a copper conductor with f in kHz

$$t = \frac{2.09}{\sqrt{f}} \text{ mm} \tag{6.13}$$

In copper therefore, for $t = 100\,\mu\text{m}$, $f = 437\,\text{kHz}$ and for $t = 10\,\mu\text{m}$, $f = 43.7\,\text{MHz}$.

Proximity Effect in Symmetrical Pairs of Conductors

The d.c. resistance of a matched pair of conductors is twice the d.c. resistance of each conductor. However this is not true for the a.c. resistance which is greater than the sum, owing to the proximity effect. This effect, which is due to the mutual repulsion of opposite direction current filaments, decreases the current density in those portions of the conductors which are closest together (see Figure 6.3). This increases the total a.c. resistance and increases the sum of the internal inductances.

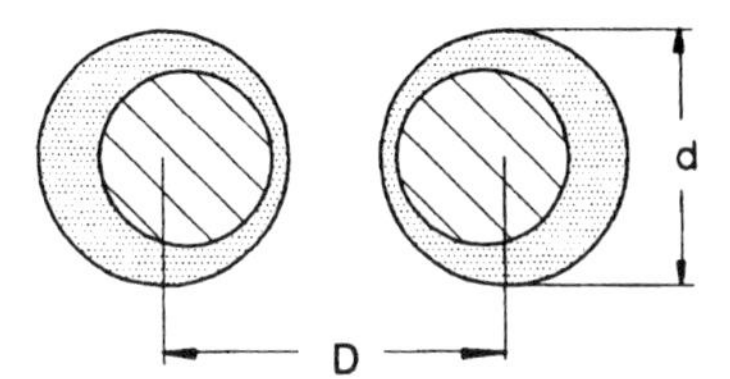

Figure 6.3
Proximity effect in an isolated pair

The proximity effect is proportional to d/D and can be calculated exactly by means of Bessel functions. But a sufficient approximation can be obtained by modifying Equations (6.8) and (6.9) as follows

$$\frac{R}{R_0} = \frac{1}{4}\left(1 + \sqrt[6]{3^6 + \frac{8 \cdot u^6}{(1 - d/D)}}\right) \pm 2\% \tag{6.14}$$

$$\frac{X}{R_0} = \frac{\omega L_0}{R_0} = \frac{0.455 \cdot u^2}{(1 + \sqrt[6]{3^6 + 8 \cdot u^6})\sqrt{1 - (d/D)^2}} \pm 10\% \tag{6.15}$$

These empirical formulae are compared with the Bessel forms in Figures 6.4 and 6.5, showing the effect of proximity. Equation (6.14) agrees with the Bessel curve to better than 2% for d/D up to 0.6. Equation (6.15) agrees with the Bessel form to better than 10%, up to $d/D = 0.6$ which when L_0 is one-tenth of the total inductance, corresponds with an acceptable accuracy of 1%. The usual range of d/D is from 0.25 to 0.6.

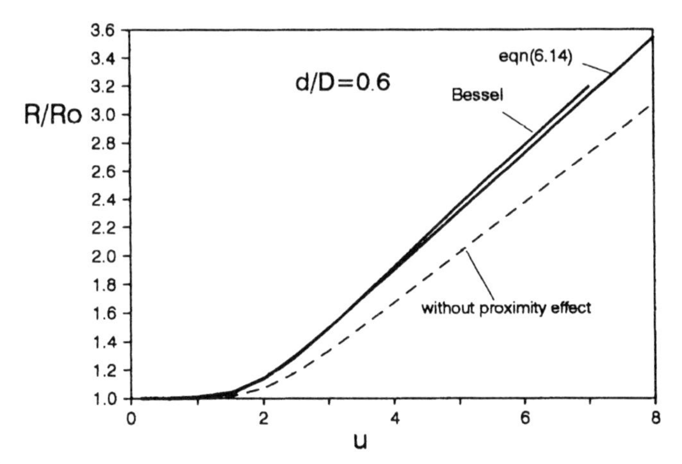

Figure 6.4
Proximity effect on a.c. resistance

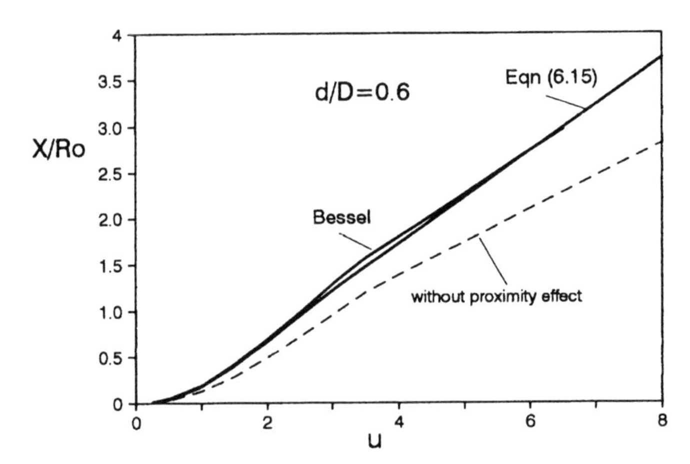

Figure 6.5
Proximity effect on reactance

Resistance and Reactance of Tubular Conductors

The asymptotic approximations to the Bessel formulae for cylindrical tubular conductors are (note the change in sign of the third term)

For $u \geq 6$
$$\frac{R}{R_s} = \frac{\sqrt{2}u}{4} - \frac{1}{4} + \frac{3\sqrt{2}}{32u} \pm 0.05\% \tag{6.16}$$

$$\frac{X}{R_s} = \frac{\omega L_0}{R_s} = \frac{\sqrt{2}u}{4} - \frac{3\sqrt{2}}{32u} \pm 0.1\%$$

In these expressions the ratios are expressed in terms of R_s which is the d.c. resistance of a solid cylindrical conductor of diameter S equal to the internal diameter of the tubular conductor. The value of u is now

$$u = \frac{S}{2}\sqrt{\frac{\mu\omega}{\rho}} \tag{6.17}$$

Resistance and Reactance of Thin Tubular Conductors

For thin tubular conductors of thickness t, there are approximate expressions [7], for all values of v,

$$\frac{R}{R_s} = \frac{S \cdot v}{8 \cdot t} \frac{\sinh v + \sin v}{\cosh v - \cos v} - \frac{1}{4} \tag{6.18}$$

and

$$\frac{X}{R_s} = \frac{S \cdot v}{8 \cdot t} \frac{\sinh v - \sin v}{\cosh v - \cos v}$$

where

$$v = t\sqrt{\frac{2\mu\omega}{\rho}} = 2\sqrt{2}\cdot\frac{t}{S}\cdot u \tag{6.19}$$

The error in these latter expressions is of the order of $(t/S)^2$ at low frequencies (e.g. 30 kHz) and diminishes gradually as the frequency increases.

Effect of Screens on Resistance of Balanced Pairs

When a balanced pair is surrounded by a cylindrical metallic screen, the signal currents in the pair induce eddy currents in the screen which represent a further source of loss in the signal. The author has found that this can be accounted for by a further increment in the a.c. resistance of the pair, which is again a function of the normalized frequency u.

$$R_E = 0.8(\sqrt{u} - \tfrac{1}{2}) \times (\text{Resistance of the screen from Equation (6.18)}) \tag{6.20}$$

Here, u takes the value pertaining to the conductors. The value of R_E is added to the resistance calculated in Equation (6.14) so that

$$R = \frac{R_0}{4}\left(1 + \sqrt[6]{3^6 + \frac{8\cdot u^6}{(1 - d/D)}}\right) + 0.8\left(\sqrt{u} - \frac{1}{2}\right)$$
$$\times (\text{Resistance of the screen from Equation (6.18)}) \tag{6.21}$$

The effect of a screen on the internal reactance of a pair is more complicated. A circular screen tends to reduce the effect of proximity, but a very close screen which follows the external profile of the insulated conductors tends to increase the reactance. In this latter case the constant in Equation (6.15) should be increased from 0.455 to 0.55

Pair in Cable Surrounded by Other Pairs

In this case, the six similar pairs surrounding the signal pair form a Faraday cage which is a virtual screen. This cage also dissipates energy in the form of induced eddy currents. To estimate the increase in a.c. resistance of the pair, a thin screen, having the same d.c. resistance as the 12 wires, is postulated to surround the pair at a diameter S, which is approximately equal to $3D$. The thickness of this fictional screen is obtained by equating the d.c. resistances, thus

$$R_0 = \frac{\pi}{t\cdot\pi S} = \frac{\rho}{12\cdot(\pi/4)d^2}$$

therefore

$$t = \frac{3d^2}{S} \tag{6.22}$$

and the a.c. resistance increment due to the screen is obtained by applying Equations (6.18) and (6.20).

For a quad in a cable surrounded by similar quads, the two other conductors in the quad dominate the effect of conductors in surrounding quads and the thickness of the fictional screen is empirically reduced by a factor of 6.

Coaxial Pair of Conductors

A coaxial pair of conductors consists of an inner conductor of diameter d surrounded coaxially by an outer tubular conductor of inner diameter D, as shown in Figure 6.6.

The a.c. resistance is the sum of the inner conductor resistance from Equation (6.14) and the outer conductor resistance from Equation (6.16) or if the outer conductor is thin, from Equation (6.18). In the latter case the total resistance is

$$R = \frac{Ri}{4}\left(1 + \sqrt[6]{3^6 + \frac{8 \cdot u^6}{(1 - d/D)}}\right) + R_D\left(\frac{Dv}{8t}\,\frac{\sinh v + \sin v}{\cosh v - \cos v} - \frac{1}{4}\right) \tag{6.23}$$

The total internal reactance is the sum of the individual conductor reactances from Equations (6.15) and (6.18), namely

$$X = \omega L_0 = \frac{0.54 \cdot u^2 \cdot R_i}{(1 + \sqrt[6]{3^6 + 8 \cdot u^6})\sqrt{1 - (d/D)^2}} + \frac{R_D \cdot D \cdot v}{8t}\,\frac{\sinh v - \sin v}{\cosh v - \cos v} \tag{6.24}$$

In these equations S has been substituted by D, R_i is the d.c. resistance of the inner conductor and R_D is the d.c. resistance of a solid conductor of diameter D. Similarly u and v of Equations (6.17) and (6.19) have S substituted by D.

High-Frequency Approximations for Coaxials

When $u \geq 6$ Equations (6.6), (6.7) and (6.16) become asymptotic to

$$\frac{R}{R_i} = \frac{\sqrt{2}u}{4} + \frac{1}{4}, \quad \frac{R}{R_S} = \frac{\sqrt{2}u}{4} - \frac{1}{4} \quad \text{and} \quad \frac{X}{R_i} = \frac{X}{R_s} = \frac{\sqrt{2}u}{4}$$

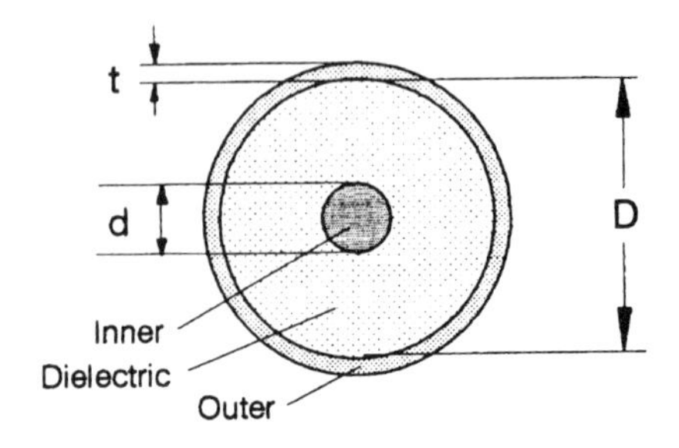

Figure 6.6
Coaxial pair

so that the total hf resistance of the coaxial becomes

$$R = \frac{R_i - R_s}{4} + \frac{\sqrt{2}}{4}(R_i \cdot u_i + R_s \cdot u_0)$$

Substituting for the values of u_i and u_0, D for S, and for the d.c. resistances gives

$$R = \frac{1}{\pi}\left(\frac{\rho_i}{d^2} - \frac{\rho_0}{D^2}\right) + \frac{\sqrt{2\mu_0\omega}}{2\pi}\left(\frac{\sqrt{\rho_i}}{d} + \frac{\sqrt{\rho_0}}{D}\right) = r + s\sqrt{\omega}$$

The first term r is always small compared with $s\sqrt{\omega}$ so that for non-magnetic conductors

$$R = 0.632\sqrt{f_{MHz}}\left(\frac{\sqrt{\rho_i}}{d} + \frac{\sqrt{\rho_0}}{D}\right) \tag{6.25}$$

Proceeding in a similar manner for the total internal inductance of the coaxial at high frequency

$$L_0 = \frac{10^{-7}}{\sqrt{f_{MHz}}}\left(\frac{\sqrt{\rho_i}}{d} + \frac{\sqrt{\rho_0}}{D}\right) \tag{6.26}$$

For d, D in m and ρ_i, ρ_0 in $\Omega \cdot$m, R will be in Ω/m and L_0 in H/m.

Comparison of Calculations and Measurements

In Figure 6.7 are shown the calculated resistance (using Equation (6.21)) and measured resistance (symbols) of a close-screened pair, with conductor diameter 0.634 mm and insulation diameter 2.435 mm. The screen thickness was 0.05 mm and fitted closely to the pair profile. The measurements have a basic accuracy of 1% and the agreement with the calculation is very good.

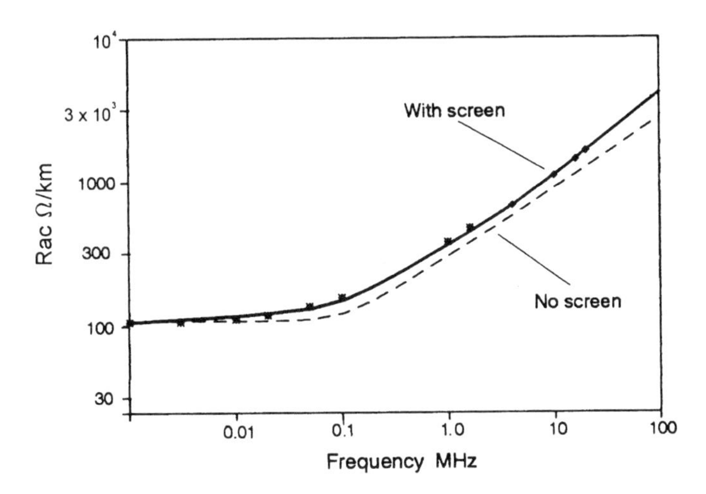

Figure 6.7
Resistance of screened pair

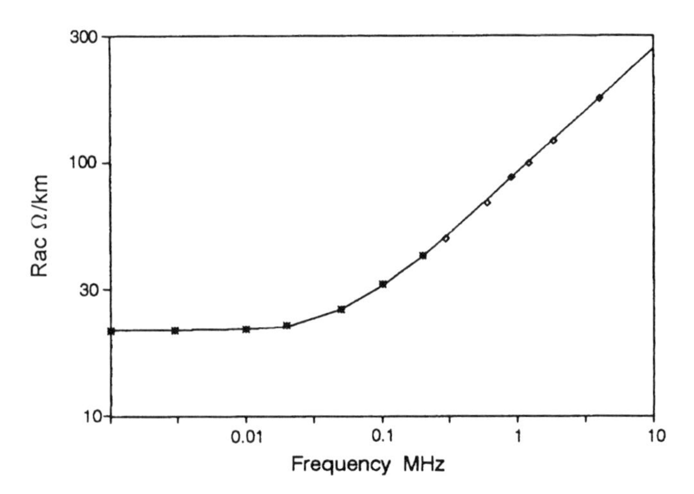

Figure 6.8
Resistance of coaxial pair

In Figure 6.8 is shown the calculated resistance (using Equation (6.23)) and measured resistances (symbols) for a coaxial pair having an inner conductor diameter of 1.19 mm and an outer conductor of diameter 4.42 mm. The thickness of the outer conductor was 0.19 mm. Again the measurements were made to a basic accuracy of 1% and the agreement is also very good.

CAPACITANCE AND INDUCTANCE BETWEEN CONDUCTORS

The capacitance, C, and inductance, L, formulae have a general form of

$$C = \frac{n\pi \cdot \varepsilon_0 K}{W} \text{ F/m} \quad \text{and} \quad L = \frac{\mu_0}{n\pi} W + L_0 \text{ H/m}$$

$$C = \frac{1000K}{k_C \cdot W} \text{ nF/km} \quad \text{and} \quad L = k_L \cdot W + L_0 \text{ mH/km}$$

where

$n = 1$ for balanced conductors and 2 for unbalanced conductors
$k_C = 36$ for balanced conductors and 18 for unbalanced conductors
$k_L = 0.4$ for balanced conductors and 0.2 for unbalanced conductors

K is the effective relative permittivity of the insulation surrounding the conductors. A balanced construction is where the potential of the two conductors is symmetrical about the earth potential and an unbalanced construction is where one conductor is at earth potential.

L_0 is the internal inductance of the conductors as calculated in the previous sections. W is a dimensionless geometric factor depending on the physical arrangement of the conductors and in 'perfect' arrangements (e.g. an isolated unscreened pair or a coaxial pair) is the same for capacitance and inductance. For other arrangements, where the electric and magnetic fields do not have the same boundaries, the geometric factor is approximately the same for the capacitance and the inductance. It is therefore convenient to deal with these two primary parameters together.

Isolated Symmetrical Pair

Dealing first with a 'perfect' symmetrical case, the isolated pair. The electric and magnetic lines of force are as sketched in Figure 6.9. The exact solution for the capacitance of this arrangement gives

$$W = \cosh^{-1}(D/d) = \ln[(D/d)\{1 + \sqrt{1 - (d/D)^2}\}]$$

Hence

$$C = \frac{1000 \cdot K}{36 \cdot \cosh^{-1}(D/d)} = \frac{1000 \cdot K}{36 \cdot \ln[D/d\{1 + \sqrt{1 - (d/D)^2}\}]} \text{ nF/km} \tag{6.27}$$

Similarly for the inductance of the pair

$$\begin{aligned} L &= 0.4 \cosh^{-1}(D/d) + L_0 \\ &= 0.4 \ln[D/d\{1 + \sqrt{1 - (d/D)^2}\}] + L_0 \text{ mH/km} \end{aligned} \tag{6.28}$$

For $D/d > 3$ these equations reduce to

$$C = \frac{1000 \cdot K}{36 \cdot \ln(2D/d)} \text{ nF/km} \tag{6.29}$$

$$L = 0.4 \ln(2D/d) + L_0 \text{ mH/km} \tag{6.30}$$

where L_0 is the internal inductance calculated from Equation (6.15). Note that the error in using (6.29) and (6.30) for $D/d = 2$ is about 5% and for $D/d = 1.5$ is about 14% (C too small, L too large).

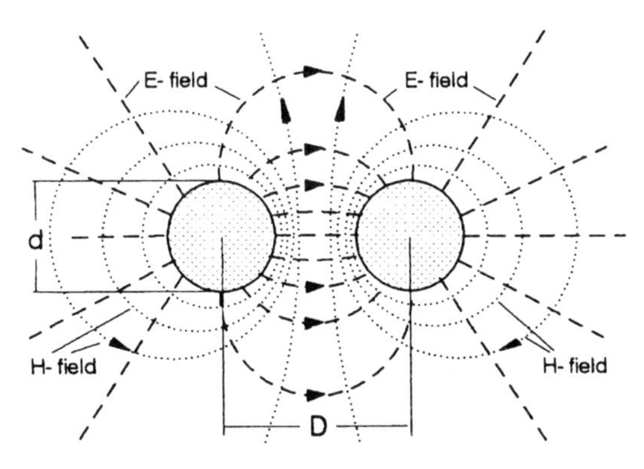

Figure 6.9
Electric and magnetic fields of isolated symmetric pair

Screened Balanced Pair

If the idea of a 'perfect' arrangement of conductors is defined as a case where the electric and magnetic fields have the same boundaries, then another 'perfect' arrangement is a balanced pair surrounded by a circular electromagnetic screen, i.e. a screen of low resistivity and high magnetic permeability, Figure 6.10. This arrangement was analysed by H. R. F. Carsten [8], among others. Carsten obtained the geometric factor as a series

$$W_S = \ln\left(\frac{2D}{d}\cdot\frac{1-h^2}{1+h^2}\right) - \left(\frac{d}{2D}\right)^2\left(1-\frac{4h^2}{1-h^2}\right)^2$$
$$-\left(\frac{d}{2D}\right)^4\left\{\frac{1}{2}\left(1+\frac{16h^2}{1-h^4}\right)^2+\left(\frac{1-4h^2}{1-h^4}\right)^2\left(1+\frac{8h^2(1+h^4)}{1-h^4}\right)\right\}-\cdots \quad (6.31)$$

where $h = D/S$ and S is the diameter of the screen. When $h = 0$ this series reduces to $W_S = \cosh^{-1}(D/d)$ as it should. As D/d gets smaller more and more terms are required to obtain an accurate answer.

For $D/d > 1.7$ and $h > 0.35$ it is sufficiently accurate (better than 3% error) to use only the first term of this series, so that

$$C = \frac{1000\cdot K}{36\cdot\ln\left(\frac{2D}{d}\cdot\frac{1-h^2}{1+h^2}\right)}\ \text{nF/km} \quad (6.32)$$

Since this is a 'perfect' arrangement we also have

$$L = 0.4\ln\left(\frac{2D}{d}\cdot\frac{1-h^2}{1+h^2}\right) + L_0\ \text{mH/km} \quad (6.33)$$

In Figure 6.11 a plot, calculated from Carsten's formula, of K/C against D/d is shown for a range of $h = D/S$. The advantage of this presentation is that the curves all run to zero at $D/d = 1$ instead of to infinity. To obtain the capacitance in nF/km, the value of the effective permittivity K is divided by the y-axis reading.

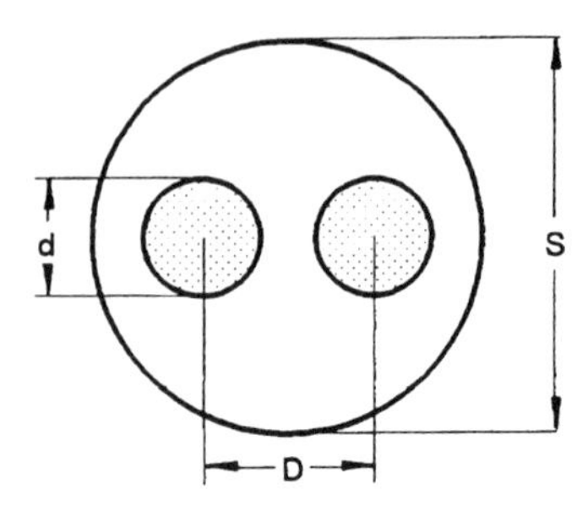

Figure 6.10
Screened balanced pair

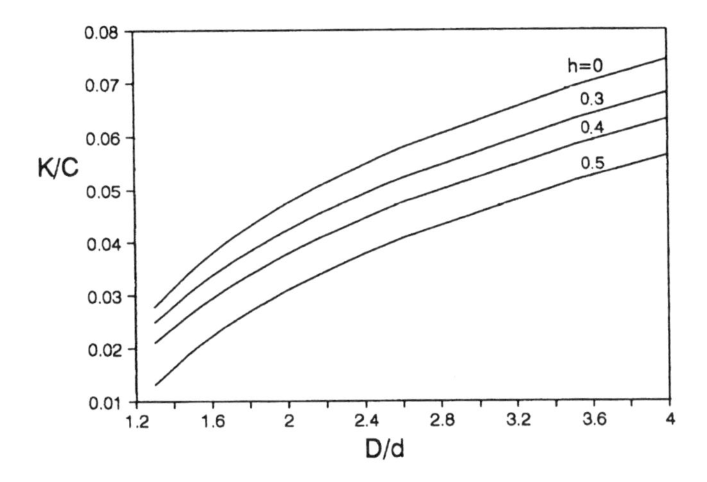

Figure 6.11
Capacitance of screened pair

Electrostatically Screened Pair

When the pair screen is not magnetic, the electric field terminates on the inner surface of the screen but the magnetic field penetrates the screen to an extent which varies with the frequency depending on the induced eddy currents. The author has found that an adequately accurate value (within 3%) of the inductance is given by

$$L = 0.4 \ln\left\{\frac{1.8D}{d} \cdot \frac{1-h^2}{1+h^2}\right\} + L_0 \text{ mH/km} \tag{6.34}$$

where

$$h = \frac{D}{S}\{1 - \exp(-u^{3/2})\} \tag{6.35}$$

u is the normalized frequency from Equation (6.2).

Multipair and Multiquad Cables

When many pairs or many quads are stranded together to form a large cable, any individual pair is surrounded by many others which are statistically at earth potential. These form a virtual screen around the pair under consideration. Carsten's analysis was extended to these cases by D. R. Robinson [9]. Because the analyses are in the form of infinite series and are tedious to apply, the author has derived empirical formulae which match the analytical values to better than 1% for $D/d \geq 1.25$. They are

$$\text{Multipair cable} \qquad C = \frac{1000 \cdot K}{36 \cdot \ln\left(\frac{2D}{d} - 0.9\right)} \text{ nF/km} \tag{6.36}$$

$$\text{Multiquad cable} \qquad C = \frac{1000 \cdot K}{36 \cdot \ln\left(\frac{2.4D}{d} - 1\right)} \text{ nF/km} \tag{6.37}$$

If these values are compared with Carsten's formulae for screened pairs or quads the effective value of h will be seen to vary from about 0.38 at $D/d = 1.5$ to 0.2 at $D/d = 4.5$ (for multipair cable), and from 0.45 to 0.35 for multiquad cables over the same range of D/d. Empirical formulae for the effective values of $S/D = 1/h$ are

$$\text{Multipair cable} \qquad \frac{S}{D} = 1.41 + 1.8\sqrt{D/d - 1} \tag{6.38}$$

$$\text{Multiquad cable} \qquad \frac{S}{D} = 2.33 + 0.88\sqrt{D/d - 1} \tag{6.39}$$

N.B. In all these formulae, D is the diameter of the conductor insulation.

Equations (6.36) and (6.37) are shown in Figure 6.12 and Equations (6.38) and (6.39) in Figure 6.13.

Cables such as these are 'imperfect' in the sense that the magnetic field penetrates the virtual screen to a greater or lesser extent depending on the frequency. The author has found that the following empirical formulae give values of inductance which agree with measurements on cables to within about 2%.

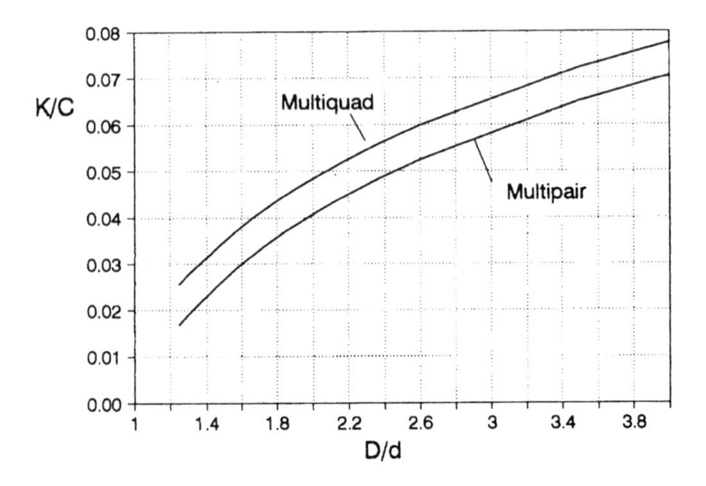

Figure 6.12
Capacitance of pairs in cables

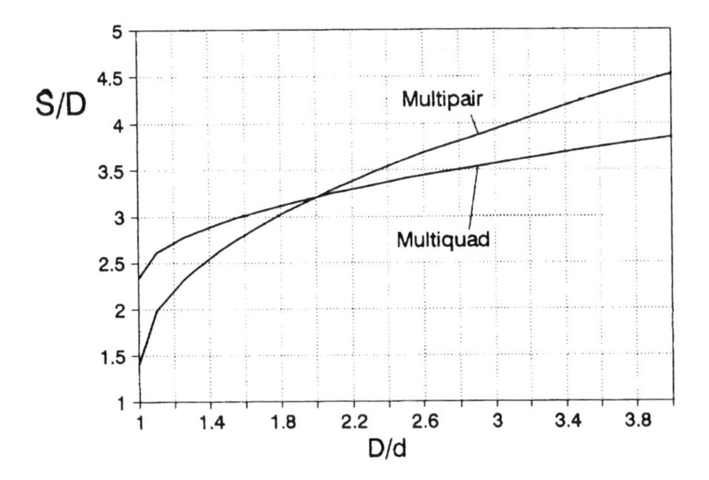

Figure 6.13
Virtual screen diameter in cables

Multipair cable $$L = 0.4\ln\left\{\frac{2D}{d}\cdot\frac{1-h^2}{1+h^2}\right\} + L_0 \text{ mH/km} \tag{6.40}$$

where $$h = \frac{0.85\{1 - 0.333\cdot\exp(-u^{3/2})\}}{1.41 + 1.8\sqrt{D/d - 1}} \tag{6.41}$$

Multiquad cable $$L = 0.4\ln\left\{\frac{3.18D}{d}\cdot\frac{1-h^2}{1+h^2}\right\} + L_0 \text{ mH/km} \tag{6.42}$$

where $$h = \frac{0.85\{1 - 0.333\cdot\exp(-u^{3/2})\}}{2.33 + 0.88\sqrt{D/d - 1}} \tag{6.43}$$

In Equations (6.39) and (6.41) the denominators are the values of S/D from (6.36) and (6.37) respectively and the numerators are similar to Equation (6.35). u is defined in Equation (6.2).

$$u = \frac{d}{2}\sqrt{\frac{\mu\mu_0\omega}{\rho}}$$

The good agreement between Equations (6.40) and (6.41), shown as lines, and measurements on actual cables, shown as symbols, is demonstrated in Figures 6.14 and 6.15.

Small Pairage Cables

The formulae given above for C and L of multipair and multiquad cables only apply if there is at least one layer of pairs or quads around the pair in question. For small pairage cables, or for the layer of pairs/quads next to a metallic cable screen, S should be

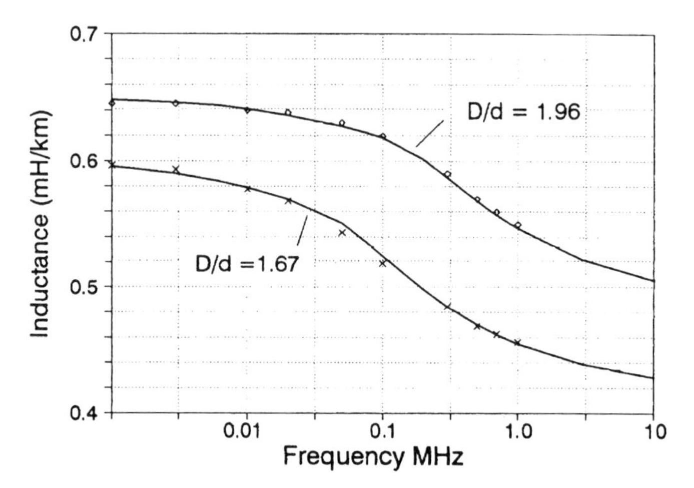

Figure 6.14
Inductance of pairs in multipair cables

determined by finding the average distance from the centre of the pair to neighbouring conductors. This can be done using a scale drawing of the cable, by drawing eight equally spaced radii from the pair centre to neighbouring pairs/quads/screens, measuring their lengths and taking the arithmetic mean in relation to the measured D. This determination of S/D is then used in Equation (6.32) for capacitance and as the denominator in either Equation (6.41) or (6.43) as appropriate for the inductance.

Coaxial Pairs

The capacitance and external inductance between the inner conductor of diameter d, and an outer coaxial conductor of internal diameter D, is analytically simple and exact.

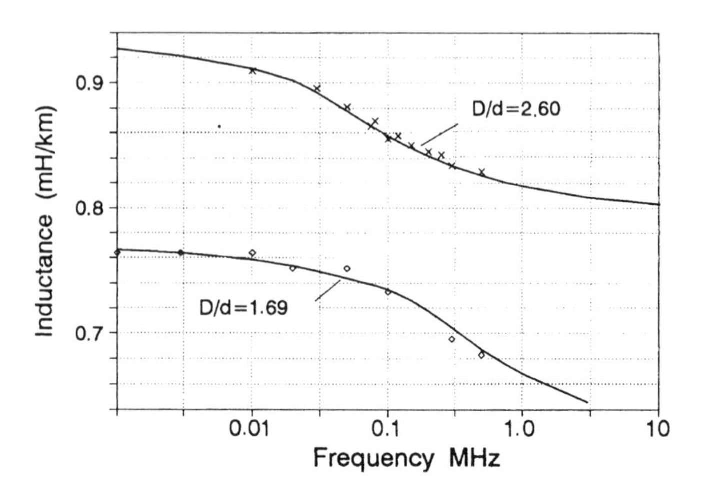

Figure 6.15
Inductance of pairs in multiquad cables

For the total inductance the internal inductance of the inner and outer conductors using Equation (6.24) must be added.

$$C = \frac{1000 \cdot K}{18 \ln(D/d)} \text{ nF/km} \tag{6.44}$$

$$L = 0.2 \ln(D/d) + X_0/\omega \text{ mH/km} \tag{6.45}$$

where X_0 is in Ω and ω is the angular frequency in krad/s.

For thin (<0.25 mm) outer conductors the internal inductance of the outer conductor is less than 2% of L at low frequencies and diminishes as the frequency rises. Very good agreement with measurements is achieved as shown in Figures 6.16 and 6.17. In Figure 6.16 the rise in measured inductance below 10 kHz is due to the magnetic field starting to penetrate the steel tapes lapped around the outer conductor for crosstalk protection. There are no steel tapes in the 2.12/9.43 polyethylene foam coaxial design.

For ease of reference the high-frequency approximations ($u > 6$) for the a.c. resistance and inductance for coaxial pairs are restated here from Equations (6.25) and (6.26)

$$R = 0.632\sqrt{f_{MHz}}\left(\frac{\sqrt{\rho_i}}{d} + \frac{\sqrt{\rho_0}}{D}\right)$$

$$L_0 = \frac{10^{-7}}{\sqrt{f_{MHz}}}\left(\frac{\sqrt{\rho_i}}{d} + \frac{\sqrt{\rho_0}}{D}\right)$$

For d, D in metres and ρ_i, ρ_0 in $\Omega \cdot$m, R will be in Ω/m and L_0 in H/m. For $u > 6$ (above 300 kHz for $d = 1$ mm copper inner), L_0 is substituted for X_0/ω in Equation (6.45).

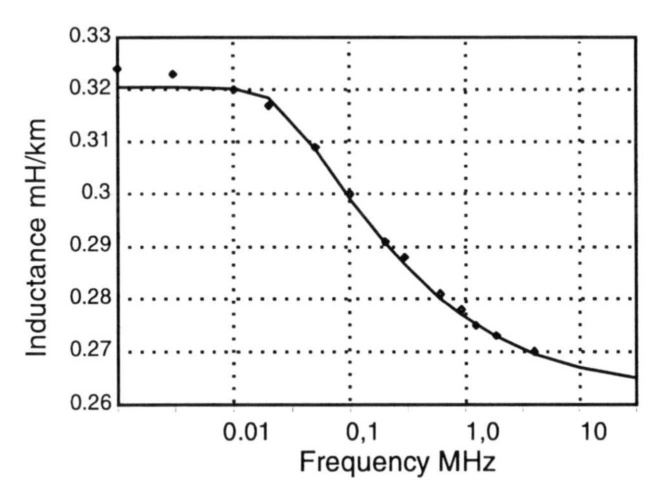

Figure 6.16
Inductance of 1.2/4.4 mm coaxial pair

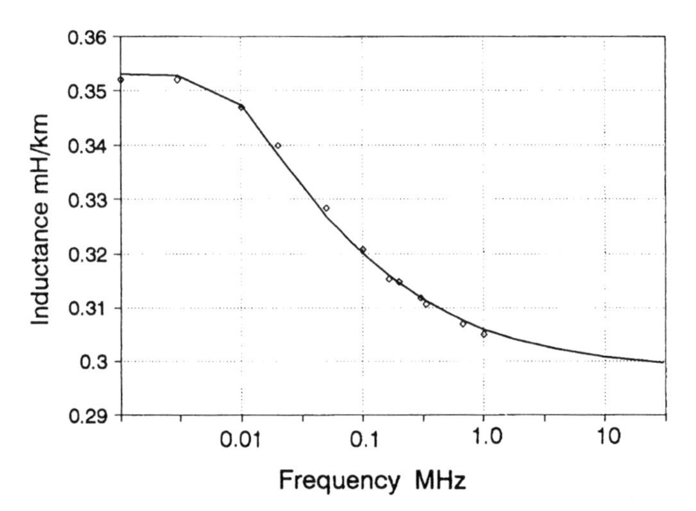

Figure 6.17
Inductance of 2.12/9.43 PEF coaxial pair

Isolated Conductor with Earth Return

For an isolated conductor suspended parallel to the surface of the earth, there is an image conductor caused by reflection in an effective, perfectly conducting, earth plane. This effective earth plane is a distance H below the physical earth surface, hence the image conductor is $2H$ below the physical conductor. See Fig. 6.18.

$$H = 330\sqrt{\frac{\rho}{f}}\,\text{metres} \tag{6.46}$$

where f is the frequency in Hz, and ρ is the earth resistivity in $\Omega \cdot$ metres. At 50 Hz with $50\,\Omega \cdot$ m earth resistivity, $H = 330$ m. Thus H is large compared with the conductor diameter up to quite high frequencies.

The capacitance of the conductor to ground is

$$C = \frac{1000K}{18\ln(4H/d)}\,\text{nF/km} \tag{6.47}$$

where d is the conductor diameter in metres. Since the electric field is most concentrated near the conductor the effective permittivity K is dominated by the air dielectric and is essentially equal to unity.

The inductance with earth return, and the mutual inductance between parallel conductors with earth return, have been calculated by A. Rosen [10] to be

$$L = 0.2\ln\left(\frac{4H}{d}\right) + 0.175\,\frac{y}{H} - j\left(0.16 - 0.175\,\frac{y}{H}\right)\,\text{mH/km} \tag{6.48}$$

$$M = 0.2\ln\left(\frac{2H}{D}\right) + 0.175\,\frac{\bar{y}}{H} - j\left(0.16 - 0.175\,\frac{\bar{y}}{H}\right)\,\text{mH/km} \tag{6.49}$$

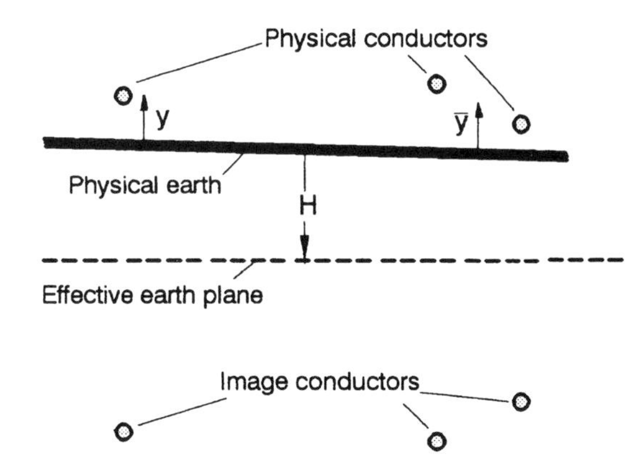

Figure 6.18
Conductors with earth return

where y is the height of the conductor above the physical earth surface and $\bar{y}$ is the mean height of the two conductors. D is the separation of the two conductors in metres. In both cases the second and third terms represent the complex inductance of the earth return path.

EFFECTIVE PERMITTIVITY

All the preceding capacitance formulae involve the multiplier K which is the effective relative permittivity of the space between the conductors. In the case of vacuum $K = 1$, and for gases nearly so. In the majority of practical constructions the conductors are first insulated with a concentrically applied dielectric, and then assembled together into pairs/quads and cables. This creates numerous air-filled interstices between the insulated conductors. In the case of coaxial constructions there is often an annular air-filled gap between the insulation and the outer conductor. Also it is fairly usual for the insulation itself to be a mixture of solid dielectric and air (string-and-tape and cellular or foam insulations). Thus the effective permittivity is the resultant permittivity of the mixed dielectrics.

The analysis of mixed dielectrics can be quite complex. In cases where the field equipotentials are concentric with the dielectric boundaries simple solutions are possible. In other cases empirical approaches are used based on the relative areas of dielectric (A_d) to the total area of the cross-section (A_t). If k is the permittivity of the solid dielectric

$$K = (k - 1)\frac{A_d}{A_t} + 1 \tag{6.50}$$

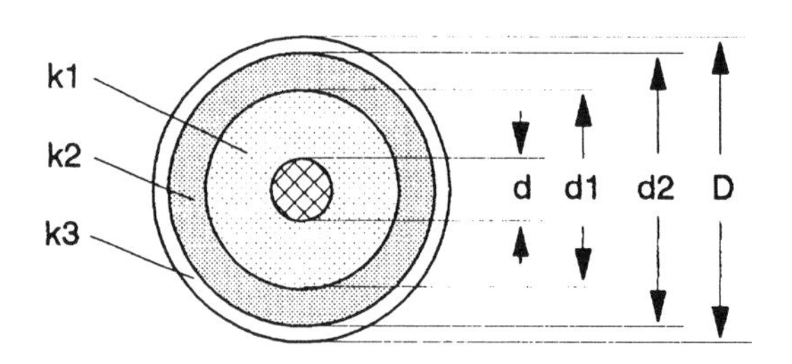

Figure 6.19
Concentric dieelectric layers

Concentric Dielectric Layers

This case, Fig. 6.19, can be analysed exactly with the following result for n layers.

$$K = \frac{\ln(D/d)}{\dfrac{1}{k_1}\ln\left(\dfrac{d_1}{d}\right) + \dfrac{1}{k_2}\ln\left(\dfrac{d_2}{d_1}\right) + \dfrac{1}{k_3}\ln\left(\dfrac{d_3}{d_2}\right) + \cdots + \dfrac{1}{k_n}\ln\left(\dfrac{D}{d_{n-1}}\right)} \tag{6.51}$$

Obviously for one layer this gives $K = k_1$.

Cellular Insulations

In the case of cellular or foam insulations it is assumed that the mixture of air and solid dielectric is homogeneous. The amount of air to solid is proportional to the percentage blow B, and the effective permittivity is

$$K = (k - 1) \cdot (1 - B/100) + 1 \tag{6.52}$$

The percentage blow is determined by the density of the unblown insulation to the density of the blown insulation. Thus for an unblown insulation $B = 0$ and for a highly blown insulation $B = 65\%$ (for example).

Multipair and Multiquad Cables

For a pair in a multipair cable, the neighbouring pair conductors form a virtual screen of diameter S. A well-compacted cable of n wires of insulation diameter D form a bunch of overall diameter

$$D_c = 1.7D\sqrt{n/2}$$

Hence in a diameter S (by equating D_c to S and rearranging) there will be a non-integer number of insulated wires

$$n = 0.692\left(\frac{S}{D}\right)^2$$

The area of solid dielectric in diameter S, to the total dielectric area in diameter S, is

$$p = \frac{n\pi/4 \cdot (D^2 - d^2)}{\pi/4 \cdot (S^2 - n \cdot d^2)} = \frac{n(D^2 - d^2)}{(S^2 - n \cdot d^2)}$$

This does not take into account the inhomogeneity of the electric field however. Assume arbitrarily that the solid dielectric between adjacent conductors of a pair, being in the most concentrated field, has twice the effect of other portions of solid dielectric, then the ratio becomes

$$p = \frac{(n+1)\{(D/d)^2 - 1\}}{\{(S/D \cdot D/d)^2 - n\}} \tag{6.53}$$

$$\text{whence} \quad K = (k - k_I) \cdot p + k_I \tag{6.54}$$

Here, k is the effective permittivity of the solid insulation and k_I is the permittivity of the interstices. This is unity in in the case of air-filled interstices but is greater than unity in the case of interstices filled, for example, with a waterproofing gel.

Proceeding in the same way for multiquad cables one arrives at

$$D_c = 1.8\sqrt{n/2} \quad \text{hence} \quad n = 0.617\left(\frac{S}{D}\right)^2$$

$$q = \frac{(n+2)\{(D/d)^2 - 1\}}{\{(S/D \cdot D/d)^2 - n\}} \tag{6.55}$$

$$\text{whence} \quad K = (k - k_I) \cdot q + k_I \tag{6.56}$$

Both p and q are graphed in Figure 6.20. using Equations (6.38) and (6.39) to obtain S/D from D/d These factors have been confirmed experimentally over a wide range of constructions to better than 1%.

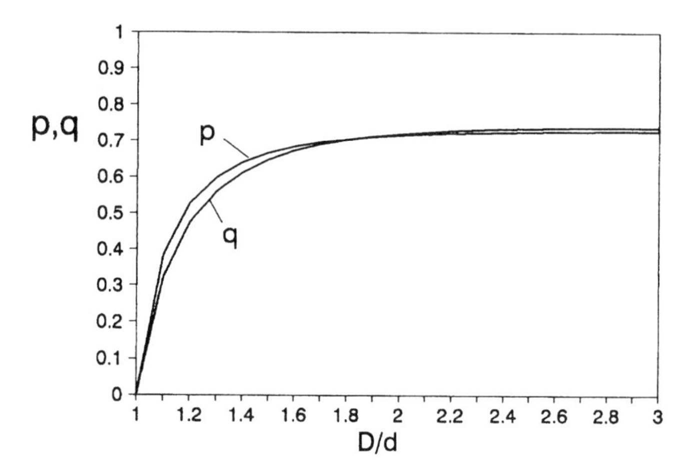

Figure 6.20
Permittivity factor for cables

Table 6.1
Permittivities and effective permittivities

Insulation	k	K
Dry paper/air		
Quad trunk	2.05	1.7
Quad carrier	1.75	1.55
Unit twin	2.2	1.84
Solid polythene	2.28	1.91
Cellular polythene		
40% blow; twin construction	1.77	1.54
60% blow; twin construction	1.51	1.36
Polyvinylchloride (PVC)	3–5	2.4–3.8
Polytetrafluoroethylene (PTFE)	2.1	1.77

Semi-air spaced Dielectrics

For coaxial semi-air-spaced insulations as in Figure 2.5, the value of p in Equation (6.54) becomes

$$p = t/T$$

where t is the thickness of the disk and T the spacing of the disks, or for the thread insulation

$$p = \theta/360$$

where θ is the angle subtended by the thread diameter at the centre conductor.

Table 6.1 gives guidelines to the values of permittivities and effective permittivities found in practice.

Table 6.2
Values of loss tangent

Insulation	$\tan \delta$
Dry paper/air	0.015
Solid polyethylene	0.0005
Cellular polyethylene	0.0013
Air-spaced polyethylene	0.0001
Polyvinylchloride	0.03
Polytetrafluorethylene	0.0003

CONDUCTANCE

The conductance between the conductors of a transmission line is the dissipative component of the capacitance and is simply calculated from the tangent of the loss angle,

$$G = \omega C \cdot \tan\delta$$
$$= 2\pi f_{kHz} \times C_{nF/km} \times \tan\delta \ \mu\text{S/km} \quad (6.57)$$

In most practical insulations δ is very small and $\tan\delta = \delta$. For semi-air-spaced dielectrics $\tan\delta$ is reduced in much the same way that the permittivity is reduced to the effective permittivity. However in cellular insulations some blowing agents used in manufacture leave residues in the insulation which increase the loss.

For paper insulations, it is usual to insulate with the paper containing approximately 6% moisture to increase the tensile elongation. In order to restore the low dielectric loss the cables must be well dried before sheathing. This is done by holding the stranded cables at about 120 °C in a vacuum of about 0.01 Torr for 3–4 h. During testing and installation it is necessary to avoid the ingress of atmospheric moisture.

Table 6.2 gives typical values of for cable insulations.

7

Transmission Parameters

In order to derive the transmission behaviour of lines from their primary parameters, and to investigate the effect of terminations, some mathematics is necessary [20].

MATHEMATICAL TREATMENT OF TRANSMISSION LINES

Consider a uniform line with primary parameters of R, L, G and C per kilometre. A short section of this line of length dx will have a series impedance of $Z_S = (R + j\omega L) \cdot dx$ and a shunt admittance of $Y_P = (G + j\omega C) \cdot dx$.

The voltage across dx will be

$$E - (E + dE) = (R + j\omega L) \cdot I \cdot dx$$

hence $$-\frac{dE}{dx} = I(R + j\omega L)$$

The current change through dx will be

$$I - (I + dI) = (G + j\omega C) \cdot E \cdot dx$$

hence $$-\frac{dI}{dx} = E(G + j\omega C)$$

and

$$\frac{d^2 I}{dx^2} = -\frac{dE}{dx}(G + j\omega C) \cdot I = (R + j\omega L)(G + j\omega C) \cdot I = \gamma^2 \cdot I$$

This is a differential equation whose solution is

$$I = a \cdot \varepsilon^{\gamma x} + b \cdot \varepsilon^{-\gamma x} \tag{7.1}$$

For a line of infinite length $I \rightarrow 0$ as $x \rightarrow \infty$ hence $a = 0$. When $x = 0, I = I_S = b$ hence $I = I_S \varepsilon^{-\gamma x}$. This represents an exponentially decaying current with a propagation constant defined by

$$\gamma = \sqrt{(R + j\omega L)(G + j\omega C)} \tag{7.2}$$

Also

$$E = \frac{-1}{(G + j\omega C)} \frac{dI}{dx} = \frac{-1}{(G + j\omega C)} \cdot I_S(-\gamma) \cdot \varepsilon^{-\gamma x}$$

$$= I_S \frac{\sqrt{(R + j\omega L)(G + j\omega C)}}{(G + j\omega C)} \tag{7.3}$$

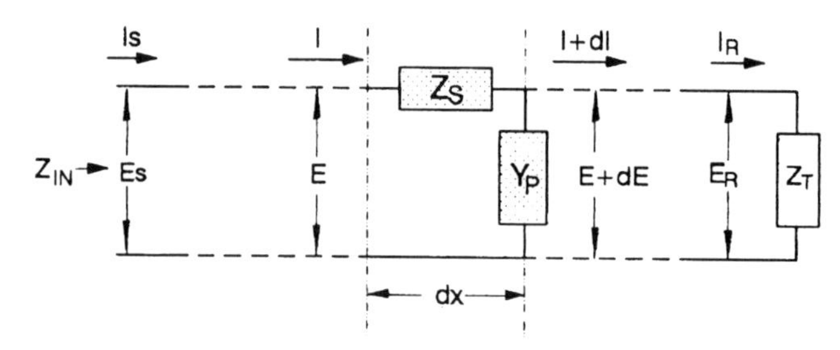

Figure 7.1
Elementary line section

When $x = 0$, $E = E_S$ which is the sent voltage. By definition the ratio E_S/I_S on an infinite length of line is the characteristic impedance

$$Z_0 = \frac{E_S}{I_S} = \sqrt{\frac{(R + j\omega L)}{(G + j\omega C)}} \tag{7.4}$$

Note that

$$\gamma Z_0 = R + j\omega L \tag{7.5}$$

and

$$\frac{\gamma}{Z_0} = G + j\omega C \tag{7.6}$$

Write Z_0 and γ in real and imaginary parts, with Z_R and Z_I being the real and imaginary parts of the characteristic impedance and α and β being the attenuation and phase constants. Substituting in (7.5) and (7.6) we have

$$(\alpha + j\beta)(Z_R + jZ_I) = R + j\omega L$$

Multiplying through and equating real and imaginary parts gives

$$R = \alpha Z_R - \beta Z \quad \text{and} \quad \omega L = \beta Z_R + \alpha Z_I \tag{7.5a}$$

Similarly

$$\alpha + j\beta = (Z_R + jZ_I)(G + j\omega C) \tag{7.5b}$$

whence

$$G = \frac{\alpha Z_R + \beta Z_I}{Z_R^2 + Z_I^2} \quad \text{and} \quad \omega C = \frac{\beta Z_R - \alpha Z_I}{Z_R^2 + Z_I^2} \tag{7.6a}$$

If $G \ll \omega C$

$$\alpha = -\omega C \cdot Z_I \quad \text{and} \quad \beta = \omega C \cdot Z_R \tag{7.6b}$$

Hyperbolic Form of Line Equations

Since

$$\sinh x = \frac{\varepsilon^x - \varepsilon^{-x}}{2} \quad \text{and} \quad \cosh x = \frac{\varepsilon^x + \varepsilon^{-x}}{2}$$

we can write

$$\varepsilon^{\gamma x} = \cosh \gamma x + \sinh \gamma x \quad \text{and} \quad \varepsilon^{-\gamma x} = \cosh \gamma x - \sinh \gamma x$$

and substituting in (7.1)

$$\begin{aligned} I &= (a+b)\cosh \gamma x + (a-b)\sinh \gamma x \\ &= A\cosh \gamma x + B\sinh \gamma x \end{aligned} \tag{7.7}$$

From (7.3)

$$E = \frac{-1}{(G+j\omega C)} \cdot \frac{dI}{dx} = \frac{-\gamma}{(G+j\omega C)}(A\sinh \gamma x + B\cosh \gamma x)$$

$$= -Z_0 \cdot (A\sinh \gamma x + B\cosh \gamma x) \tag{7.8}$$

For $x = 0$, from (7.7) and (7.8). $I_S = A$ and $E_S = -B \cdot Z_0$, thus at any point of the line

$$E = E_S \cosh \gamma x - I_S Z_0 \sinh \gamma x \tag{7.9}$$

$$I = I_S \cosh \gamma x - \frac{E_S}{Z_0}\sinh \gamma x \tag{7.10}$$

For a line of finite length l, the received current and voltage are I_R and E_R. By definition the terminating impedance is

$$Z_T = \frac{E_R}{I_R} = \frac{E_S \cosh \gamma l - I_S Z_0 \sinh \gamma l}{I_S \cosh \gamma l - \dfrac{E_S}{Z_0}\sinh \gamma l}$$

Cross-multiply and re-arrange to extract the input impedance $Z_{in} = E_S/I_S$

$$Z_{in} = Z_0 \cdot \frac{Z_T \cosh \gamma l + Z_0 \sinh \gamma l}{Z_0 \cosh \gamma l + Z_T \sinh \gamma l} \tag{7.11}$$

$$= Z_0 \cdot \frac{Z_T + Z_0 \tanh \gamma l}{Z_0 + Z_T \tanh \gamma l} \tag{7.12}$$

Special Cases

(i) When $Z_T = Z_0$ or when $\gamma l \geq 3$ then $Z_{in} = Z_0$ (7.13)

(ii) When the line is short circuited $Z_T = 0$ then $Z_{in} = Z_{sc} = Z_0 \cdot \tanh \gamma l$ (7.14)

(iii) When the line is open circuited $Z_T = \infty$ then $Z_{in} = Z_{oc} = Z_0 \cdot \coth \gamma l$ (7.15)

Case (i) shows that if the line is terminated in an impedance which matches Z_0 in magnitude and phase, or if the total attenuation of the line is large, then the input impedance will be Z_0. Cases (ii) and (iii) are used when measuring Z_0 and γ using impedance bridges and by the method of resonances.

Techniques for Dealing with Complex Quantities

All the impedances and propagation constants are complex, having real and imaginary parts. Some mathematical techniques for dealing with them are given below. Having measured Z_{sc} and Z_{oc}, from (7.14) and (7.15) we have

$$Z_0 = \sqrt{Z_{sc} \cdot Z_{oc}} \tag{7.16}$$

hence

$$|Z_0| = \sqrt{|Z_{sc}||Z_{oc}|} \quad \text{and} \quad \angle Z_0 = \tfrac{1}{2}(\angle Z_{sc} + \angle Z_{oc}) \tag{7.17}$$

So that

$$Z_0 = Z_R + jZ_I = |Z_0|\{\cos(\angle Z_0) + j\sin(\angle Z_0)\} \tag{7.18}$$

Also from (7.14) and (7.15)

$$\tanh \gamma l = \tanh(\alpha l + j\beta l) = \sqrt{\frac{Z_{sc}}{Z_{oc}}} = A + jB \quad \text{(say)} \tag{7.19}$$

Using hyperbolic identities it may be shown that

$$\tanh 2\alpha l = \frac{2A}{1 + A^2 + B^2} = C \quad \text{(say)} \tag{7.20}$$

and

$$\tan 2\beta l = \frac{2B}{1 - (A^2 + B^2)} \tag{7.21}$$

Now by definition

$$\tanh 2x = \frac{\varepsilon^{2x} - \varepsilon^{-2x}}{\varepsilon^{2x} + \varepsilon^{-2x}} = \frac{\varepsilon^{4x} - 1}{\varepsilon^{4x} + 1}$$

hence from (7.20)

$$C = \frac{\varepsilon^{4\alpha l} - 1}{\varepsilon^{4\alpha l} + 1} \quad \text{leading to} \quad \varepsilon^{4\alpha l} = \frac{1 + C}{1 - C}$$

Thus

$$\alpha l = \frac{1}{4}\ln\left(\frac{1 + C}{1 - C}\right) \quad \text{and} \quad \beta l = \frac{1}{2}\tan^{-1}\left(\frac{2B}{1 - A^2 - B^2}\right) \tag{7.22}$$

The input impedance of a terminated line can be dealt with in the same way. From (7.11)

$$\frac{Z_{in}}{Z_0} = \frac{Z_T \cosh \gamma l + Z_0 \sinh \gamma l}{Z_0 \cosh \gamma l + Z_T \sinh \gamma l} = \frac{\cosh \gamma l + \dfrac{Z_0}{Z_T} \sinh \gamma l}{\dfrac{Z_0}{Z_T} \cosh \gamma l + \sinh \gamma l} \tag{7.23}$$

Define

$$\frac{Z_0}{Z_T} = \tanh(p + jq) = A + jB \tag{7.24}$$

Then as in (7.20) and (7.21)

$$\tanh 2p = \frac{2A}{1 + A^2 + B^2} = C \quad \text{and} \quad \tan 2q = \frac{2B}{1 - (A^2 + B^2)}$$

whence

$$p = \frac{1}{4} \ln\left(\frac{1 + C}{1 - C}\right) \quad \text{and} \quad q = \frac{1}{2} \tan^{-1}\left(\frac{2B}{1 - A^2 - B^2}\right) \tag{7.25}$$

Substituting (7.24) in (7.23) and using hyperbolic identities results in

$$\frac{|Z_{in}|}{|Z_T|} = \frac{\sqrt{\sinh^2(\alpha l + p) + \cos^2(\beta l + q)}}{\sqrt{\sinh^2(\alpha l + p) + \sin^2(\beta l + q)}} \tag{7.26}$$

and

$$\angle\left(\frac{Z_{in}}{Z_0}\right) = \tan^{-1}\{\tanh(\alpha l + p) \cdot \tan(\beta l + q)\}$$

$$- \tan^{-1}\{\coth(\alpha l + p) \cdot \tan(\beta l + q)\} \tag{7.27}$$

If the transmission parameters are to be calculated from the primary parameters R, L, G and C per km then from (7.4)

$$Z_0 = \sqrt{\frac{R + j\omega L}{G + j\omega C}}$$

$$|Z_0| = \sqrt[4]{\frac{R^2 + \omega^2 L^2}{G^2 + \omega^2 C^2}} \tag{7.28}$$

$$\angle Z_0 = \frac{1}{2}\left\{\tan^{-1}\left(\frac{\omega L}{R}\right) - \tan^{-1}\left(\frac{\omega C}{G}\right)\right\} \tag{7.29}$$

and from (7.2)

$$\gamma = \sqrt{(R + j\omega L)(G + j\omega C)}$$

$$|\gamma| = \sqrt[4]{(R^2 + \omega^2 L^2)(G^2 + \omega^2 C^2)} \tag{7.30}$$

$$\angle\gamma = \frac{1}{2}\left\{\tan^{-1}\left(\frac{\omega L}{R}\right) + \tan^{-1}\left(\frac{\omega C}{G}\right)\right\} \tag{7.31}$$

From (7.30) and (7.31) the attenuation and phase constants are calculated as

$$\alpha = |\gamma| \cdot \cos(\angle\gamma) \quad \text{nepers/km} \tag{7.32}$$

$$\beta = |\gamma| \cdot \sin(\angle\gamma) \quad \text{radians/km} \tag{7.33}$$

In all the formulae the attenuations are in nepers; to convert to dB multiply by 8.686.

Reflections

When a line is not terminated in its characteristic impedance, a signal travelling down the line is reflected partially (or completely if the line is open- or short-circuited) and the reflection travels back to the input. Similarly there will be a reflection when two lines of different impedance are connected together. To find the size of the reflection replace the line by a generator of e.m.f. E with impedance Z_0 and replace the termination Z_T by Z_0 in series with Z_R. Then further replace Z_R by the impedance-less generator with e.m.f. $-IZ_R$. The two circuits in Figure 7.2 are identical.

We have the following relations

$$Z_T = Z_0 + Z_R \qquad E = I(2Z_0 + Z_R)$$

$$I = I_1 + I_2 + \frac{E - IZ_R}{2Z_0}$$

$$I_1 = \frac{E}{2Z_0} \qquad I_2 = \frac{-IZ_R}{2Z_0}$$

$$\frac{I_2}{I_1} = \frac{-IZ_R}{E} = \frac{-Z_R}{2Z_0 + Z_R} = \frac{Z_T - Z_0}{Z_0 + Z_T} \tag{7.34}$$

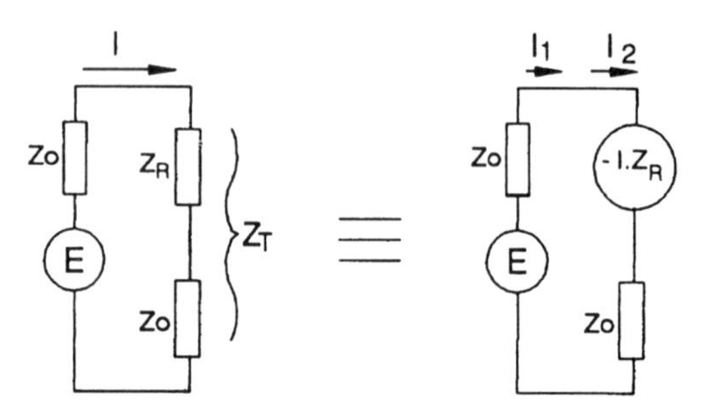

Figure 7.2
Equivalent circuit of mis-terminated line

This is the current reflection coefficient. Since $E = IZ_0$ the same ratio gives the voltage reflection coefficient also.

The return loss is defined as $$RL = 20\log_{10}\left|\frac{Z_T - Z_0}{Z_0 + Z_T}\right| \text{dB} \tag{7.35}$$

The voltage transmission coefficient $$= 1 + \frac{Z_T - Z_0}{Z_0 + Z_T} = \frac{2 \cdot Z_T}{Z_0 + Z_T} \tag{7.36}$$

Wavelength and Velocity of Propagation

The wavelength λ, of the signal is defined as the distance between two points where the voltage or current are in the same phase (i.e. the phase has retarded by 2π radians between the points). Since the phase constant is β radians/km the distance between these points is $\lambda = 2\pi/\beta$. The phase velocity of propagation is the product of wavelength and frequency

$$V_P = \lambda \cdot f = \frac{2\pi \cdot f}{\beta} = \frac{\omega}{\beta} \tag{7.37}$$

When $G \ll \omega C$ and $\omega \to \infty$, from (7.6b) $\beta = \omega C \cdot Z_R = \omega C\sqrt{L/C} = \omega\sqrt{LC}$. Thus

$$V_P \to \frac{1}{\sqrt{LC}} \qquad \text{as} \quad \omega \to \infty$$

$$= \sqrt{\frac{36 \ln W_c}{K \cdot 10^{-6} 0.4 \cdot 10^{-3} \ln W_l}} = \frac{300}{\sqrt{K}}\sqrt{\frac{W_c}{W_l}} \text{Mm/s} \tag{7.38}$$

where K is the effective permittivity, W_c is the geometric factor for capacitance and W_l is the geometric factor for the inductance of the line. This is the velocity of propagation of a single frequency continuous sine wave signal. Real signals are composed of the superposition of many such sine waves, harmonically related in Fourier series. The propagation velocity of a distinctive point of the resultant signal is called the group velocity V_G, which is slower than the phase velocity due to the dispersion

$$V_G = V_P - \lambda\frac{dV_P}{d\lambda} = \frac{d\omega}{d\beta} \tag{7.39}$$

APPROXIMATE FORMS

We have seen in (7.6b) that if $G \ll \omega C$ it may be neglected resulting in simplification of the formulae to

$$\alpha = -\omega C \cdot Z_I \quad \text{and} \quad \beta = \omega C \cdot Z_R$$

Substituting in (7.5a) gives

$$R = 2\alpha \cdot Z_R \quad \text{and} \quad L = C \cdot |Z_0|^2 \tag{7.40}$$

Similarly since $\omega L \gg R$ as $\omega \to \infty$

$$Z_0 = \sqrt{\frac{R + j\omega L}{j\omega C}} = \sqrt{\frac{L}{C}} \tag{7.41}$$

and

$$\gamma = \sqrt{(R + j\omega L) \cdot (G + j\omega C)} = j\omega\sqrt{LC}\sqrt{\left(1 + \frac{R}{j\omega L}\right) \cdot \left(1 + \frac{G}{j\omega C}\right)}$$

If $R \ll \omega L$ and $G \ll \omega C$ this may be expanded by the binomial theorem to give

$$\gamma = \alpha + j\beta = j\omega\sqrt{LC}\left\{1 + \frac{R}{2j\omega L} + \frac{G}{2j\omega C}\right\} = \frac{R}{2}\sqrt{\frac{C}{L}} + \frac{G}{2}\sqrt{\frac{L}{C}} + j\omega\sqrt{LC}$$

Thus

$$\alpha = \frac{R}{2}\sqrt{\frac{C}{L}} + \frac{G}{2}\sqrt{\frac{L}{C}} = \frac{R}{2Z_0} + \frac{GZ_0}{2} \text{ nepers/km} \tag{7.42}$$

and

$$\beta = \omega\sqrt{LC} \text{ radians/km} \tag{7.43}$$

Loaded Lines

By adding inductance to the line, ωL may be made greater than R even at low frequencies, so that the approximate expressions above may be used. This loading may be done continuously by winding a close helix of magnetic wire around each individual conductor (as in some early submarine cables) or by connecting lumped inductances (loading coils) in series with the pair at regular intervals, e.g. 88 mH/2000 yds (88 mH/1.14 km). The advantages of loading are to reduce the attenuation (G being negligible) and to achieve a constant impedance, attenuation and phase velocity with frequency.

With lumped loading, as in any LC low-pass filter, there is a cut-off frequency f_{co} where the attenuation suddenly rises steeply

$$f_{co} = \frac{1}{\pi S\sqrt{LC}} \tag{7.44}$$

where S is coil spacing in km and L and C are the total inductance and capacitance per km.

Coaxial Pair Approximations

At very high frequencies, as in (7.41),

$$Z_0 = Z_\infty = \sqrt{\frac{L}{C}} = \sqrt{\frac{(\mu_0/2\pi)\ln(D/d)\,\mathrm{H/m}}{(2\pi\varepsilon_0 K)/\{\ln(D/d)\}\,\mathrm{F/m}}}$$

$$= \frac{1}{2\pi}\sqrt{\frac{\mu_0}{K\varepsilon_0}}\ln\left(\frac{D}{d}\right) = \frac{60}{\sqrt{K}}\ln\left(\frac{D}{d}\right) \qquad (7.45)$$

At moderately high frequencies e.g. $> 60\,\mathrm{kHz}$

$$Z_0 = \sqrt{\frac{R + j\omega(L + L_0)}{j\omega C}} = \sqrt{\frac{L}{C}}\cdot\sqrt{1 + \frac{R + j\omega L_0}{j\omega L}}$$

Since $R + j\omega L_0 \ll j\omega L$ at these frequencies

$$Z_0 = \sqrt{\frac{L}{C}}\cdot\left(1 + \frac{R + j\omega L_0}{2\cdot j\omega L}\right) = Z_\infty + \frac{R + j\omega L_0}{2j\omega\sqrt{LC}} \qquad (7.46)$$

At these higher frequencies the resistance and internal inductance of the inner and outer conductors may be obtained from (6.7) and (6.16) utilizing the first terms only. Combined in series and substituting for u_i and u_o and the d.c. resistances, these give

$$R + j\omega L_0 = \left(\frac{1}{d} + \frac{1}{D}\right)\sqrt{\frac{\mu_0 \rho f}{\pi}}(1 + j) \qquad (7.47)$$

Also

$$\sqrt{LC} = \sqrt{(\mu_0/2\pi)\ln(D/d)(2\pi\varepsilon_0 K)/\ln(D/d)} = \sqrt{\mu_0\varepsilon_0 K} \qquad (7.48)$$

Substituting in (7.46) we get

$$Z_0 = Z_\infty + \left(\frac{1}{d} + \frac{1}{D}\right)\cdot\frac{1}{2\pi}\cdot\sqrt{\frac{\rho}{2\varepsilon_0 K\cdot\omega}}(1 - j)$$

$$= Z_\infty + \frac{A}{\sqrt{\omega}}(1 - j) \qquad (7.49)$$

where

$$A = \left(\frac{1}{d} + \frac{1}{D}\right)\cdot\frac{1}{2\pi}\cdot\sqrt{\frac{\rho}{2\varepsilon_0 K}} \qquad (7.50)$$

$\mu_0 = 4\pi\times 10^{-7}\,\mathrm{H/m}$ and $\varepsilon_0 = (1/36\pi)\times 10^{-9}\,\mathrm{F/m}$. d, D are the inner and outer conductor diameters in metres and K is the effective permittivity. L_0 is the total internal inductance of the conductors. The frequency f is in Hz.

Similarly for the attenuation from (7.42). Substituting for R, Z_∞ and G gives

$$\alpha = \frac{\left(\frac{1}{d}+\frac{1}{D}\right)\sqrt{\frac{\mu_0 \rho f}{\pi}}}{\frac{2}{2\pi}\sqrt{\frac{\mu_0}{K\varepsilon_0}}\ln\left(\frac{D}{d}\right)} + \frac{\omega C \tan\delta}{2}\cdot\frac{1}{2\pi}\sqrt{\frac{\mu_0}{K\varepsilon_0}}\ln\left(\frac{D}{d}\right)$$

Substituting for C from (7.45) and splitting the resistivity between inner and outer conductors gives

$$\alpha = \sqrt{\frac{\varepsilon_0 K\omega}{2}}\frac{1}{\ln\left(\frac{D}{d}\right)}\left(\frac{\sqrt{\rho_i}}{d}+\frac{\sqrt{\rho_0}}{D}\right) + \frac{1}{2}\sqrt{\mu_0\varepsilon_0 K}\cdot\omega\cdot\tan\delta\,\text{nepers/km} \quad (7.51)$$

$$= \frac{4.51\sqrt{Kf}}{\ln\left(\frac{D}{d}\right)}\left(\frac{\sqrt{\rho_i}}{d}+\frac{\sqrt{\rho_0}}{D}\right) + 9.61\sqrt{K}\cdot f\cdot\tan\delta\,\text{dB/100 m} \quad (7.52)$$

with f in MHz. ρ_i and ρ_0 are in $\Omega\cdot$m. These approximate expressions were most useful in the days before PCs and spreadsheets. Now they serve to make more clear the dependencies of the secondary parameters on the physical make-up of the coaxial.

Effect of Dimensional Variations in Coaxial Pairs

By partial differentiation of Equation (7.45), the effect of dimensional variations in the coaxial may be summarized by

$$\frac{\Delta Z_\infty}{Z_\infty} = \frac{-1}{\ln(D/d)}\frac{\Delta d}{d} + \frac{1}{\ln(D/d)}\frac{\Delta D}{D} - \frac{1}{2}\frac{\Delta K}{K} \quad (7.53)$$

Eccentric Coaxial Pairs

If the axis of the inner conductor is displaced from the axis of the outer conductor by a small amount y, Schelkunoff [7] has shown that

$$Z_\infty = \frac{60}{\sqrt{K}}\ln x\left(1 - \frac{e^2x^2}{(x^2-1)\ln x}\right) \quad (7.54)$$

where $x = D/d$ and the eccentricity $e = 2y/D$. Thus an eccentric inner conductor reduces the impedance by the fraction

$$\frac{\Delta Z_\infty}{Z_\infty} = \frac{e^2x^2}{(x^2-1)\ln x} \quad (7.55)$$

and the capacitance and inductance are changed by the fractions

$$\frac{\Delta C}{C} = 2\,\frac{e^2x^2}{(x^2-1)\ln x} \quad \text{and} \quad \frac{\Delta L}{L} = -2\,\frac{e^2x^2}{(x^2-1)\ln x} \tag{7.56}$$

The attenuation is increased by the fraction

$$\frac{\Delta\alpha}{\alpha} = \left\{1 + \frac{2e^2}{x} + \frac{e^2x^2}{(x^2-1)\ln x}\right\} \tag{7.57}$$

Elliptic Confocal Pairs

If the line is formed from confocal elliptical conductors, where the foci are distance f apart, d and D are the major axes of the inner and outer conductors respectively and $x = D/d$, then Shebes [11] has shown that

$$Z_\infty = \frac{60}{\sqrt{K}} \ln x + \frac{15f^2}{\sqrt{K}\,D^2}\,(x^2-1) \tag{7.58}$$

PRACTICAL APPLICATIONS OF THEORY

To illustrate the application of the formulae developed in Chapters 6 and 7 we will apply them to the design and measurement of a local area network (LAN) cable containing two foil-screened twisted pairs. The cable is required to work up to 20 MHz. A sketch of the cable is shown in Figure 7.3.

The requirements are

Conductors: Plain copper wire 22 AWG (0.635 mm diam.)
max. d.c. resistance 57.1 Ω/km at 20 °C.

Insulation: Cellular polyethylene

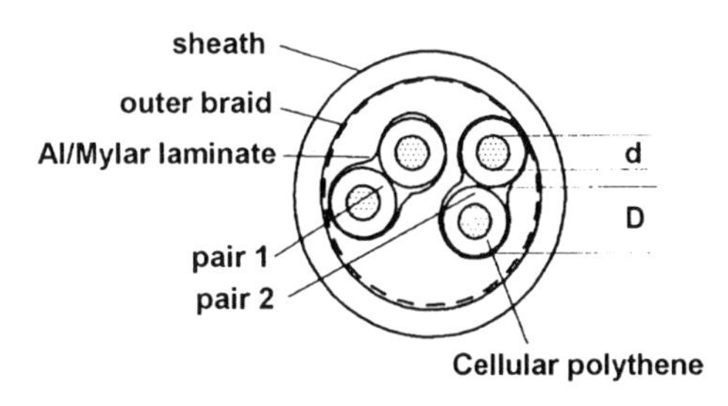

Figure 7.3
Local area network cable

Twisted pairs to be individually close-screened with aluminium/Mylar laminate (foil thickness 50 μm). Thus $h = D/S = 0.5$

Characteristic impedance

150 Ω ± 10% from 3 to 20 MHz
155 Ω ± 10% at 28.4 kHz
270 Ω ± 10% at 9.6 kHz

Attenuation: Less than

45 dB/km at 16 MHz
22 dB/km at 4 MHz
5 dB/km at 28.4 kHz
3 dB/km at 9.6 kHz

Since the impedance rises at low frequency we aim for a high frequency impedance of 145 Ω to give more tolerance at low frequency. From Equation (7.4)

$$Z_0 = \sqrt{\frac{R + j\omega L}{G + j\omega C}} \rightarrow \sqrt{\frac{L}{C}} \quad \text{at high frequency.}$$

From Equations (6.34) and (6.36) when u is large and for $h = 0.5$

$$Z_0 = \sqrt{\frac{36 \times 0.4 \times 10^{-3} \times \ln(1.08D/d) \times \ln(1.2D/d)}{1000 \times K \times 10^{-9}}} = \frac{120}{\sqrt{K}} \cdot \ln\left(\frac{1.14D}{d}\right)$$

For cellular polythene at 60% blow, Equation (6.52) gives the effective permittivity

$$K = (2.28 - 1)(1 - 60/100) + 1 = 1.512$$

Hence

$$\ln\left(\frac{1.14D}{d}\right) = \frac{145}{120}\sqrt{1.512} = 1.486$$

and

$$\varepsilon^{1.486} = 4.419 = \frac{1.14D}{d}$$

Then for $d = 0.635$ mm, $D = 2.46$ mm and from Equation (6.32), $C = 27.32$ nF/km.

A cable was made to this design and although within specification the impedance was lower than designed (138.7 Ω) and the capacitance was higher at 30.05 nF/km. Careful measurements showed

Conductor diameter $d = 0.634$ mm

Insulation diameter $D = 2.435$ mm

The cellular insulation had the correct blow of 60% based on mass and volume but there was an 0.1 mm skin of solid polythene next to the conductor.

From Equation (6.51)

$$K = \frac{\ln(2.435/0.634)}{\frac{1}{2.28}\ln\left(\frac{0.834}{0.634}\right) + \frac{1}{1.512}\ln\left(\frac{2.435}{0.834}\right)} = 1.623$$

corresponding to an effective blow of 51%. Using this with the measured d and D gives good agreement with the measurements.

$$C = \frac{1000K}{36\ln(1.2D/d)} = \frac{1000 \times 1.623}{36 \times \ln(1.2 \times 2.435/0.634)} = 29.5\,\text{nF/km} \quad \text{(measured 30.05)}$$

$$Z_0 = \frac{120}{\sqrt{1.623}}\sqrt{\ln\left(\frac{1.2 \times 2.435}{0.634}\right) \times \ln\left(\frac{1.08 \times 2.435}{0.634}\right)} = 138.9\,\Omega \quad \text{(measured 138.7)}$$

Impedance Measurements

The open- and short-circuit impedances were measured over a frequency range of 1 kHz to 1.6 MHz on a 10 m sample of the cable using an S & H balanced impedance bridge (R217). A 10 m length was chosen to avoid resonance effects and also to ensure that the average of physical measurements at each end of the sample would be typical of the whole length. Equations (7.16) to (7.22) were used in a straightforward spreadsheet to calculate all the cable parameters, Table 7.1.

The continuation of the spreadsheet to calculate the primary and secondary (transmission) parameters is shown in Table 7.2.

Commentary on Measurements

The S & H impedance bridge has a basic accuracy of 1% up to 1.6 MHz but this degrades at high frequencies for the minor vector component and for small phase

Table 7.1
Spreadsheet for calculating parameters from Z_{oc} and Z_{sc}

1 Freq. kHz	2 R_{sc} Ω	3 L_{sc} mH	4 G_{oc} mS	5 C_{oc} nF	6 $\lvert Z_{sc}\rvert$ Ω	7 $\lvert Y_{oc}\rvert$ mS	8 $\angle Z_{sc}$ rad	9 $\angle Y_{oc}$ rad	10 $\lvert\tanh \gamma l\rvert$	11 $\angle\tanh \gamma l$	12 A	13 B	14 C
1	1.07	0.0091	0	0.301	1.07	0.002	0.053	1.568	0.0014	0.8106	0.0010	0.0010	0.0020
3	1.07	0.0092	0	0.301	1.09	0.006	0.159	1.570	0.0025	0.8646	0.0016	0.0019	0.0032
10	1.12	0.0087	0	0.301	1.24	0.019	0.455	1.571	0.0049	1.0128	0.0026	0.0041	0.0051
20	1.19	0.0081	0	0.301	1.57	0.038	0.707	1.571	0.0077	1.1388	0.0032	0.0070	0.0064
50	1.37	0.0075	0	0.301	2.72	0.095	1.045	1.571	0.0160	1.3081	0.0042	0.0155	0.0083
100	1.57	0.0072	0	0.301	4.76	0.189	1.235	1.571	0.0300	1.4028	0.0050	0.0296	0.0100
1000	4.01	0.0063	0.005	0.308	39.8	1.935	1.470	1.568	0.2775	1.5192	0.0143	0.2771	0.0266
1600	5.39	0.0063	0.015	0.320	63.6	3.217	1.486	1.566	0.4522	1.5261	0.0202	0.0355	0.0335

Table 7.2
Continuation of spreadsheet

15 Freq. kHz	16 $\|Z_0\|$ Ω	17 $\angle Z_0$ deg	18 Att. dB/km	19 Phase rad/km	20 R Ω/km	21 L mH/km	22 G mS/km	23 C nF/km	24 $\|\gamma\|$ N	25 $\angle\gamma$ deg	26 $\|\gamma Z_0\|$ Ω	27 $\angle\gamma Z_0$ deg	28 $\|\gamma/Z_0\|$ S	29 $\angle\gamma/Z_0$ deg	30 V_P Mm/s
1	753	−43	0.85	0.10	107	0.91	0.05	30.10	0.14	46.44	107	3.0	0.000	89.85	61
3	438	−40	1.40	0.19	107	0.92	0.05	30.10	0.25	49.54	109	9.1	0.001	89.95	100
10	257	−32	2.23	0.41	112	0.87	0.01	30.10	0.49	58.03	124	26.1	0.002	90.00	153
20	204	−25	2.80	0.70	119	0.81	0.07	30.10	0.77	65.25	157	40.5	0.004	90.00	180
50	170	−15	3.62	1.55	137	0.75	0.05	30.10	1.60	74.95	272	59.9	0.009	90.00	203
100	159	−10	4.35	2.96	157	0.71	−0.09	30.09	3.00	80.38	476	70.8	0.019	90.00	212
1000	143	−3	11.6	27.04	381	0.61	−1.96	30.05	27.1	87.18	3882	84.4	0.189	90.00	232
1600	141	−2	14.6	42.44	475	0.59	−20.6	30.06	42.5	87.74	5971	85.4	0.302	90.00	237

angles. The difficulty is illustrated by the values in column 22 for G. The error can be traced back to the measurement of G_{oc} in column 4. Here the bridge is incapable of resolving the angle $\pi/2 - \delta$ from $\pi/2$. This makes a considerable difference to the cosine of $\angle(\gamma/Z_0)$ but not to the sine of this angle. The only parameter seriously affected is G.

In column 30 the phase velocity V_P is becoming asymptotic to 240 Mm/s as the frequency rises. Applying Equation (7.39) gives

$$V_P = \frac{300}{\sqrt{K}} \cdot \sqrt{\frac{\ln(1.2D/d)}{\ln(1.08D/d)}} = \frac{300}{\sqrt{K}} \cdot \sqrt{\frac{\ln 4.6088}{\ln 4.1479}} = \frac{300}{\sqrt{K}} \times 1.0364$$

Thus

$$K = \left\{\frac{300}{240}\right\}^2 \times 1.0364 = 1.619$$

This is in good agreement with the value of K calculated from the physical measurements on the insulated core (namely 1.623). The worksheet programme outlined above is perfectly general and can be applied to sc/oc measurements made on any type of conductive transmission line.

Comparison of Measured and Calculated Parameters

We now calculate the primary parameters from Equations (6.14), (6.15), (6.18), (6.20), (6.32), (6.34) and (6.35), and the secondary parameters using Equations (7.2), (7.4) and (7.28)–(7.33) starting from $d = 0.634$ mm, $D = 2.435$ mm, $K = 1.623$, $R_O = 108\ \Omega$/loop km, $t = 0.05$ mm, $h = 0.5$ and $\tan\delta = 0.0013$ as shown in Table 7.3

Capacitance

It has been seen that the formulae of Chapter 6 enable the capacitance and permittivity to be calculated to better than 2% in most cases. This example confirms this so long as the physical dimensions are known to better than 1%. Since the capacitance depends

Table 7.3
Spreadsheet for calculating primary and secondary parameters

1 Freq. kHz	2 R_{ac} Ω/km	3 L mH/km	4 G mS/km	5 $\|Z_0\|$ Ω	6 $\angle Z_0$ deg	7 α dB/km	8 β rad/km	9 V_P Mm/s	10 h'	11 $\angle Z_s$ rad/km	12 $\angle Y_P$ rad/km	13 R/R_0	14 $\|Z_s\|$ kΩ/km	15 $\|Y_P\|$ S/km	16 ω Mrad/s	17 u cond	18 v scrn	19 R_{sac} Ω/km
1	107	0.92	0	761	−43	0.84	0.1	61	0.05	0.05	1.57	1.00	0.11	0	0.01	0.22	0.04	36.5
10	118	0.88	0	265	−32	2.29	0.4	152	0.22	0.44	1.57	1.00	0.13	0	0.06	0.68	0.12	36.5
20	122	0.84	0	209	−25	2.80	0.7	178	0.31	0.71	1.57	1.00	0.16	0	0.13	0.96	0.17	36.5
50	132	0.77	0.01	172	−14	3.44	1.5	203	0.42	1.07	1.57	1.02	0.28	0.01	0.31	1.53	0.27	36.5
100	150	0.71	0.02	160	−9	4.14	2.9	215	0.48	1.25	1.57	1.12	0.47	0.02	0.63	2.16	0.38	36.5
1 M	363	0.64	0.24	145	−3	11.1	26.8	235	0.50	1.48	1.57	2.79	3.87	0.19	6.28	6.82	1.19	36.9
1.6 M	447	0.60	0.39	143	−2	13.8	42.5	237	0.50	1.50	1.57	3.46	6.09	0.30	10.0	8.62	1.50	37.5
4 M	684	0.59	0.96	142	−1	21.6	105	239	0.50	1.52	1.57	5.32	14.9	0.74	25.1	13.6	2.37	42.5
10 M	1.1 k	0.58	2.41	141	−1	35.7	261	241	0.50	1.54	1.57	8.27	36.7	1.85	62.8	21.6	3.75	64.3

only on the dimensions it will be constant with frequency if the permittivity is also constant with frequency. This is so for most types of cable insulation, however paper insulations if slightly moist can show a decrease in permittivity of about 1% as the frequency rises to about 50 kHz followed by a slight rise thereafter.

Resistance

Figure 7.4 shows the measured and calculated values of resistance with frequency. An agreement of better than 3% is achieved even with the complicated effects of the screen. The data are plotted on log–log scales and the effect of the screen on the calculated values is also shown. Without the screen the resistance becomes asymptotic to a square-root law with frequency.

With the screen, significant extra losses are seen. The calculation is quite sensitive to the screen thickness used. The best agreement was obtained with $t = 0.050$ mm (the design value) although the measured value was 0.055 ± 0.005 mm. Also included on the graphs are values of resistance calculated from sweep frequency measurements to 20 MHz.

Inductance

In Chapter 6 the good agreement of inductance calculations using the various empirical relationships was demonstated. Figure 7.5 shows that good agreement, within 0.03 mH/km or 3.7%, is also achieved in this case despite the complicating effect of the screen.

The 4% scatter is partly due to measurement error (about 2%) and also about 2% due to the empirical relation for proximity effect, as discussed in Chapter 6.

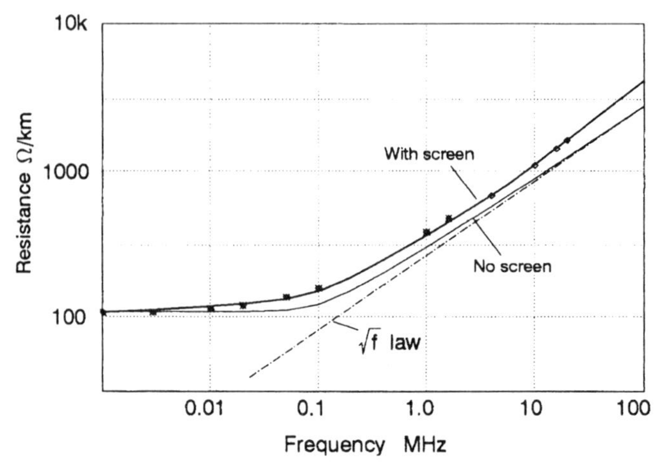

Figure 7.4
Resistance of screened twin cable

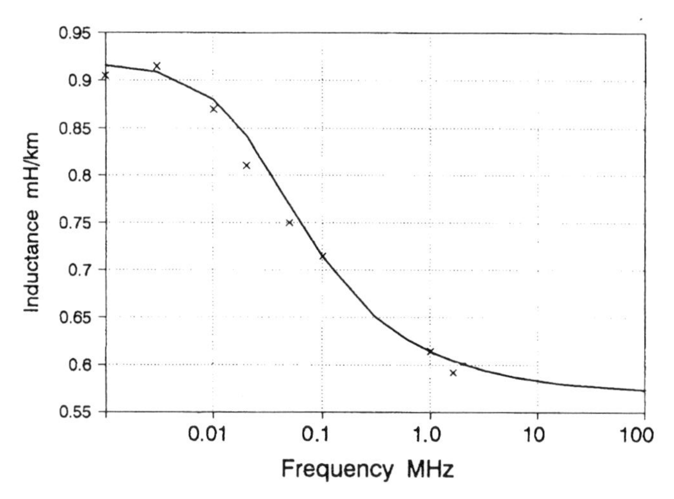

Figure 7.5
Inductance of screened twin cable

Conductance

The difficulties with conductance measurement have been discussed earlier. The expected conductance variation with frequency for a constant loss-angle of $\delta = 0.0013$ is shown in Figure 7.6 The linear variation with frequency in this log–log plot is to be expected from Equation (6.57)

$$G = \omega C \cdot \tan \delta$$

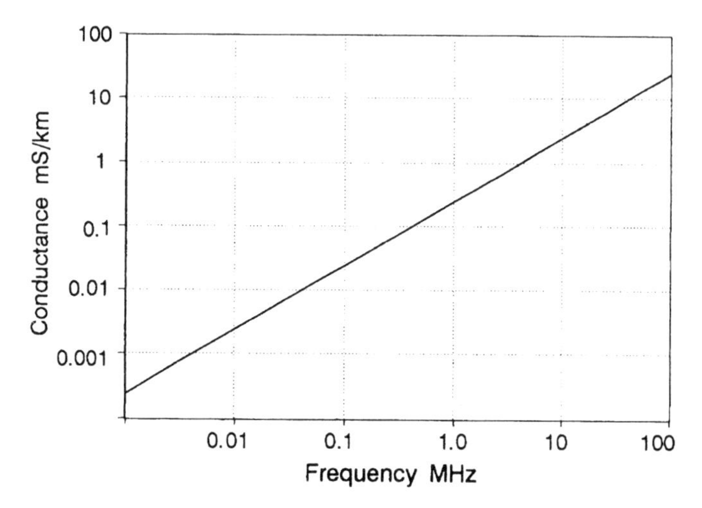

Figure 7.6
Conductance of screened twin

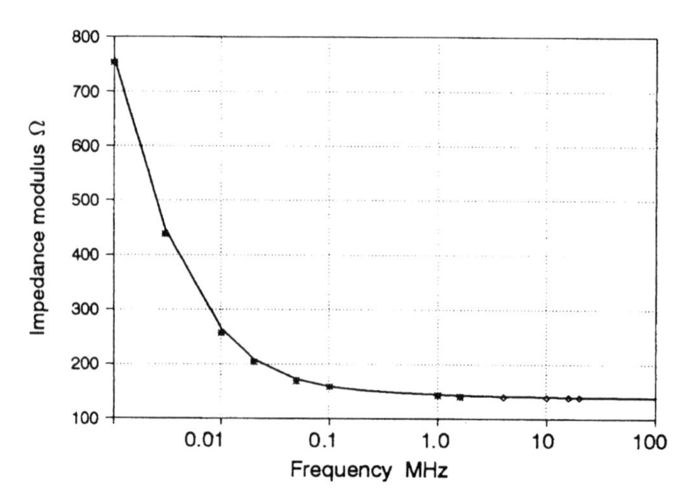

Figure 7.7
Impedance modulus of screened twin

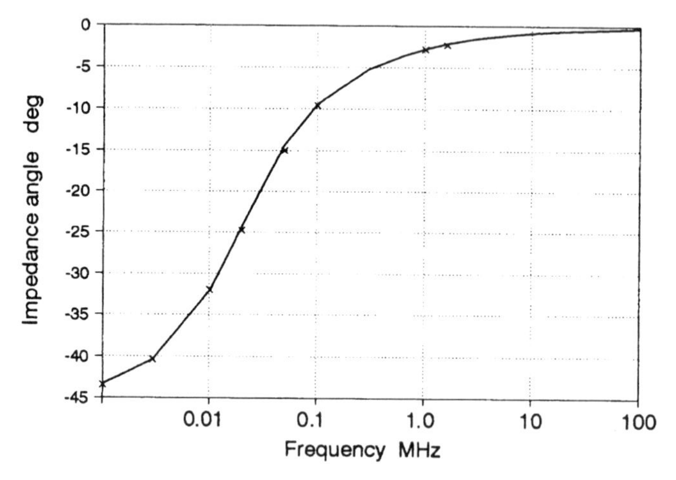

Figure 7.8
Impedance angle of screened twin

Impedance Modulus and Angle

Figures 7.7 and 7.8 show the measured and calculated values of the impedance with frequency. Above 1.6 MHz the values of the modulus were calculated from swept-frequency return loss measurements up to 20 MHz. Good agreement is seen to within 2% and 0.5° respectively.

Attenuation

The measured and calculated attenuations are shown in Figure 7.9 on log–log scales. This plot shows how the attenuation becomes asymptotic to square-root law at both

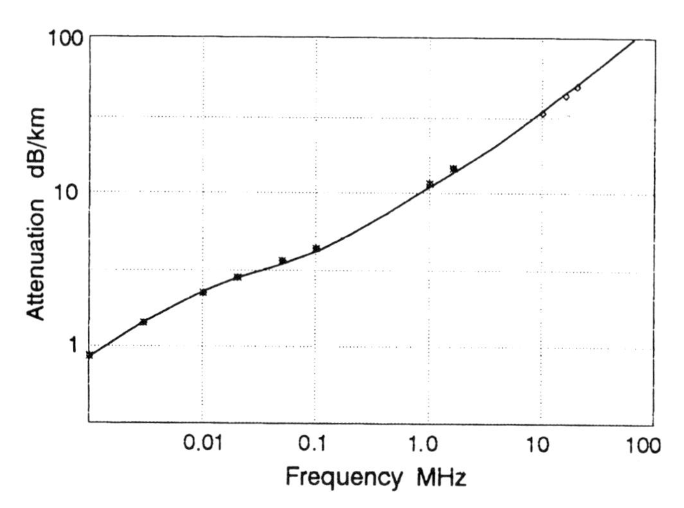

Figure 7.9
Attenuation coefficient of screened twin

low and high frequencies with a transition region between 5 kHz and 100 kHz. Above 3 MHz the effect of screen losses becomes apparent. The attenuation values above 1.6 MHz were taken from a swept-frequency insertion loss measurement up to 20 MHz. This form of measurement is sensitive to the termination impedance and it would be expected that above 1 MHz, where the pair impedance falls below the 150 Ω terminating resistance, that the insertion loss (measured on 500 m) would be about 1.3 dB/km lower than the calculated attenuation. This is about what is observed. Otherwise the measured values agree with the calculated values to within 0.5 dB/km.

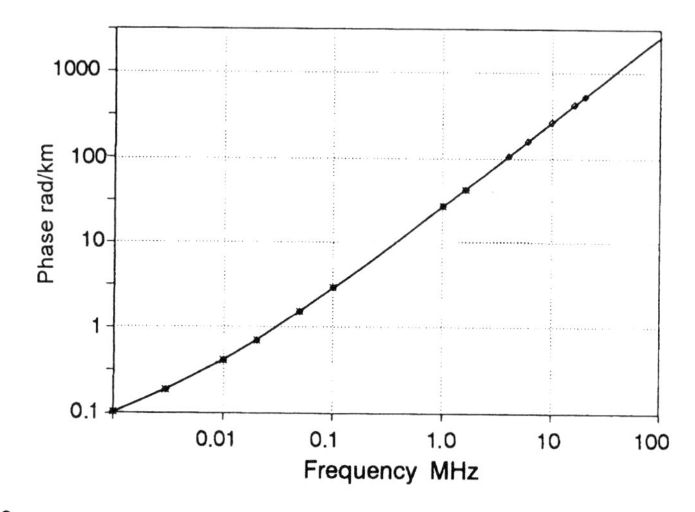

Figure 7.10
Phase constant of screened twin

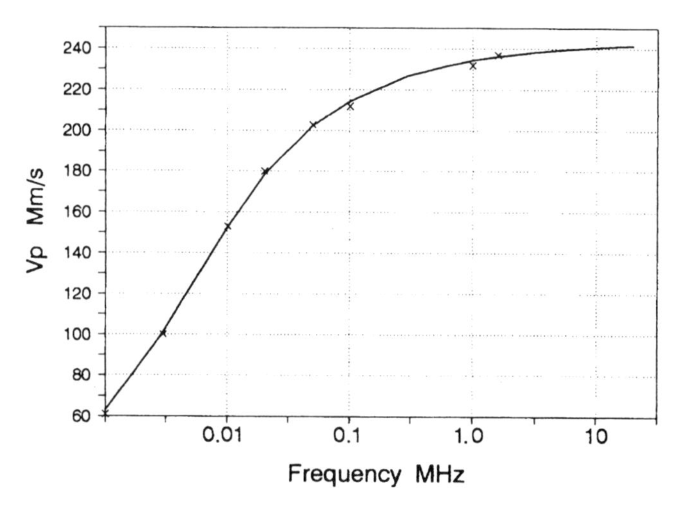

Figure 7.11
Phase velocity of screened twin

Phase Constant and Phase Velocity

Figures 7.10 and 7.11 also show good agreement in β and V_P measurements. Notice that β becomes asymptotic to a linear law as the frequency rises and consequently $V_P = \omega/\beta$ becomes asymptotic to a constant value of 240 Mm/s. This detailed survey of measurements and calculations on screened twin cable serves to show that excellent agreement to within 3% can be achieved. However the accuracy of measurement must be good, especially of the physical dimensions.

MIS-TERMINATED LINES

In this section we look at the effect of mis-termination on the input impedance of the same type of cable as in the previous section (but with a slightly lower h.f. impedance, 138.5 Ω). Figure 7.12 shows the input impedance versus frequency when a 100 m length of line is terminated in a 150 Ω resistor (the system impedance) and in a nominally correct termination of a 138.5 Ω resistor. Even in the latter case there are significant ripples due to the characteristic impedance having an imaginary component which is not terminated. If the line were terminated in its true characteristic impedance (e.g. by using a long length of identical cable terminated in a 138.5 Ω resistor) the input impedance would be the smooth mean curve in the figure, which is of course the modulus of the characteristic impedance. Although the swing of the input impedance is $\pm 8\,\Omega$ with the 150 Ω termination, the return loss is better than -28 dB. This would generally be considered acceptable to the system.

The frequency of the impedance ripple, $\Delta f = 1.2$ MHz, is the same for both terminations shown, or for any mismatched termination on this length of this type of cable. Thus if this frequency of ripple is observed on this type of cable one knows that

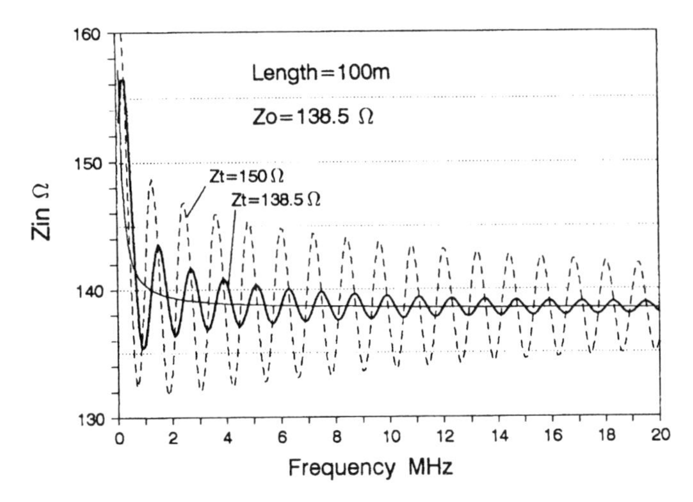

Figure 7.12
Input impedance of screened twin with resistive terminations

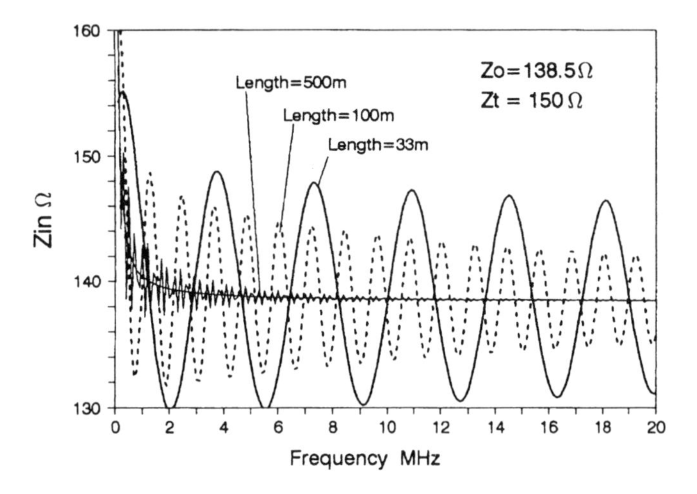

Figure 7.13
Effect of length on input impedance

there must be a reflective fault 100 m down the line. Figure 7.13 shows the effect of varying the line length, and it will be seen that the ripple frequency is inversely proportional to the length. Notice also that the amplitude of the ripple decreases as the length increases (for the same mismatch) owing to the line attenuation increasing with length.

Mathematically, from Equation (7.26) we have

$$\frac{|Z_{in}|}{|Z_0|} = \sqrt{\frac{\sinh^2(\alpha l + p) + \cos^2(\beta l + q)}{\sinh^2(\alpha l + p) + \sin^2(\beta l + q)}}$$

This has maxima at frequencies for which

$$\cos^2(\beta l + q) = 1 \quad \text{and simultaneously} \quad \sin^2(\beta l + q) = 0 \quad \text{i.e. for } (\beta l + q) = n\pi$$

For adjacent maxima $(\beta_2 - \beta_1)l = \{(n+1)\pi - q_2\} - (n\pi - q) = \pi - (q_1 - q_2)$. It may be shown that $(q_1 - q_2)$ is less than 1% of π (and is zero for oc or sc termination). Hence

$$\Delta\beta = \frac{\pi}{l} = \frac{\Delta\omega}{V_P} \quad \text{thus} \quad \Delta f = \frac{V_P}{2l} = \frac{3000}{\sqrt{K} \cdot 2l} \tag{7.59}$$

The phase velocity for this line is known to be 240 Mm/s within 1% above 3 MHz, so that by measuring the ripple frequency Δf, the distance to the fault or termination can be determined to within 1%. Alternatively V_P and hence K may be determined.

Impedance Simulator

Even if a resistive termination is adjusted to equal the line impedance the resultant ripple may be too large for measurement or diagnostic purposes. A better match can be achieved by using an impedance simulator which is an RC network matching both the real and imaginary parts of the line impedance over a wide frequency range. Generally speaking a two-branch simulator as shown in Figure 7.14 can match the impedance at only one frequency. With more branches the match can be made at several frequencies. In the figure the capacitor was adjusted for best match at 1.5 MHz and the variable resistor was adjusted to match $|Z_0|$ at 8 MHz. The resulting simulator impedance matches the line impedance within $\pm 0.6\,\Omega$ over the whole band on this 100 m length.

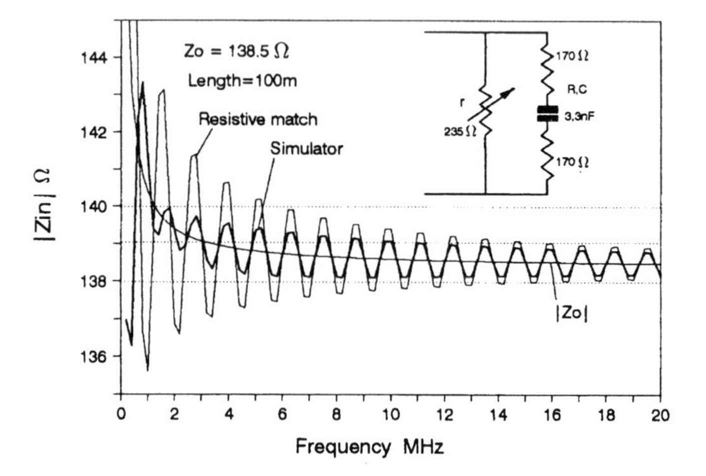

Figure 7.14
Input impedance with simulator termination

For longer lengths the match improves as the attenuation increases; for shorter lengths the ripple amplitude increases but even at 30 m the ripple is less than $\pm 1\,\Omega$ over the band. The input impedance of the two-branch simulator is

$$Z_R = \frac{r\{rR + R^2 + 1/(\omega C)\}^2}{(R + r)^2 + 1/(\omega C)^2} \tag{7.60}$$

$$Z_I = \frac{-r^2/(\omega C)}{(R + r)^2 + 1/(\omega C)^2} \tag{7.61}$$

where Z_R and Z_I are the real and imaginary parts of the simulator impedance. The variable resistor can be adjusted to match the real part of the line impedance while still maintaining a good match to the imaginary part of the impedance.

IMPEDANCE IRREGULARITIES

Multiple Reflections

In the same way that mis-termination can cause ripples on the input impedance trace of a uniform line, so does an impedance fault on the transmission line. The frequency of the ripple will be related to the distance down the line from the test end. It is not unknown for an input impedance trace on a network analyser to resemble that shown in Figure 7.15. The problem then is to interpret this curve to try to find the physical causes. In fact this curve is simulated by the superposition of the four sinusoids shown in Figure 7.16. These sinusoids are the individual responses to impedance faults at distances

$$x = 50\,\text{m} \qquad x = 25\,\text{m} \qquad x = 9\,\text{m} \qquad x = 1.6\,\text{m}$$

Although all the faults are of approximately the same size, the more distant ones are subject to more attenuation and therefore produce smaller deviations at the test end.

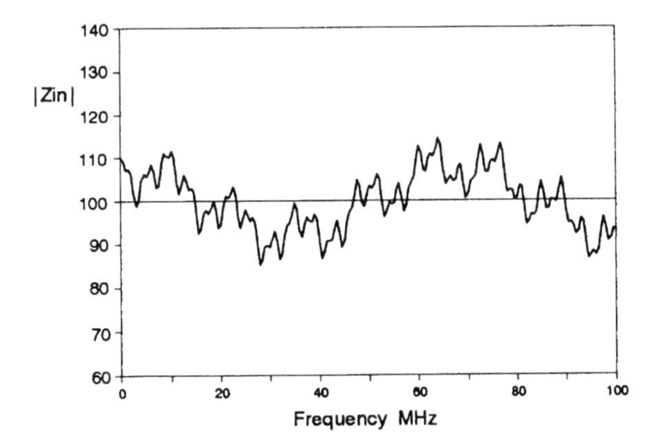

Figure 7.15
Simulated input impedance of twin with discrete faults

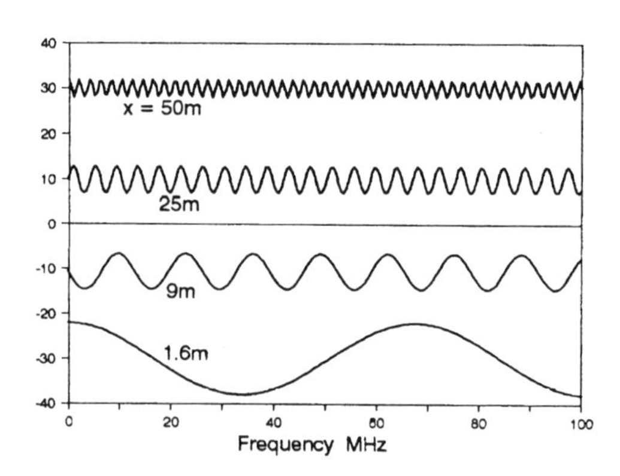

Figure 7.16
Sinusoidal components of simulated input impedance

The size of the fault can be estimated from Equations (7.42), (6.32) and (6.34) at high frequency ($L_0 = 0$)

$$Z_0 = \sqrt{\frac{L}{C}} = \frac{120}{\sqrt{K}} \cdot \ln\left(\frac{2D}{d} \cdot \frac{1-h^2}{1+h^2}\right) \tag{7.62}$$

For a 100 Ω line with $K = 1.9$ and $h = 0.3D/d = 1.89$. For $Z_0 = 105\,\Omega$ $D/d = 2.00$ i.e. a +5.9% change. This could be due to the conductor diameter changing from 0.5 mm to 0.472 mm (a reduction of 28 μm). This could be caused by a severe snatch on a small pairage cable having minimal tensile strength.

If this cable was 100 m in length and was tested from the opposite end, the reflections from the faults would be more attenuated particularly at the higher frequencies. Also the ripple frequencies would increase and a very different impedance trace would result. In fact if these were the only faults on the length it is unlikely that they would be detected.

Periodic Reflections

A more complex situation arises if there are small impedance faults spaced at regular distances down the cable as illustrated in Figure 7.17. These faults result in reflections all of similar amplitude. They will arrive at the test end with constant phase

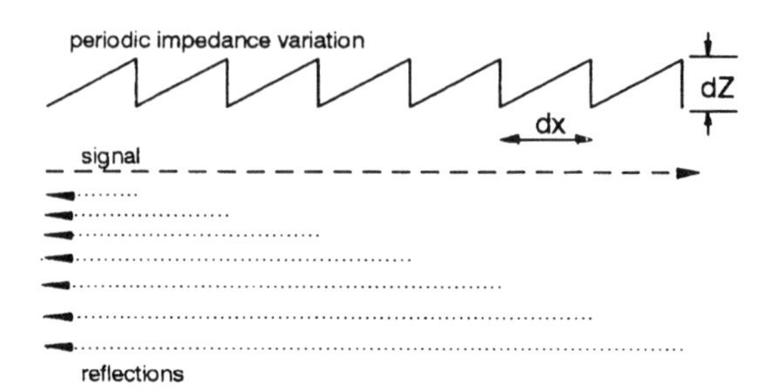

Figure 7.17
Periodic reflections on line

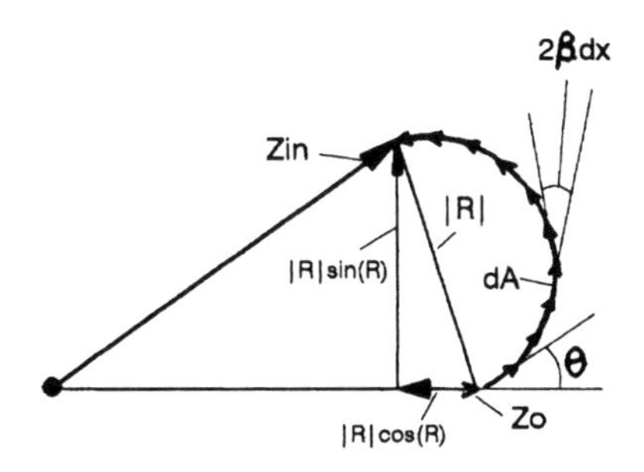

Figure 7.18
Addition of periodic vectors

differences where they add together vectorially to give a resultant vector R as illustrated in Figure 7.18. The modulus and angle of this resultant are given by

$$|R| = \sqrt{\frac{\delta A^2 \sin^2(N\beta\delta x)}{\sin^2(\beta\delta x)}} \tag{7.63}$$

$$\angle R = (N-1)\beta\delta x + \theta \tag{7.64}$$

where the number of reflections is N. The magnitude of this vector shows principal maxima (from the denominator of (7.63)) of magnitude $N \cdot \delta A$ at frequencies where $\beta \cdot \delta x = 0, \pi, 2\pi \ldots$ and subsidiary maxima of much smaller size when $N\beta \cdot \delta x = 0, \pi, 2\pi \ldots$ due to the numerator. This is illustrated in Figure 7.19 for two values of N. The principal maxima are caused by the individual vectors all coming into phase alignment when the phase change from one reflection to the next and back is $0, 2\pi, 4\pi \ldots$, or alternatively when

$$\delta x = \frac{\lambda}{2} = \frac{V_P}{2f_0} \tag{7.65}$$

θ in Equation (7.64) is the phase difference between the first reflected vector and $|Z_{in}|$. Notice that $|R|$ is always positive, but the size of the vector addition of Z_{in} and R depends on the sign of $\cos(\angle R)$which is always positive at the principal maxima if N is

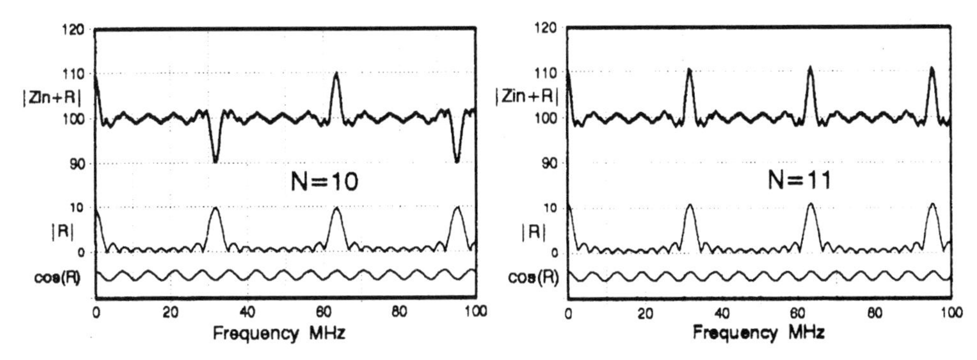

Figure 7.19
Addition of periodic refections for $N = 10$ and 11, $\delta A = 1\,\Omega$

an odd integer, but is alternately positive and negative if N is an even integer. The size of the principal maximum is $N \times \delta A$.

In practice for cables having attenuation and phase varying with frequency the situation is somewhat different. N is not constant with frequency and the principal maxima are not exactly harmonic. Owing to attenuation the reflected vectors at the test end diminish in size exponentially (even with constant δA) each one being $2\alpha \cdot \delta x$ dB smaller than the preceding one. Mathematically, the effective value of N for this geometric series is the integral number of reflections which have an amplitude greater than $1/e$ of the first. This is equivalent to saying that beyond 8.686 dB of attenuation down the line the reflected vectors do not influence the resultant vector. Thus

$$N = \frac{8.686}{\alpha \cdot \delta x} + 1 \tag{7.66}$$

Figure 7.20 shows the effect of periodic reflection points spaced at 1.33 m down a 100 Ω, 0.5 mm conductor line, each point having an impedance change of 0.4 Ω. Since for this line $V_p = 210$ Mm/s from Equation (7.66) the principla maxima should occur at 0 MHz, 79 MHz, 158 MHz.... The attenuation at 80 MHz is about 20 dB/100 m so that only reflections closer than $8.686 \times 100/20 = 43$ m should contribute to the maximum at 79 MHz, thus N is about $1 + 43/1.33 = 33$ and the size of the peak should be about $33 \times 0.4 = 13.2\,\Omega$. Thus once a peak has been identified as due to periodic reflections, a quick calculation using (7.65) gives the spacing and (7.66) gives N. Then the size of the peak enables the size of the individual reflections to be estimated.

From Equation (7.62), an 0.4 Ω fault could be due to a wire diameter change of −0.6% (−3 μm in 0.5 mm). This is a very small change and is unlikely to be found by physical inspection, so that correcting the manufacturing process needs to be done on a deductive basis. The most likely, but not the only, cause of such periodic faults at such short intervals would be a regular snatch on a pair during manufacture. This could , for example, be due to a sloppy driving pin on a 500 mm diameter spool.

Reversing the test end makes δA negative but does not change the sign of $|R|$ (by definition). Thus the sign of the principal maximum at 79 MHz which depends on the

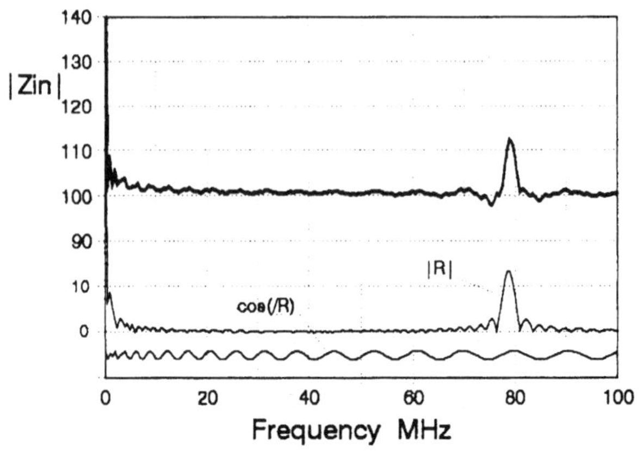

Figure 7.20
Input impedance of screened twin with periodic reflections

sign of cos ($\angle R$) should not change. It is assumed that θ is very small or zero. However N is sensitive to δx and small changes in the periodicity could easily change it from even to odd which, as we have seen, reverses the sign of alternate principal maxima. So the effect of reversing test ends is difficult to predict.

In practice nice clean traces such as this are rarely found owing to other irregularities on the line adding their contribution to the reflections. Also the presence of two or more series of periodic reflections can give complex traces, particularly if they produce several maxima in the frequency band (δx large).

Nature of Impedance Profile

So far the impedance profile has been considered to be a periodic series of step changes. Since the impedance of the pair does not change significantly from end to end, these must be connected by a slow return to the baseline which does not return any significant reflections to the input. It is conceivable that this situation can occur in practice, however the variety of impedance profiles due to manufacturing errors can lie between this and a rectangular pulse type of profile such as that shown in Figure 7.21. This will produce two sets of periodic reflections both with the same δx but displaced in phase by $2\beta \cdot w + \pi$ as shown in the figure. The vector addition of these two series with the input impedance vector produces a resultant, Figure 7.22, which behaves in much the same way as that due to a sawtooth profile.

The resultant varies as the impedance pulse width, w, changes from 0 to δx and as the length Dx changes. The effect of Dx is to rotate the arcs of both series of reflections by the same amount relative to Z_{in}, namely by $\theta = 2\beta \cdot Dx$ thus changing the resultant. The effect of changing the test end, i.e. replacing Dx by Dx' where $Dx = 0.25\,dx$ and $Dx' = 0.5\,dx$, is shown in Figure 7.22. Thus what could be an acceptable input impedance from one end, might be unacceptable when measured from the other end.

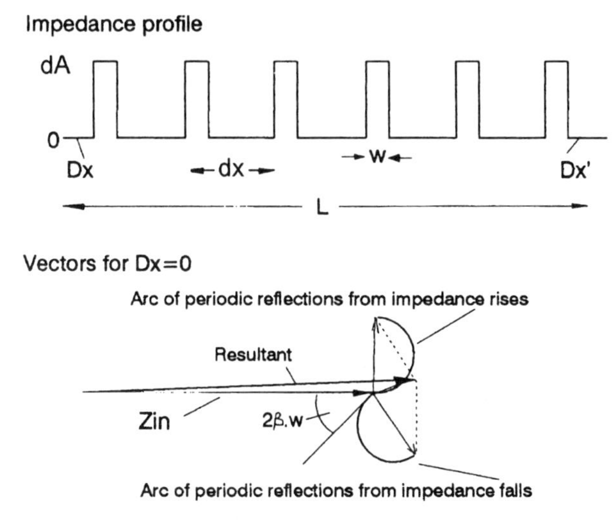

Figure 7.21
Addition of periodic reflections for rectangular impedance deviations

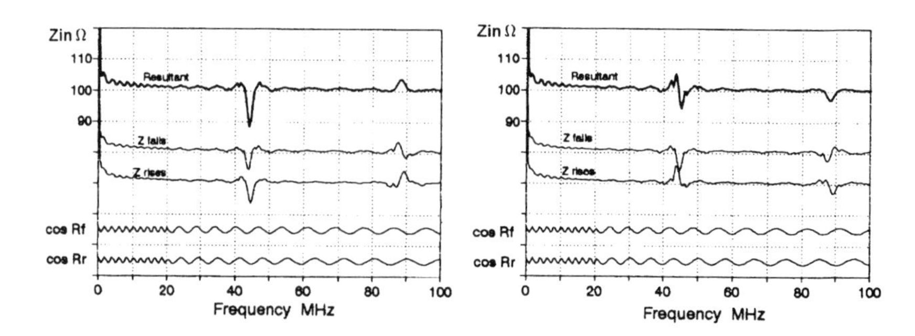

Figure 7.22
Input impedance of screened twin with periodic rectangular impedance deviations. dA = 0.4 Ω spaced at dx = 2.37 m for Dx = 0.25 dx and 0.5 dx

OPTIMUM DESIGN OF CABLES

We have been shown how to design cables for given primary or transmission parameters but no guidance has been given on choosing the optimum parameters. If all parameters are specified there is no room to manoeuver, but if some design freedom is available various options may be chosen, such as designing for minimum overall diameter, or minimum attenuation or minimum cost.

Balanced Pair Local Area Cables

Stimulated by the volatile and high price of copper in the 1960s, Spencer [12] investigated the problem of optimizing the design of both copper conductor and aluminium conductor local area cables in the telephone network. This enlarged on earlier work by Madsen [13]. His design parameters were to achieve the lowest-cost solution for cables having a fixed attenuation coefficient, taking into account material, duct and manufacturing costs.

The attenuation coefficient at low frequencies is given by

$$\alpha = \sqrt{\frac{\omega C \cdot R}{2}}\,\text{N/km} \tag{7.67}$$

$$C \cdot R = 2\alpha^2/\omega \tag{7.68}$$

$$R = 8\rho/\pi d^2 \tag{7.69}$$

where R is the resistance in Ω/km, C the capacitance in F/km, ρ is the resistivity and $\omega = 2\pi \times$ frequency in Hz. If the attenuation coefficient is fixed at a given frequency, R and C may be varied as long as the CR product remains constant. For multipair cables from Equation (6.36)

$$C = \frac{1000K}{36\ln(2x - 0.9)} = \frac{51}{\ln(2x - 0.9)}\,\text{nF/km} \tag{7.70}$$

where $x = D/d$ and K has been fixed at a value of 1.84 for solid polythene insulation. Thus from Equations (7.68) and (7.69)

$$d = \sqrt{\frac{4\rho\omega \cdot k}{\pi\alpha^2 \ln(2x - 0.9)}} \quad (7.71)$$

with $k = 51.0\,\text{nF/km}$.

The cable core cross-sectional area is given by $2\sqrt{3} \cdot N \cdot (xd)^2$ in accordance with Equation (4.4), and taking the sheath thickness to be directly proportional to the core diameter, i.e. $t = t_s \cdot D_0$ with $t_s = 0.1$, the material plus duct cost *per pair* per km is given by

$$d^2\left[\frac{\pi}{2}p_1 + \frac{\pi}{2}(x^2 - 1)p_2 + 8\sqrt{3}x^2 t_s p_3 + 2\sqrt{3}(1 + 4t_s)x^2 p_4\right] \quad (7.72)$$

where p_1, p_2, p_3, and p_4 are the volumetric costs of conductor, insulation, sheath material and duct space respectively. From Equations (7.71) and (7.72), the material and duct cost per pair per km is

$$\frac{4\rho\omega k\left[\frac{\pi}{2}p_1 + \frac{\pi}{2}(x^2 - 1)p_2 + 8\sqrt{3}x^2 t_s p_3 + 2\sqrt{3}(1 + 4t_s)x^2 p_4\right]}{\pi\alpha^2 \ln(2x - 0.9)} \quad (7.73)$$

It can be shown that this is a minimum when

$$x[(2x - 0.9)\ln(2x - 0.9) - x] = \frac{p_1 - p_2}{p_2 + 8.82t_s p_2 + 2.2(1 + 4t_s)p_4} \quad (7.74)$$

This shows that the value of x for minimum material and duct cost is

- dependent on the volumetric costs of the materials and the value of the duct space occupied.
- independent of the chosen resistivity, permittivity, attenuation and frequency.

Since the value of x and the capacitance factor k determine the capacitance, the optimum capacitance is fixed solely by the combination of materials used for the cable and their costs.

When Spencer published his analysis in 1969 the sterling prices of the materials were as shown in Table 7.4 (1994 prices are also shown with material densities and resistivities).

The 1969 duct costs were taken by Spencer to be an average between existing and 'new work' costs and to obtain the 1994 duct costs *the author* has applied the same average factor of 5 by which the material costs have increased over the 25-year period (corresponding to 6.6% annual inflation).

Manufacturing costs for the cables are more difficult to include in the analysis since they are not a linear function of pair count. *The author* has found that the manufacturing costs for this type of cable are proportional to $\sqrt{N/x}$. For 100 pair cables the manufacturing costs are approximately 40% and 60% of the material costs

Table 7.4
Material costs, densities and resistivities

Material	1969	1994	Density kg/m^3	Resistivity $\Omega \cdot$m
Copper (£/kg)	0.45	1.73	8.89×10^3	0.01724
Aluminium (£/kg)	0.20	1.24	2.70×10^3	0.0282
Polythene insulant (£/kg)	0.17	0.982	0.918×10^3	–
Polythene sheath (£/kg)	0.204	0.89	0.920×10^3	–
Duct space (£/mm^2/km)	0.098	0.49	–	–

for copper and aluminium conductor cables respectively. These percentages increase for smaller pair counts. By taking a range of values of capacitance, the values of x, diameter and cost for 100 pair copper conductor cables and aluminium conductor cables can be calculated.

Comparisons of the costs and diameters are made relative to the values for the existing copper/polythene cable specification at a capacitance of 53 nF/km. The attenuation coefficient used was that for existing 0.5 mm copper cables, viz. 1.68 dB/km (0.193 N/km). The comparisons are shown in Figure 7.23. The figures on the curves are the capacitances in nF/km. The optimum capacitance in this case is that giving minimum cost. It will be seen that aluminium conductor cables are about 20% cheaper than copper cables in this comparison. The optimum capacitance for the aluminium cables is 70nF/km. However the cables are 22% greater in diameter than the present copper cable specification.

It will also be seen that the optimum capacitance for the copper cables is about 40 nF/km, which would give a cost reduction of 5% but increase their diameter by 12%, making them only 9% smaller than the optimum aluminium cables. The

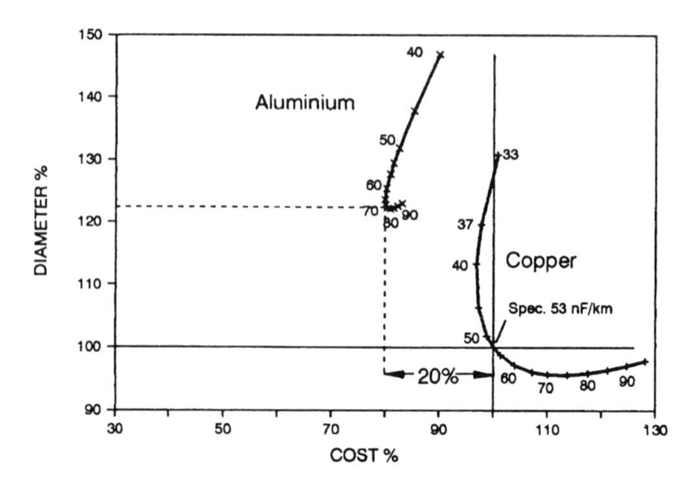

Figure 7.23
Comparison of total 1969 cost of copper and aluminium conductor 100-pair cables versus diameter with fixed attenuation coefficient of 1.68 dB/km (reproduced by permission of IEE)

minimum size of copper cable would be given by a capacitance of 75 nF/km but the cost would increase by 32%.

When duct costs are eliminated, for the case of directly buried cable, the cost advantage of the aluminium cables is increased to 30%, the optimum capacitance reduces to 67 nF/km and the diameter penalty increases slightly to 23%. For copper the optimum capacitance also reduces, to 35 nF/km, but the diameter is now only 2% less than the aluminium cable. The cost saving for copper cables at this capacitance is now about 10%.

The cost penalties for departing from the optimum capacitances can be read from these curves in any particular case. For small departures the cost penalties are also small. The optimum capacitances calculated for this analysis are dependent on the ratios of the polythene and duct costs to the price of the metal conductors Thus the optimum design values are not dependent on the value of money. However if one of the costs changes out of proportion to the other costs the optimum design will change. This is most likely to be the cost of the metal conductors and will result in cost penalties as the actual design departs from the new optimum. Metal cost variations of moderate magnitude only produce small cost penalties, so it is reasonable to stabilize designs for long periods of time. For larger conductor cost changes considerable changes occur.

It is interesting to review these designs in the light of 1994 material costs. These are included in Table 7.4. It will be seen that the ratio of copper to polythene costs has changed from 2.65 in 1969 to 1.76 in 1994. For aluminium the change is smaller, from 1.18 to 1.27.

Applying the 1994 costs gives the results in Figure 7.24. Although the optimum capacitance for aluminium cables is not very much different, that for copper cable is significant. The cost savings for aluminium cables are no longer so compelling as in 1969. In view of other difficulties associated with handling aluminium conductors both in manufacture and installation/maintainance, aluminium conductors are not presently favoured.

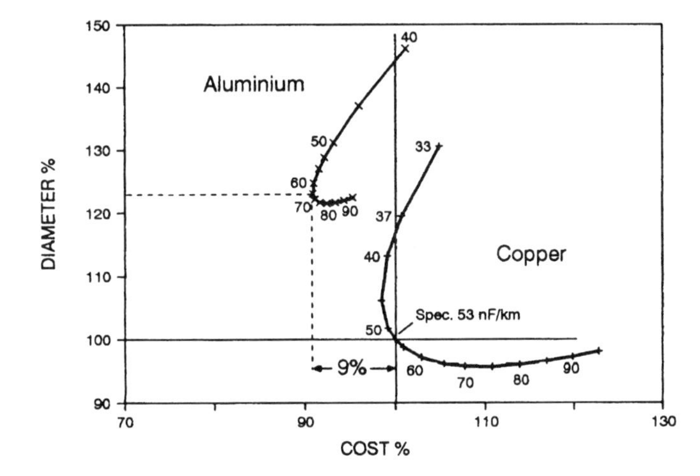

Figure 7.24
Comparison of total 1994 cost of copper and aluminium conductor 100-pair cables versus diameter with fixed attenuation coefficient of 1.68 dB/km (reproduced by permission of IEE)

Effect on Service Area

Taking the optimum capacitance of the aluminium cables as 65 nF/km, this corresponds to an optimum x of 1.993 and to a polythene insulation diameter of 0.997 mm on a conductor diameter 0.5 mm. The attenuation at 1.59 kHz from Equation (7.67) is 2.65 dB/km and the loop resistance from Equation (7.69) is 287 Ω/km. The planning rules for the local network were set by Spencer to limiting values of 1000 Ω loop resistance and 10 dB attenuation. Thus this cable could serve subscribers out to a distance from the exchange of 3.5 km for resistance and 3.8 km for attenuation. This is a reasonably balanced condition which could serve 95% of subscribers in an urban situation.

For a more extended area a larger conductor diameter is required. If we choose 0.8 mm and still maintain the optimum capacitance we find that the resistance limit is 8.9 km and the attenuation limit is 6.0 km which is not a balanced condition and is therefore wasteful of conductor material. Let us try to obtain this balanced condition for all sizes of conductor.

If L_R and L_α are the resistance and attenuation limits then

$$L_R = 1000 \cdot \frac{\pi d^2}{8\pi} \quad \text{and} \quad L_\alpha = \frac{10}{8.686\sqrt{\dfrac{\omega CR}{2}}} \tag{7.75}$$

For these to be equal

$$C \cdot R = 2\left(\frac{8\rho}{8.686 \cdot \omega \cdot \pi d^2}\right)^2 \quad \text{or} \quad C = \frac{2 \times 8\rho}{\pi d^2}\left(\frac{1}{8.686\omega}\right)^2 \tag{7.76}$$

For $\omega = 10^4$ (1.59 kHz), $C = 0.675\rho/d^2$ nF/km so that C varies with d and ρ. The impedance of the pair is

$$|Z_0| = \sqrt{\frac{R}{\omega C}} = \sqrt{\frac{8\rho}{\pi d^2} \cdot \frac{\pi d^2 (8.686 \cdot \omega)^2}{\omega \times 2 \times 8\rho}}$$

$$= 8.686 \cdot \sqrt{\frac{\omega}{2}} = 614.3\,\Omega \quad \text{at} \quad \omega = 10^4 \tag{7.77}$$

Thus we have a constant impedance condition in which the impedance is independent of both d and ρ. The only conductor size which will satisfy *this* 'balanced' condition *and* the optimum capacitance condition is given by

$$d = \sqrt{\frac{0.675\rho}{C}} \tag{7.78}$$

$$= 0.541 \text{ mm for aluminium with } C_{opt} = 65 \text{ nF/km}$$

$$= 0.509 \text{ mm for copper with } C_{opt} = 45 \text{ nF/km}$$

No other conductor size can satisfy both conditions. For a larger or smaller conductor size therefore a compromise is required.

Optimizing Coaxial Pair Design

Coaxial pair designs can optimized for a number of criteria such as

- Minimum diameter for a required attenuation
- Minimum dielectric area
- Minimum conductor cross-section with various configurations
- Maximum voltage rating
- Maximum power rating

The general method will be illustrated by three cases.

Minimum diameter of coaxial pair for required attenuation

From Equation (7. 53) the attenuation of a coaxial pair at high frequency is given by

$$\alpha = \frac{4.58\sqrt{Kf}}{\ln(D/d)}\left(\frac{\sqrt{\rho_i}}{d} + \frac{\sqrt{\rho_0}}{D}\right) + 9.1\sqrt{Kf}\tan\delta \text{ dB/100 m}$$

For stranded inner conductors and for braided outer conductors two form factors K_i and K_0 need to be introduced (see Chapter 10). Also since the dielectric loss factor does not affect the optimum proportioning of the diameters it may be ignored for this analysis. Thus

$$\alpha = \frac{4.58\sqrt{Kf}}{\ln(D/d)}\left(\frac{K_i\sqrt{\rho_i}}{d} + \frac{K_0\sqrt{\rho_0}}{D}\right) \tag{7.79}$$

Let $D/d = x$ and $K_0\sqrt{\rho_0}/K_i\sqrt{\rho_i} = g$ then at a given frequency

$$\alpha = \text{const.}\ \frac{K_i\sqrt{\rho_i}}{D}\left(\frac{x+g}{\ln x}\right)$$

whence

$$D = \frac{\text{const.}\ K_i\sqrt{\rho_i}}{\alpha}\left(\frac{x+g}{\ln x}\right) \tag{7.80}$$

Differentiating this with respect to x gives

$$\frac{dD}{dx} = \text{const.}\ \frac{\ln x - (x+g)/x}{(\ln x)^2}$$

For D to be a minimum, $dD/dx = 0$ thus

$$x_{opt}\{\ln(x_{opt} - 1\} = g \tag{7.81}$$

The optimum diameter ratio is therefore a function of g and dependent on the form and resistivity of the conductors. For equal resistivities, a solid inner and a cylindrical tube outer conductor, $x_{opt} = 3.59$.

For a stranded inner conductor and a braided outer conductor

$K_i = 1.25$ and $K_0 \approx 2.3$ thus for equal resistivities $g \approx 1.84$ and $x_{opt} \approx 4.2$

For a solid inner conductor and a braided outer conductor

$K_i = 1.0$ and $K_0 \approx 2.3$ thus for equal resistivities $g \approx 2.3$ and $x_{opt} \approx 4.55$

From these optimum ratios, $D_{\min}$ can be calculated from Equation (7.80) for a given attenuation and effective permittivity K.

Maximum voltage rating of coaxial pair

If the electric field at breakdown for the insulation is E V/mm then the maximum voltage which can be applied is

$$V = \frac{Ed}{2} \ln x, \text{ which may be re-written as } D = \frac{2V}{E} \cdot \frac{x}{\ln x} \text{ mm} \tag{7.82}$$

By differentiation, D will be a minimum for a given voltage breakdown V when

$$\ln x = 1 \text{ or for } x_{opt} = 2.72$$

Maximum power rating of coaxial pair

For a low duty cycle pulse transmission, the optimum design is usually that for maximum voltage rating using the peak value of the pulses. The peak voltage will be least when the coaxial is properly terminated in its characteristic impedance. In this case the power will be

$$P = V^2/2Z_0 \quad \text{whence} \quad D = 120 \cdot K^{-4} \cdot \frac{\sqrt{P}}{E} \cdot \frac{x}{\sqrt{\ln x}} \tag{7.83}$$

Differentiating this gives a maximum P, limited by breakdown, when $\ln x = 0.5$ or $x_{opt} = 1.65$.

For maximum power rating with thermal limitations, the simplest solution for minimum temperature of the external surface of the coaxial is given by the minimum attenuation for a given diameter D, given above. A more difficult problem is to design for minimum temperature of the inner conductor to prevent dielectric softening or deterioration. For a discussion of this consult Dummer and Blackband [21].

8

Crosstalk

Experience using parallel open-wire lines for telegraphy and telephony showed that couplings between circuits enabled the signal on one circuit to be picked up on a neighbouring circuit. This is known as crosstalk. Circuits formed from single wires with earth return were particularly bad in this respect owing to the large inductive loop formed by the earth return. The use of two wires per circuit produced considerable improvement if both circuits were well balanced with respect to earth.

Further improvements were made by transposing the wires of one pair with respect to the others (Figure 8.1), so that over the length of the line equal positive and negative inductions would cancel out. For many circuits, complex transposition schemes were worked out which were successful enough to enable carrier frequency transmissions up to 100 kHz or more to be used.

Transpositions were made at double-insulators on the mounting poles. For such transpositions to be effective they should be made at separations not exceeding about a twentieth of a wavelength at the maximum working frequency, e.g. for 100 kHz working the wavelength is 3 km and the transpositions should be closer than 150 m. On a good pole route the poles would be spaced approximately 80 to 100 m apart.

The economic and security benefits of enclosing many pairs within one protective sheath to form a cable required similar methods of reducing crosstalk. These took the form of individual twist lengths (or lay lengths) for each of the highly balanced pairs, to form a lay scheme, analagous to the transposition schemes of open-wire lines. In time

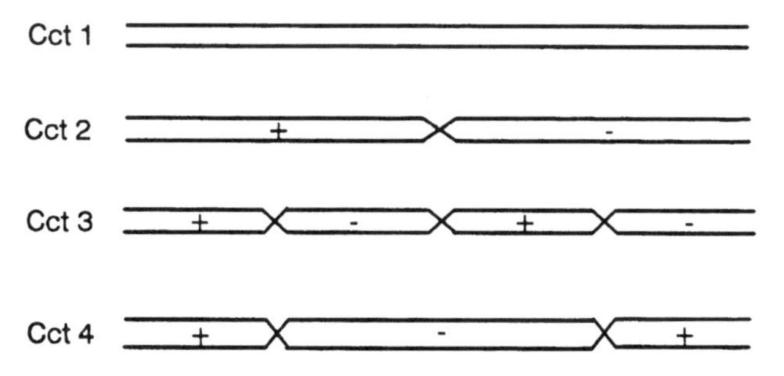

Figure 8.1
Transposition of open wire lines

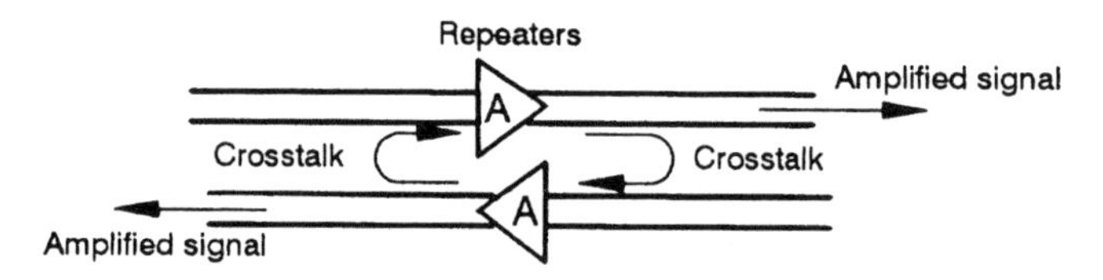

Figure 8.2
Effect of crosstalk at repeaters

these lay schemes were sufficiently successful to enable the transmission of carrier frequencies up to 550 kHz and later the transmission of pulse-code-modulations up to 2 Mbit/s. over 6 km lengths. On shorter cables, video and data signals up to 100 MHz can be carried on balanced pair cables.

For long-distance carrier telephony, coaxial pair constructions superseded balanced pairs and were worked up to frequencies of 4, 12 and even 60 MHz, until they in turn were superseded by optical fibre cables. The use of repeater amplifiers on long distance analogue transmissions made the control of crosstalk vital to prevent uncontrolled oscillation between different directions of transmission (Figure 8.2).

COUPLINGS BETWEEN BALANCED CIRCUITS

Electrostatic Couplings

The electrostatic couplings between terminated pairs are shown in Figure 8.3. The physical arrangement in (a) can be redrawn as an equivalent bridge circuit (b). By star-delta transformations and a little algebra the voltage appearing across the diagonal CD, i.e. the crosstalk from AB to CD, can be shown to be

$$e = E \cdot \frac{j\omega(x_1 x_3 - x_2 x_4)}{j\omega(x_1 + x_2)(x_3 + x_4) + (2/Z_2)(x_1 + x_2 + x_3 + x_4)} \tag{8.1}$$

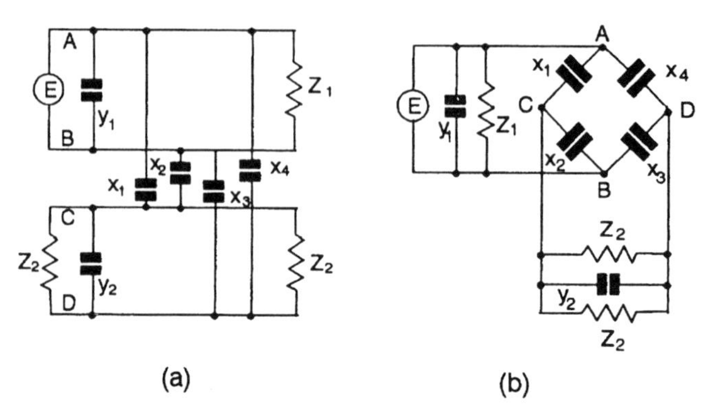

Figure 8.3
Electrostatic couplings between balanced pairs

If it is assumed, as is generally the case, that x_1, x_2, x_3 and x_4 are very similar in value and equal to x_m with small deviations, then

$$(x_1 \cdot x_3 - x_2 \cdot x_4) \cong x_m\{(x_1 + x_3) - (x_2 + x_4)\}$$
$$= x_m \cdot C_{ub} \tag{8.2}$$

where $C_{ub} = (x_1 + x_3) - (x_2 + x_4)$ and is called the capacitance unbalance. Applying the same approximation to the denominator of Equation (8.1) gives

$$(x_1 + x_2)(x_3 + x_4) \cong 4x_m^2 \quad \text{and} \quad (x_1 + x_2 + x_3 +_4) \cong 4x_m$$

$$e = E \cdot \frac{j\omega C_{ub}}{4 \cdot j\omega x_m + 8/Z_2} \tag{8.3}$$

$$\frac{e}{E} = \frac{j\omega C_{ub} Z_2}{8} \tag{8.4}$$

since the first term in the denominator of Equation (8.3) is negligible with respect to the second term. Thus the crosstalk is independent of x_m and only depends on the value of C_{ub}. By including the earth capacitances of the wires in the derivation of Equation (8.1), it can be shown that the pair/pair crosstalk is also substantially independent of the earth unbalances.

Magnetic Couplings

The transmitted current I_1 flowing in the disturbing circuit causes an induced voltage e to appear in the disturbed circuit, which drives a crosstalk current I_2.

$$e = -j\omega M \cdot I_1$$

$$I_2 = \frac{e}{2Z_2} = -\frac{j\omega M \cdot I_1}{2Z_2} \tag{8.5}$$

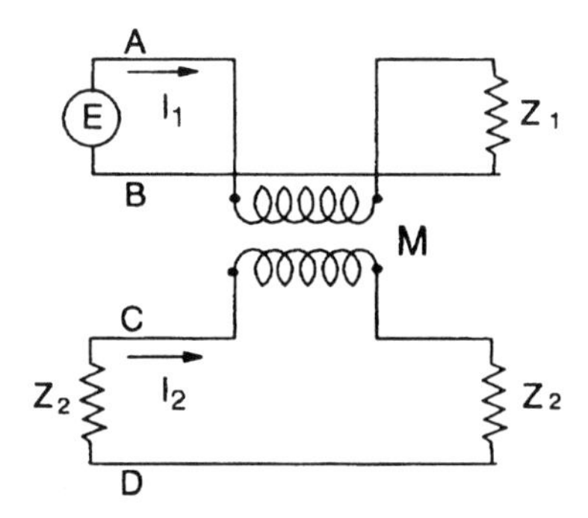

Figure 8.4
Mutual inductance coupling between balanced pairs

Crosstalk Attenuation

Crosstalk is defined in terms of the power transfer between circuits. The crosstalk attenuation is the logarithm of the power ratio between the disturbing and disturbed circuits expressed in nepers or decibels. From Equations (8.4) and (8.5)

$$A_c = \ln\left|\sqrt{\frac{Z_2}{Z_1}}\cdot\frac{E}{e}\right| = \ln\sqrt{\left|\frac{Z_2}{Z_1}\right|}\,\frac{8}{\omega C_{ub}|Z_2|}$$

$$= \ln\frac{8}{\sqrt{|Z_1\cdot Z_2|}\,\omega C_{ub}}\ \text{nepers}$$

$$= 20\log\frac{8}{\sqrt{|Z_1\cdot Z_2|}\,\omega C_{ub}}\ \text{dB} \tag{8.6}$$

$$A_m = \ln\left|\sqrt{\frac{Z_1}{Z_2}}\,\frac{I_1}{I_2}\right| = \ln\sqrt{\left|\frac{Z_1}{Z_2}\right|}\,\frac{2|Z_2|}{\omega M}$$

$$= \ln 2\,\frac{\sqrt{|Z_1 Z_2|}}{\omega M}\ \text{nepers}$$

$$= 20\log 2\,\frac{\sqrt{|Z_1 Z_2|}}{\omega M}\ \text{dB} \tag{8.7}$$

Note that 1 neper = 8.686 dB. Equating Equations (8.6) and (8.7) gives the mutual inductance which will give the same crosstalk attenuation as a given capacitance unbalance.

$$\frac{8}{\sqrt{|Z_1 Z_2|}\,\omega C_{ub}} = 2\frac{\sqrt{|Z_1 Z_2|}}{\omega M}$$

whence

$$M = \frac{C_{ub}}{4}\cdot|Z_1 Z_2| \tag{8.8}$$

Required Orders of Magnitude

To get some idea of the order of magnitude of required crosstalk attenuation consider Figure 8.2. If the repeaters have a gain of A dB, it is required to have a stability margin of 20 dB, and the crosstalk paths indicated each have the same attenuation X dB, then $2A + 2X = -20$ dB, or $X = -(10 + A)$ dB. For a repeater gain of 50 dB therefore, X must be better than -60 dB. For circuits having an impedance of 150 Ω Equation (8.6) gives a required capacitance unbalance of <53 pF at 160 kHz, which by Equation (8.8) is equivalent to a mutual inductance of $<0.2\,\mu$H.

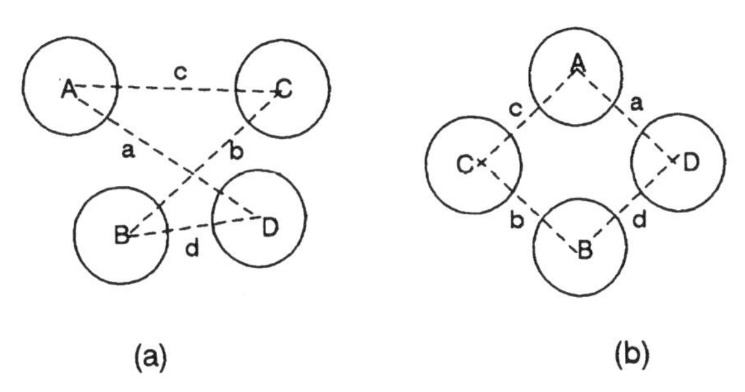

Figure 8.5
Dimensional unbalances in pairs and quads

Electric and Magnetic Couplings Between Pairs

It has been shown [14, 15] that the electric and magnetic couplings between pairs of wires AB and CD, can be calculated by the formulae

$$C_{ub} = \frac{72}{K} C_1 C_2 \cdot \ln\left(\frac{a \cdot b}{c \cdot d}\right) \text{ pF/km} \tag{8.9}$$

$$M = 0.2\mu_r \cdot \ln\left(\frac{a \cdot b}{c \cdot d}\right) \text{ mH/km} \tag{8.10}$$

where C_1 and C_2 are the pair capacitances in nF/km, K is the permittivity and μ_r is the relative permeability between the pairs. a, b, c, and d are the wire separations shown in Figure 8.5.

Quad Construction

If the wires of two pairs are arranged on the corners of a square as in Figure 8.5(b) then $a = b = c = d$ and both the capacitive unbalance and the mutual inductance are zero. If the square becomes a lozenge shape so that $a \cdot b = c \cdot d$, the couplings are also zero. However in practice there will be manufacturing imperfections and the aim is to achieve only small random variations from the ideal so that statistically a near-perfect balance is obtained. If there is a systematic increase of say 0.5% in $a \cdot b$ compared with $c \cdot d$ then Equations (8.9) and (8.10) show that a capacitance unbalance of about 200 pF/500 m would be expected on a 43 nF/km quad ($K = 1.7$) compared with a specification of 75 pF/500 m max. and in addition, a mutual inductance between the pairs of 0.5 mH/500 m. Hence the mechanical accuracy of manufacture needs to be very good for high-quality quads.

If there were no systematic displacement, and the value of say a/d, varied randomly between 1.005 and 0.995 with a correlation length of about 5 m, then the unbalance would be 1.95 pF/5 m and this would accumulate according to the square root of length to 19.5 pF/500 m. This is about the level achieved in good quad manufacture. By the same reasoning there would be a mutual inductance of about 50 nH/500 m.

Within-quad unbalances are observed to vary with the lay length of the quads. This is probably due to the longer lay quads being more unstable in cross-section than the

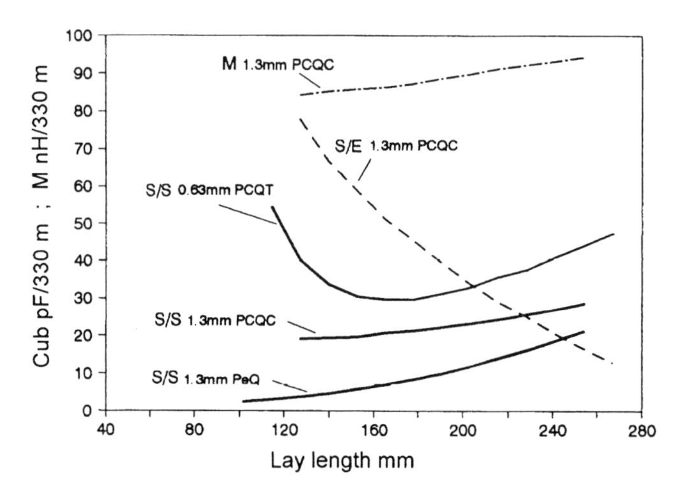

Figure 8.6
Variation of in-quad unbalances with lay length (reproduced by permission of BICC Cables Ltd)

shorter lay quads. In very short lay paper insulated quads the insulation tends to distort and this also degrades the unbalance, whereas a solid polythene insulated quad unbalance continues to decrease as the lay is shortened. Also the heavier conductors tend to produce a more uniform quad. The graphs in Figure 8.6 summarizes experiences in this respect.

The lines represent the smoothed means of the results for many cables. The actual means have a spread about the lines of about 10–20%. Notice the increase of side/side unbalance of the 0.63 mm (PCQT) paper insulated quads at short lays, compared with the harder paper–string/paper tape insulation of the 1.3 mm carrier quads (PCQC). This is even more marked in the polythene insulated carrier quads (PeQ). The strength of the heavier conductors is a factor in the overall unbalance level. The side-to-earth unbalances decrease as the lay length increases.

Near-end and Far-end Crosstalk

The crosstalk measured at the end of the disturbed line furthest from the generator is known as the far-end crosstalk (FEXT). If it is referred to the received signal at the same end it is known as the far-end signal-to-crosstalk attenuation (FES/N) or the equal level far-end crosstalk (ELFEXT). The crosstalk measured at the same end as the generator is called the near-end crosstalk (NEXT). These are indicated in Figure 8.7

$$NEXT = 20 \cdot \log\left(\frac{E}{e_n}\right) \tag{8.11}$$

$$FEXT = 20 \cdot \log\left(\frac{E}{e_f}\right) \tag{8.12}$$

$$ELFEXT = FES/N = 20 \cdot \log\left(\frac{e'}{e_f}\right) = FEXT - \alpha_l \tag{8.13}$$

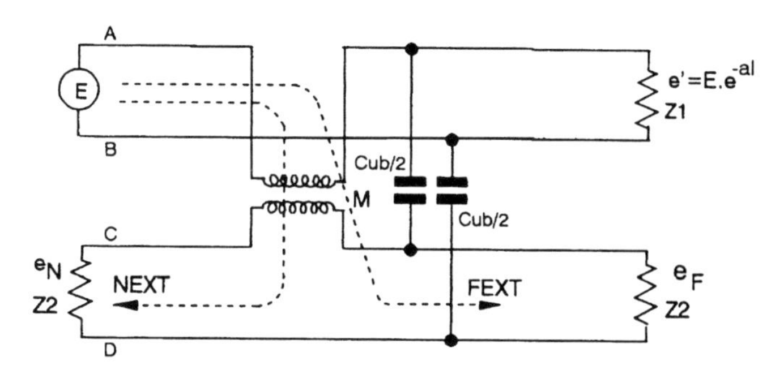

Figure 8.7
Far-end and near-end crosstalk

When magnetic and electric coulplings coexist between the pairs, as is the case for within quad couplings and for couplings between adjacent (unscreened) pairs or quads, they may be combined into a resultant magnetic coupling by using Equation (8.8). For within quad couplings the magnetic and electric couplings are systematic so that

$$M_n = \frac{|Z^2|}{4} \cdot C_{ub} + M \tag{8.14}$$

$$M_f = \frac{|Z^2|}{4} \cdot C_{ub} - M \tag{8.15}$$

whereas, although pair-to-pair electric and magnetic couplings correlate in magnitude, their phases are random and the resultant magnetic coupling is given by

$$M_r = \sqrt{\left(\frac{|Z_1 Z_2|}{4} \cdot C_{ub}\right)^2 + M^2} \tag{8.16}$$

It will be seen that within-quad FEXT attenuation is greater than NEXT attenuation. All the above theory applies to cable lengths shorter than about one tenth to one twentieth of a wavelength, i.e. for lengths of the order of 100–200 m at 100 kHz and of the order of 1–2 m at 10 MHz. For longer lengths, propagation effects cause phase changes in the measured couplings which make them increasingly complex. So that while Equation (8.16) will hold, Equations (8.14) and (8.15) will not.

Pair and Quad Lay Schemes

When combining many twisted pairs or quads within the same cable, it was soon noticed that if the lay lengths were equal, high magnetic couplings were produced. Even when the lays were different, high couplings were sometimes observed. These were named 'critical combinations' of lays. Baranov [15] published a definitive (and complex) mathematical analysis of lay schemes in 1943. She also verified her results using measurements of mutual inductance between rigidly defined pairs made by laying conductors in helical grooves machined into ebonite rods of about 5 to 10 m length. We may summarize her conclusions as follows.

Baranov's Rules

1. The mutual inductance coupling is the sum of terms which are non-accumulating periodic functions of length (the period $= p_{nm}$).
2. Some of these terms degenerate into linear functions of length when the composite lay $p_{nm} = \infty$, i.e. when
$$\frac{1}{p_{nm}} = \frac{n}{p_1} + \frac{m}{p_2} = 0$$
where n, m may take values of $-1, 1, 3, 5, \ldots$. These are the critical combinations.
3. For pairs twisted in the *same direction*, only the lay combination $p_1 : p_2 = 1:1$ is critical. In this case the coupling is proportional to length and to $1/p$.
4. For pairs twisted in *opposite directions*, all odd-to-odd combinations of n and m, i.e. $1:1$, $1:3$, $1:5$, $3:5, \ldots$ are critical ratios and give high mutual inductances linearly proportional to length, but independent of the lay length.

In applying her investigation to real manufacturing lengths of cables, Baranov was able to confirm approximately the magnitudes of critical couplings. For non-critical combinations however, instead of a non-accumulating periodic variation with length, there was an approximate square-root law dependence on length. This she ascribed to the summation of randomly distributed residual couplings and derived the formula

$$|\bar{M}| = \frac{4}{\pi}\left(\frac{r}{a}\right)^2 \cdot \frac{p_{11}^{3/2}}{k} \cdot \sqrt{L} \tag{8.17}$$

where k is a factor typifying the quality of manufacture, about 180 mm in her examples, and identified as a correlation length. The wire separation is $2r = D$ (the insulation diameter) and a is the separation between pair axes.

N. T. Pearson [16], following Baranov's work, carried out a statistical survey of the mutual inductances measured on a large number of factory lengths of 12- and 14-quad carrier cables. The mean moduli of the within layer mutual inductances measured at 5 kHz on 330 m lengths, for adjacent, 1-between, 2-between and 3-between combinations, were corrected for spacing a. Pearson then plotted $|\bar{M}| \times a^2/p_{1,1}$ against $a \times p_{1,-1}/p_{1,1}$ on log–log scales and obtained a good correlation as shown in Figure 8.8.

Since the slope of this line is one-half, i.e. a square-root law

$$\begin{aligned}
|\bar{M}| \times \frac{a^2}{p_{1,1}} &= K'\sqrt{a \cdot p_{1,-1}/p_{1,1}} \\
&= K'\sqrt{a \times p_{1,-1} \times p_{1,1}} \\
&= K'\sqrt{\frac{a}{\left(\frac{1}{p_1} - \frac{1}{p_2}\right)\left(\frac{1}{p_1} + \frac{1}{p_2}\right)}} \\
&= K'\sqrt{\frac{a \cdot p_1^2}{(1 - (p_1/p_2)^2)}}
\end{aligned}$$

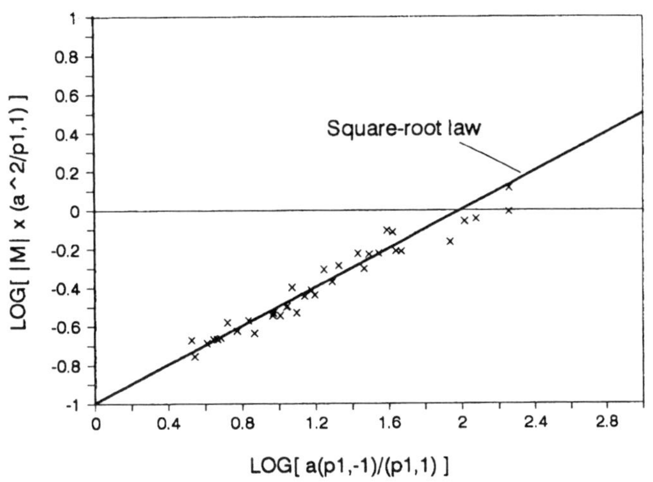

Figure 8.8
Pearson's correlation of inductance couplings between quads (reproduced by permission of BICC Cables Ltd)

Replacing a by (a/r) as in Baranov's work, this gives

$$|\bar{M}| = k \cdot \left(\frac{r}{a}\right)^{3/2} \cdot \frac{p_1}{\sqrt{1-(p_1/p_2)^2}} \tag{8.18}$$

where k is a constant typifying the quality of manufacture and will vary according to the ability of the insulated conductor to withstand the forces of twinning or quadding and subsequent cabling operations.

Equation (8.18) goes to infinity when the lays are equal, which is the critical condition for same-direction lays. From Baranov's work the critical condition results in a large but finite value of M approximately equal to 100 μH/km. Equation (8.18) can be modified to accomodate this by adding a small constant term of 5×10^{-6} to the denominator. Also k may be evaluated from the BICC results quoted by Pearson giving finally

$$|\bar{M}| = 1.9\left(\frac{r}{a}\right)^{3/2} \cdot \frac{1}{\left[5 \times 10^{-6} + \left(\frac{1}{p_1}\right)\sqrt{1-\left(\frac{p_1}{p_2}\right)^2}\right]} \text{ nH/km} \tag{8.19}$$

where p_1 is in mm and $p_1 < p_2$. The correction for length theoretically should be according to a square-root law, but from the BICC results on factory lengths, a mixed linear-square-root law appears more appropriate (as in BPO specifications) indicating the coexistence of systematic and random effects. Thus Equation (8.19) should be multiplied by $(L + \sqrt{L})/2$ with L in km.

The correlation of Equation (8.18) was checked by Pearson over about 200 cables using the same lay scheme. Good correlation was found in all down to a population of

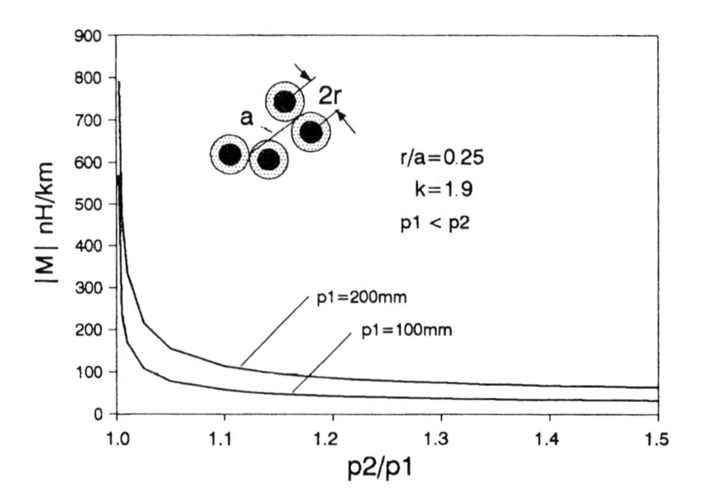

Figure 8.9
Mutual inductance between twisted pairs (reproduced by permission of BICC Cables Ltd)

10 cables. The correlation was apparent even in a population of only one cable. The theory was also checked over more than six different lay schemes, again with good agreement to Equation (8.18).

Equation (8.19) is graphed in Figure 8.9 for $L = 1$ km, $p_1 = 100$ mm and 200 mm and $r/a = 0.25$. The mutual inductance rises very rapidly as the critical condition is approached. Any lay ratio greater than 1.05 gives a good reduction from the critical value of 100 000 nH/km. Expressed as a voltage ratio the reduction is of the order of 56 dB for $p_1 = 100$ mm. Increasing the lay ratio to 1.2 only improves the reduction by 5 dB.

Quality Factor k

The value of k derived from the BICC results applies only to 1.3 mm conductors with paper–string/paper tape insulation and quadded at fairly low speed (100 r.p.m.) on detorsioning quadders. For 0.63 mm conductors with solid poythene insulation twisted at 1000 r.p.m. on non-detorsioning double twist twinners, k rises to about 5.0–5.3. It is reasonable to assume that 0.9 mm polythene twins would have k intermediate to this at about 3. A small sample (7×6 lay combinations) of 0.5 mm polythene pairs of very short average lay (12 mm) gave $k = 15$, for longer lays this may improve to about 10. By extrapolation, k for 0.4 mm conductors would be 23. These approximate to the empirical formula

$$k = 0.55d^{-4} + 1.7 \tag{8.20}$$

The inverse fourth power law is reminiscent of the bend strength of circular beams, in this case the bend strength of the insulated conductors.

Pearson gives the correlation between the capacitance unbalance of adjacent, unscreened, pairs and their mutual inductance as shown in Figure 8.10. This would be expected to be linearly proportional to the effective permittivity (in this case 1.45). Hence

$$C_{ub} = 0.32 \times 10^{-3} \cdot K_{eff}|\bar{M}| \quad \text{F (for } M \text{ in H)} \tag{8.21}$$

Applying Equation (8.16) this gives the resultant coupling in terms of mutual inductance as

$$M_r = \sqrt{\left(\frac{Z_1 Z_2}{4} \cdot 0.32 \times 10^{-3} \cdot K_{eff} M\right)^2 + M^2}$$

$$= M\sqrt{1 + \left(\frac{Z_1 Z_2}{4} \cdot 0.32 \times 10^{-3} \cdot K_{eff}\right)^2} \tag{8.22}$$

This only applies to unscreened adjacent pairs; for screened pairs, or pairs separated by another pair, the capacitance coupling is very small and tends to zero. Thus for pairs of equal impedance of 150 Ω and for permittivity 1.45

$$M_r = M\sqrt{1 + (2.61)^2} = M \times 2.80$$

The effect of placing an electrostatic screen around the pairs would be to reduce the resultant by a factor of 2.80, equivalent to a crosstalk reduction of 8.93 dB. This would apply when the impedances were approaching Z_∞, say at about 100 kHz. For voice frequencies the impedances would be much higher, say 500 Ω, in which case the reduction on screening would also be higher at 29 dB.

For non-adjacent pairs/quads, the intervening unit acts as an electrostatic screen, also the separation increases by at least a facter of two, reducing M by a factor of 2.83 (Equation (8.19)) or 9 dB. So that even if the disturbed and disturbing pairs/quads had exactly the same lay there would be a reduction, at voice frequencies, of 38 dB, on the

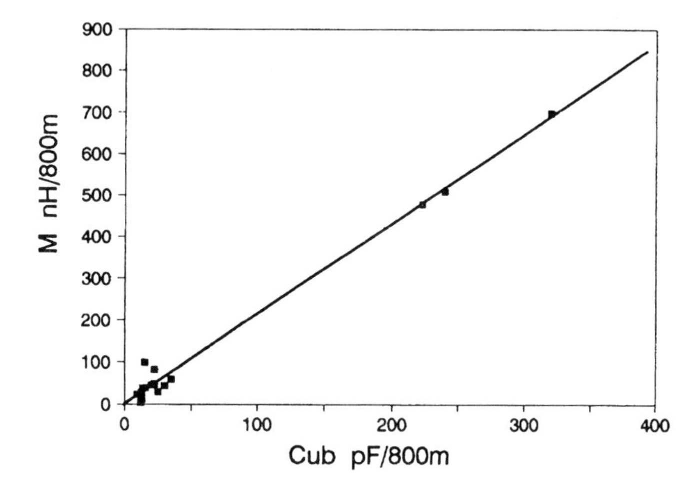

Figure 8.10
Mutual inductance versus capacitance unbalance for adjacent pairs (reproduced by permission of BICC Cables Ltd)

100 μH/km expected for this case. This is equivalent, Equation (8.8), to a 42 pF/km unbalance, a very acceptable figure for voice communication. This effect is made use of in quad trunk cables where alternate quads in a layer have identical lays (to reduce the number of different lays required in a large cable).

Again it must be stressed that all this applies only when the cable is short enough to avoid transmission effects at the frequency considered. Having evaluated the crosstalk on the short length, the effect of increasing the length should be considered as a further exercise.

Numerical Examples for Short Lengths

In calculating the electric and magnetic couplings to predict the measured values of crosstalk, the length of cable should be less than a tenth of the wavelength if the measured values are to correspond to the prediction. This means that the length should not be greater than about 250 m for a frequency of 100 kHz.

In the case of a 43 nF/km quad with ±0.5% random variation of conductor positions, the capacitive and magnetic couplings from Equations (8.9) and (8.10) would be about 1.95 pF/5 m or 14 pF/250 m, and 5 nH/5 m or 35 nH/250 m. For a 0.9 mm conductor and insulation with $K_{eff} = 1.7$ calculating the the impedance and attenuation of the pairs, and applying Equations (8.7), (8.14) and (8.15) to find the within-quad crosstalk on 250 m gives the values in Table 8.1. Such a quad would be suitable for carrier telephony up to about 300 kHz. although the measurements may differ from the predicted values above 100 kHz.

If the quad lay scheme has a mean value of lay = 100 mm and a mean lay ratio of 1.10 then Equation (8.19), using $k = 3.0$ and $r/a = 0.25$, gives the mean value of the magnetic coupling between adjacent quads as 67.8 nH/km or 25.4 nH/250 m. Applying Equations (8.22) and (8.7) gives the adjacent quad crosstalk on a 250 m length. Notice that the NEXT and FEXT are equal according to Equation (8.16). For quads separated by one intermediate quad, r/a reduces to 0.125 giving the magnetic coupling as 24 nH/km, and the capacitive coupling is zero. The between-quad couplings and crosstalk are given in Table 8.2 This lay scheme gives couplings low enough for carrier working to 500 kHz.

Table 8.1
Within-quad couplings and crosstalk on short length of 0.9 mm quad cable

f kHz	Z_0 Ω	αl dB	M_n μH	NEXT dB	M_f μH	FEXT dB	ELFEXT dB
1	449	0.18	0.75	106	0.68	107	106
3	264	0.28	0.28	100	0.21	103	102
10	167	0.38	0.13	92	0.06	99	98
30	144	0.44	0.11	83	0.04	92	92
100	138	0.63	0.1	73	0.03	83	82
300	134	1.04	0.1	63	0.03	74	73
1000	132	1.86	0.1	53	0.03	64	62

Table 8.2
Between-quad couplings and crosstalk for adjacent and one-between quads on short length of 0.9 mm quad cable

f kHz	Z_0 Ω	αl dB	M_r (adj Q) μm	F/NEXT dB	ELFEXT dB	M (lbtwn) μm	F/NEXT dB	ELFEXT dB
1	449	0.18	0.7	106	106	0.009	144	144
3	264	0.28	0.24	101	101	0.009	130	130
10	167	0.38	0.1	95	94	0.009	115	115
30	144	0.44	0.08	86	86	0.009	105	104
100	138	0.63	0.07	76	76	0.009	94	93
300	134	1.04	0.07	67	66	0.009	84	83
1000	132	1.86	0.07	56	54	0.009	73	71

ACCUMULATION OF CROSSTALK WITH LENGTH

In Figure 8.11 consider a short element dx of the interfering lines at a distance x from the generator. A coupling impedance shown as $Z_k \cdot dx$ exists between the lines at this point (equivalent to ωM of Equation (8.7), and the current in the disturbing line at x is $(E/Z_0) \cdot \varepsilon^{-\gamma x}$. This generates a voltage de in the disturbed line. The lines are considered to be similar, with impedance Z_0 and propagation constant $\gamma = \alpha + j\beta$. The voltage de travels to the near end to appear as de_n and to the far-end as de_f

$$de = Z_k \cdot \frac{E}{Z_0} \cdot \varepsilon^{-\gamma x} \cdot dx \tag{8.23}$$

$$de_n = Z_k \cdot \frac{E}{Z_0} \cdot \varepsilon^{-\gamma x} \cdot \varepsilon^{-\gamma x} \cdot dx \tag{8.24}$$

$$de_f = Z_k \frac{E}{Z_0} \cdot \varepsilon^{-\gamma x} \cdot \varepsilon^{-\gamma(L-x)} \cdot dx \tag{8.25}$$

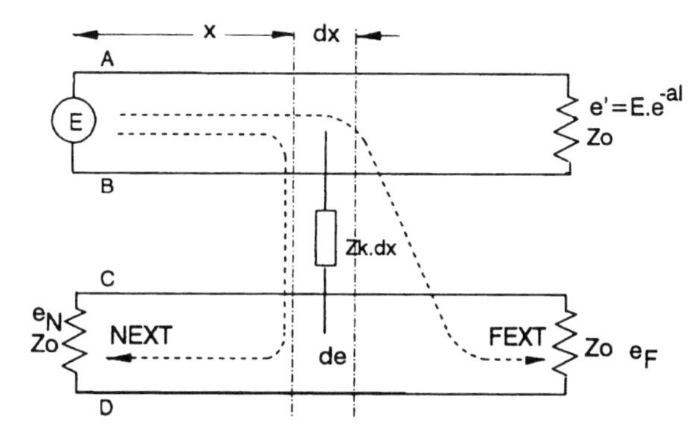

Figure 8.11
Crosstalk accumulation with length

Random Accumulation of NEXT

The coupling Z_k in balanced pairs arises by random residual couplings accumulating, so these elementary crosstalk voltages will also be in random phase relation. Their addition will therefore be a quadratic sum. If the elementary crosstalk attenuation at x is denoted as $A_e(x) = \ln(2Z_0/Z_k \cdot dx)$ then

$$|e_n|^2 = |E|^2 \cdot \int_0^L |\varepsilon^{-2\gamma x}|^2 \cdot \varepsilon^{-2A_e(x)} \cdot dx$$

$$= |E|^2 \cdot \varepsilon^{-2A_e(x)} \int_0^L \varepsilon^{-4\alpha L} \cdot dx$$

$$= |E|^2 \cdot \varepsilon^{-2A_e(x)} \cdot \frac{1 - \varepsilon^{-4\alpha L}}{4\alpha}$$

and the resultant NEXT attenuation is

$$A_N = \ln\left|\frac{E}{e_n}\right| = A_e + \frac{1}{2} \ln \frac{4\alpha}{1 - \varepsilon^{-4\alpha L}} \tag{8.26}$$

Random Accumulation of FEXT

To deal with FEXT in the same way, first consider that the line attenuations are different, viz. α_1 and α_2.

$$|e_f|^2 = |E|^2 \cdot \int_0^L |\varepsilon^{-\gamma_1 x} \cdot \varepsilon^{-\gamma_2 (L-x)}|^2 \cdot \varepsilon^{-2A_e(x)} \cdot dx$$

$$= |E|^2 \cdot \varepsilon^{-2A_e(x)} \cdot \varepsilon^{-2\alpha_2 L} \cdot \int_0^L \varepsilon^{-2(\alpha_2 - \alpha_1)x} \cdot dx$$

$$= |E|^2 \cdot \varepsilon^{-2A_e(x)} \cdot \varepsilon^{-2\alpha_2 L} \cdot \left\{\frac{1 - 2(\alpha_2 - \alpha_1)}{1 - \varepsilon^{-2(\alpha_2 - \alpha_1)L}}\right\}$$

and the resultant FEXT attenuation is

$$A_F = \ln\left|\frac{E}{e_f}\right| = A_e + \alpha_2 L + \frac{1}{2} \ln\left\{\frac{2(\alpha_2 - \alpha_1)}{1 - \varepsilon^{-2(\alpha_2 - \alpha_1)L}}\right\} \tag{8.27}$$

As $\alpha_1 \to \alpha_2$ the denominator in Equation (8.27) $\to 2(\alpha_2 - \alpha_1)L$, using the exponential series. Then

$$A_F = A_e + \alpha L - \tfrac{1}{2} \ln L \tag{8.28}$$

Omitting the second term converts the FEXT to ELFEXT (or FES/N). The ELFEXT attenuation decreases as the square root of length because of the third term. The FEXT attenuation first decreases and then increases as the attenuation increases. For Equations (8.26) and (8.28) to hold, the unit of length for both α and L must be the length dx used for A_e.

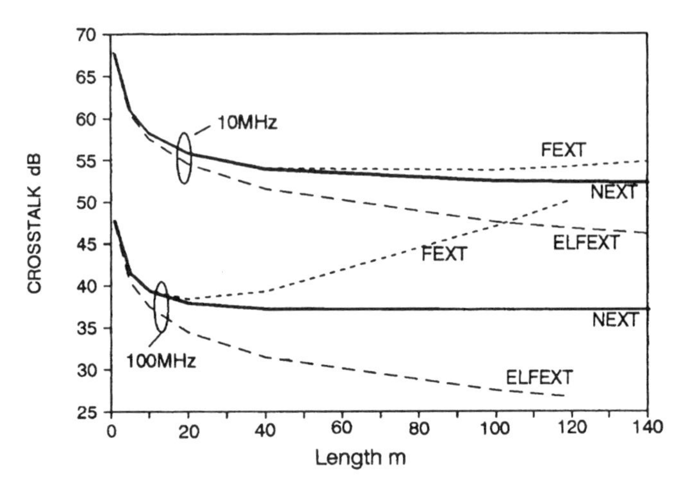

Figure 8.12
Pair crosstalk variation with length

Numerical Example

Apply these formulae to two adjacent 0.5 mm polythene pairs with impedance of 100 Ω and attenuation of 0.62 dB/100 m at 10 MHz, and having lay lengths of 20 mm and 24 mm respectively. Calculating the elementary coupling on 0.5 m of length from Equation (8.22) gives $M_r = 45 \times 1.8$ nH/km or 9.4 nH/0.5 m. The calculated crosstalks versus length at 10 MHz and 100 MHz are shown in Figure 8.12. The effect of making the elementary length 5 m instead of 0.5 m is to make the predictions about 0.5 dB pessimistic in this case.

The NEXT attenuation becomes asymptotic to $A_e + 0.5 \ln 4\alpha$ as the length increases, but the ELFEXT attenuation continues to fall as the square root of length. The FEXT at first falls as the length increases and then rises as the line attenuation increases.

This design would satisfy the requirements for NEXT for UTP-5 LAN cables. To improve these crosstalk figures by 6 dB, the mean lay length of the pairs, after cabling, should be halved while maintaining the same lay ratio.

Systematic Accumulation of Crosstalk

For completeness, consider the systematic accumulation of the elementary crosstalks. This would be the case where the magnitude and phase of the coupling impedance at each of the elementary lengths were sufficiently uniform. This happens, for example, in the case of two coaxial pairs with their outer conductors in contact throughout the length. Then Equations (8.24) and (8.25) can be directly integrated to obtain the resultant NEXT and FEXT.

$$e_N = \frac{1}{2}\frac{EZ_k}{Z_0} \cdot \int_0^L \varepsilon^{-2\gamma x} \cdot dx = \frac{EZ_k}{4\gamma Z_0}(1 - \varepsilon^{-2\gamma L})$$

The NEXT attenuation is

$$A_N = \ln\left|\frac{4Z_0\gamma}{Z_k}\right| - \ln|1 - \varepsilon^{-2\gamma L}| \tag{8.29}$$

$$= \ln\left|\frac{4Z_0\gamma}{Z_k}\right| - \ln\sqrt{1 + \varepsilon^{-4\alpha L} - 2\varepsilon^{-2\alpha L}\cos 2\beta L} \tag{8.30}$$

Similarly

$$e_F = \frac{1}{2}\frac{EZ_k}{Z_0}\cdot\int_0^L \varepsilon^{-\gamma x}\cdot\varepsilon^{-\gamma\cdot(L-x)}\cdot dx = \frac{1}{2}\frac{EZ_k}{Z_0}\cdot L\cdot\varepsilon^{-\gamma L}$$

The FEXT attenuation is

$$A_F = \ln\left|\frac{2Z_0}{Z_k\cdot L}\right| + \alpha L \tag{8.31}$$

The ELFEXT attenuation is the first term of Equation (8.31) and varies inversely with length. The NEXT Equation (8.30) has a first term that is constant with length, but the second term causes ripples as the length varies, the ripple length decreasing as the frequency increases. The ripple amplitude decreases as the length and attenuation increase, so that the NEXT becomes asymptotic to the first term for long lengths. This is illustrated for coaxials type 163 with the outer conductors in contact through the length and constant Z_k, in Figure 8.13

Reflected Crosstalk and Crosstalk via Third Circuits

Other sources contributing to the overall crosstalk between circuits arise from reflections at improperly terminated line ends, or at mismatched joints in the length.

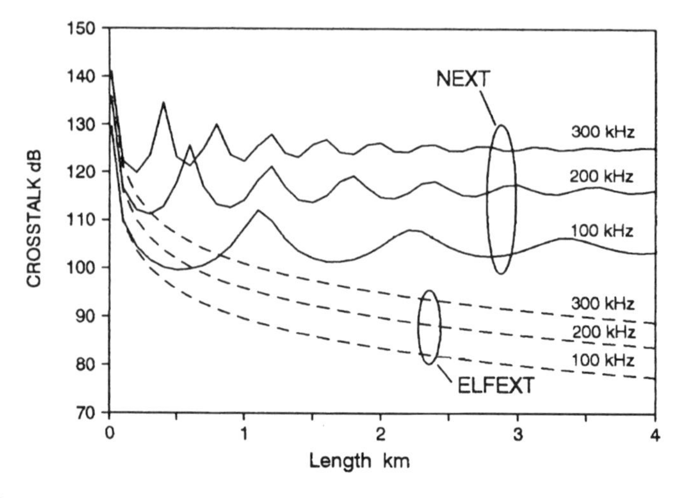

Figure 8.13
Coaxial crosstalk variation with length

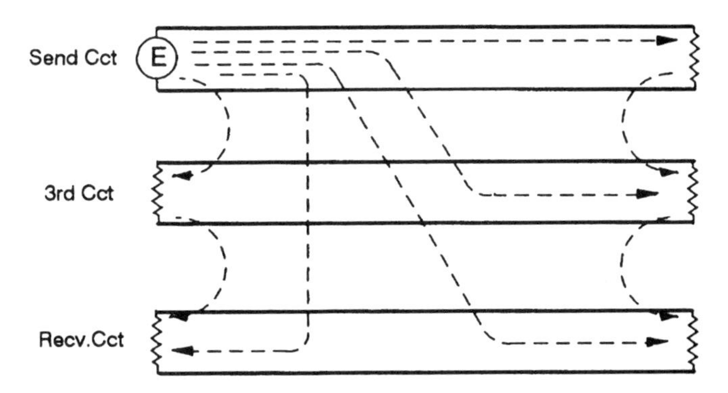

Figure 8.14
Third circuit and reflected crosstalk paths

Since these reflections are frequency dependent they give rise to ripples in the crosstalk frequency behaviour. This is particularly true for crosstalk that arrives via third circuits, Figure 8.14, which circuits are often ignored and left unterminated during use or measurement. The third circuit may not be another pair but could be, for example, a phantom circuit or an earth circuit. To control this source of crosstalk, all direct and third circuits should be properly terminated in both magnitude and phase to keep reflections below −30 dB (i.e. within 6% of the line impedances). This is also required of joints between lines, which requires the manufacturing tolerance on impedance to be within 3% in successive lengths.

Bourseau and Jarrosson [17] showed that third circuit crosstalk was primarily responsible for an asymmetric component of far end crosstalk, (i.e. the crosstalk coupling between two lines is different depending on which is used as the disturbing line) but a secondary cause which is systematic with quad lay length also plays a part (shorter lays have less AFEC) and produces a systematic component of AFEC on factory lengths. The practice in balanced pair carrier circuits was to use lumped capacitive balancing at the receive end of the cable to improve the far-end crosstalk as shown in Figure 8.15. In the presence of AFEC there was a limit to the degree of balance obtained with lumped capacitors, as shown in the figure. AFEC is a

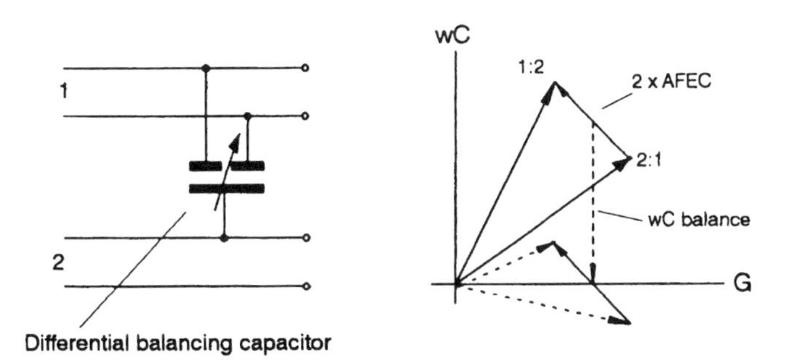

Figure 8.15
Asymmetric far-end coupling and capacitance balancing

considerable limitation on the exploitation of balanced pair cables to high frequencies (it increases as the square of the frequency) and can only be controlled by reduction of third circuit crosstalk and by transposition of wires and pairs (cross-jointing) at joints between cables.

Control of Crosstalk During Installation

The acceptance of the mixed linear/square-root law of coupling versus length of Equation (8.19) implies that the coupling contains both systematic and random components. As the length is increased by jointing, the systematic component increases more rapidly than the random component and will tend to dominate in the resultant coupling on longer lengths. This may be countered by crossing the wires within a pair at joints. This may be done according to a fixed pattern (scheduled joints), or in accordance with measurements of the actual complex couplings existing in the pairs to be jointed, after installation of the cable. For carrier cables these couplings are measured near the highest frequency to be transmitted and it is essential that the pairs (and the principal third circuit, which is the self-phantom of the quad) are well terminated in magnitude and phase during the measurement, so that the results are meaningful (to 0.05 pF) after jointing. Also, to reverse the send and receive pairs it is necessary to use relays with the relay couplings balanced out. In cables in which the phantom circuits are to be utilized, these cross-joints must also take account of side-phantom couplings. For within quad couplings there are eight combinations [19] of wire and pair crosses each with a specific effect on coupling and AFEC.

SCREENS

Schelkunoff's Theory [7]

Consider a cylindrical screen surrounding a number of circuits. The net electromagnetic field from these circuits will have a cylindrical wavefront and will be the superposition of an electric field, E, and a magnetic field, H. E and H will be perpendicular to each other and to the direction of propagation. Let the region where the field propagates have a permeability of μ, a permittivity of ε and a conductivity σ. Schelkunoff defined the impedance and propagation constants of the region as follows.

The radial propagation of the field will follow an exponentially decreasing law, $\varepsilon^{-\gamma\rho}$, with ρ being the radius of the wavefront and the intrinsic propagation constant

$$\gamma = \sqrt{j\omega\mu(\sigma + j\omega\varepsilon)} \tag{8.32}$$

For a dielectric, the conductivity is zero so that

$$\gamma = j\omega\sqrt{\mu\varepsilon} \tag{8.33}$$

which, being purely imaginary, means that there is no attenuation of the field travelling in a dielectric. For a conductor, the permittivity is zero so that

$$\gamma = \alpha + j\beta = \sqrt{j\omega\mu\sigma} \tag{8.34}$$

Consequently the wave has an attenuation constant

$$\alpha = \sqrt{\frac{\omega\mu\sigma}{2}} \tag{8.35}$$

The radial impedance for a conductive region is defined as

$$Z_\rho = \sqrt{\frac{j\omega\mu}{\sigma}} \tag{8.36}$$

This is the same for both E and H fields, and is independent of radius. However in a dielectric region the impedance at a radius ρ is

$$Z_{\rho H} = j\omega\mu\rho \quad \text{for the magnetic field} \tag{8.37}$$

$$Z_{\rho E} = \frac{1}{j\omega\varepsilon\rho} \quad \text{for the electric field} \tag{8.38}$$

In Figure 8.16 the incident wave E_0, H_0 approaches the boundary between regions 1 and 2 which have different radial impedances. There will therefore be a reflected wave, E_r, H_r and a transmitted wave E_t, H_t.

Because of the continuity of the tangential components

$$E_0 + E_r = E_t \quad \text{and} \quad H_0 + II_r = H_t \tag{8.39}$$

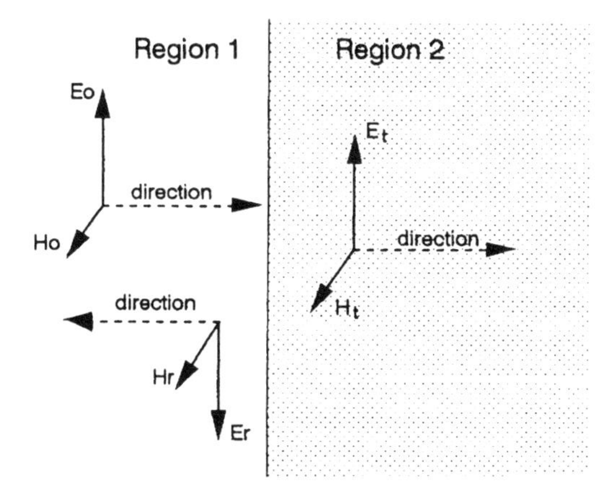

Figure 8.16
Reflection of electromagnetic wave at boundary of two materials

The radial impedances are given by

$$\frac{E_0}{H_0} = Z_{\rho 1}, \quad \frac{E_r}{H_r} = -Z_{\rho 1}, \quad \frac{E_t}{H_t} = Z_{\rho 2} \tag{8.40}$$

$$\text{Let} \quad K = \frac{Z_{\rho 1}}{Z_{\rho 2}}, \qquad \text{then} \qquad K(H_0 - H_r) = H_t \tag{8.41}$$

$$H_t = \frac{2K}{K+1} H_0 \tag{8.42}$$

$$E_t = \frac{2}{K+1} E_0 \tag{8.43}$$

Thus at a single boundary the electric and magnetic fields are modified differently. However for a conductive screen of finite thickness there are two boundaries and there will also be a reflection loss when the wave exits the screen. At the exit K becomes its reciprocal and the coefficients of Equations (8.42) and (8.43) are interchanged. To obtain the total loss by the two reflections, the coefficients are multiplied and the result is the same for both electric and magnetic fields

$$\frac{H_t}{H_0} = \frac{E_t}{E_0} = \frac{2K}{K+1}\frac{2}{K+1} = \frac{4K}{(K+1)^2} \tag{8.44}$$

or, as an attenuation

$$A_r = 20 \log \left| \frac{(K+1)^2}{4K} \right| \text{ dB} \tag{8.45}$$

However it must be remembered that K is different for the electric and magnetic fields, Equations (8.37) and (8.38). To this reflection loss is added the attenuation of the field as it propagates through the conductive screen, Equation (8.35).

If the screen is very thin there will be some interaction of the fields owing to multiple reflections. If there are multiple layers of screens, the situation is treated in the same way as the connection of transmission lines with different impedances and propagation coefficients.

Typical Values

For an air dielectric, $\mu = \mu_0 = 4\pi \times 10^{-7}$ H/m and $\varepsilon = \varepsilon_0 = \{1/36\pi\} \times 10^{-9}$ F/m. At a radius of 5 mm (0.005 m) and a frequency of 100 kHz

$$Z_{\rho E} = \frac{1}{j\omega\rho\varepsilon} = \frac{1}{j} \times 36\,\text{M}\Omega \quad \text{and} \quad Z_{\rho H} = j\omega\mu\rho = j \times 3.95\,\text{m}\Omega$$

For copper at 100 kHz $\mu = \mu_0 = 4\pi \times 10^{-7}$ H/m and $\sigma = 5.8 \times 10^7$ S/m.

$$Z_\rho = \sqrt{\frac{j\omega\mu}{\sigma}} = \sqrt{j} \times 0.116\,\text{m}\Omega \quad \text{and} \quad \alpha = \sqrt{\frac{\omega\mu\sigma}{2}} = 41.5\,\text{dB/mm}$$

For steel at 100kHz $\mu = 100\mu_0 = 4\pi \times 10^{-5}$ H/m and $\sigma = 10^7$ S/m.

$$Z_\rho = \sqrt{\frac{j\omega\mu}{\sigma}} = \sqrt{j} \times 2.81\,\text{m}\Omega \quad \text{and} \quad \alpha = \sqrt{\frac{\omega\mu\sigma}{2}} = 173\,\text{dB/mm}$$

(Note that $\sqrt{j} \cdot x$ is a vector with modulus x and an angle of 45°.)

Thus for the E-field at an air/copper boundary of 5 mm radius,

$$K_E = \frac{36\,\text{M}\Omega}{0.116\,\text{m}\Omega} = 3.1 \times 10^{11}$$

and the E-field reflection loss across a thin air/copper/air screen from Equation (8.45) is

$$A_{rE} = 218\,\text{dB}$$

For the H-field at an air/copper boundary of 5mm radius,

$$K_H = \frac{3.95\,\text{m}\Omega}{0.116\,\text{m}\Omega} = 34$$

and the H-field reflection loss across the same thin air/copper/air screen, from Equation (8.45), is

$$A_{rH} = 19\,\text{dB}$$

So the E-field is almost entirely reflected by even a thin metal screen, whereas the reflection loss of the H-field is much smaller.

Figure 8.17 shows the variation with frequency for the E-field double-reflection loss for copper at 1, 3 and 10 mm radii, and for steel at 10 mm radius. The H-field double-reflection losses at 1, 3 and 10 mm radii are shown by the heavy lines in Figures 8.18–8.20, and attenuation losses for copper, aluminium and steel screens of 10, 30 and 100 μm thickness are shown by the dashed lines.

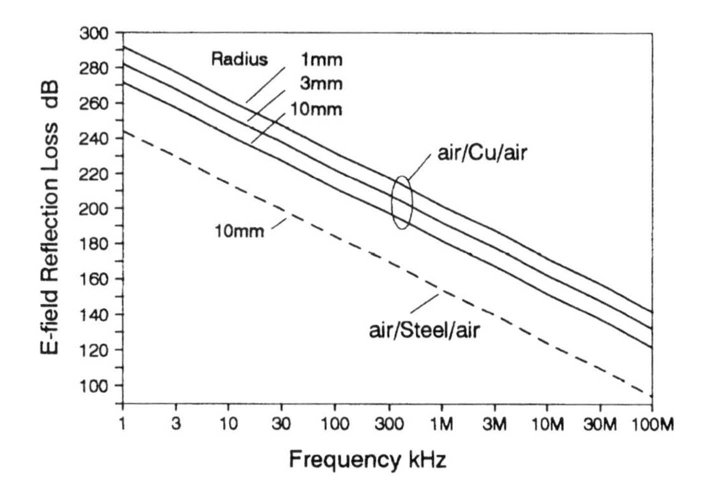

Figure 8.17
Electric field reflection loss at metallic boundary

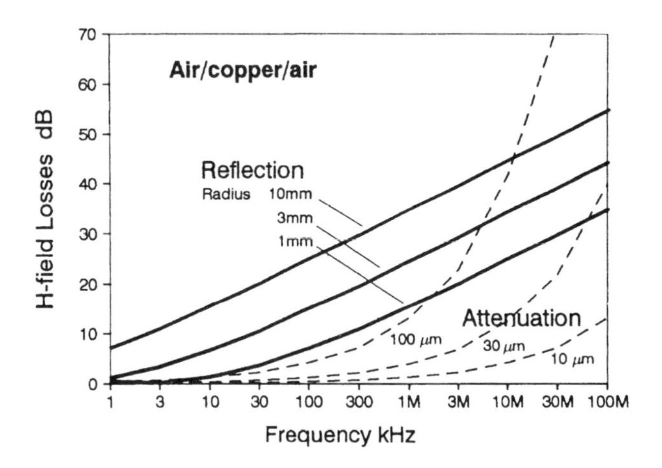

Figure 8.18

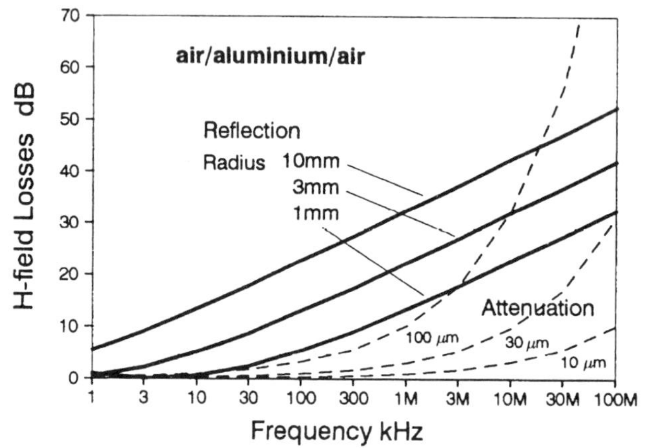

Figure 8.19
Magnetic field reflection loss and attenuation for aluminium screen

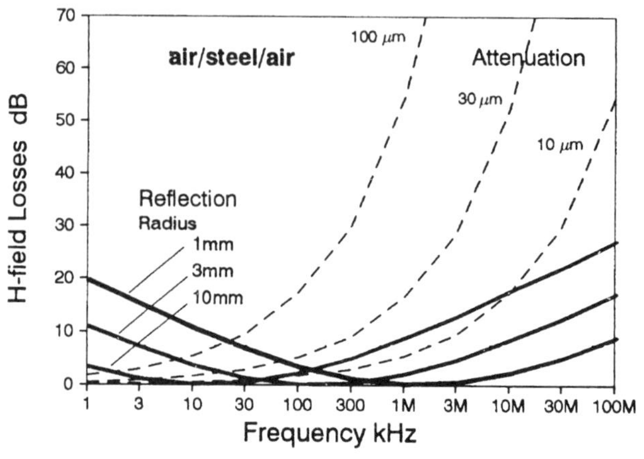

Figure 8.20
Magnetic field reflection loss and attenuation for steel screen

Notice the different shape of the steel double-reflection losses in Figure 8.20. This is due to the permeability of the steel making its intrinsic impedance higher than the H-field impedance of air at the lower frequencies. Then because of the linear increase of the latter with frequency, it crosses the steel impedance at a frequency depending on the radius. At this frequency the reflection loss is zero. As the frequency rises further the H-field air impedance becomes larger than the steel intrinsic impedance and the reflection loss builds up again. This effect can be used to increase the screen effectiveness at lower frequencies by the use of composite screens such as air/copper/steel/air.

Effect of Screens on Pair Couplings

In the discussion after Equation (8.22) it was noted that the presence of an electrostatic screen effectively removes the capacitive coupling associated with the mutual inductance coupling. From Figure 8.18 it will be seen that even as high as 100 MHz the E-field reflection loss is greater than 100 dB so this is a justifiable assumption even if the 'screen' is only another pair/quad.

If the screen is an aluminium or copper foil however there is also an H-field reflection loss as the frequency rises. At 100 kHz this loss is about 10 dB at a radius of 2–3 mm. Since it is usual when screening to screen both pairs this loss will be doubled to about 20 dB, so the mutual inductance coupling will be reduced to one-tenth of its value without the screens. Even at 3 kHz the H-field loss would be 2×5 dB and M would be reduced to one-third. If however the 'screen' were another pair/quad, the magnetic field would penetrate to a greater or lesser extent depending on frequency and the reduction in mutual inductance would not be as large.

Example of Screened Pair Crosstalk

Take as an example the two screened pair LAN cable shown in Figure 8.21. The details of the design are as follows

Conductor diameter $d = 0.63$ mm	Lay lengths p_1 and $p_2 = 60$ mm and 70 mm
Insulation diameter $D = 2.4$ mm	Screen: longitudinal Al/Mylar (9 μm Al)
Insulation: cellular polythene	Screen radius taken as a mean value = 1.5 mm
Effective permittivity $K = 1.63$	Two screened pairs twisted together
HF Impedance: $Z_0 = 143\,\Omega$	Overall screen and sheath

(For other characteristics see pages 111–120)

Applying Schelkunoff's theory above, and the effect of the E-field reflection loss on M_r in Equation (8.22), the expected screening for each screen due to H-field reflections and attenuation is shown in Figure 8.22 together with the total screening effect for two

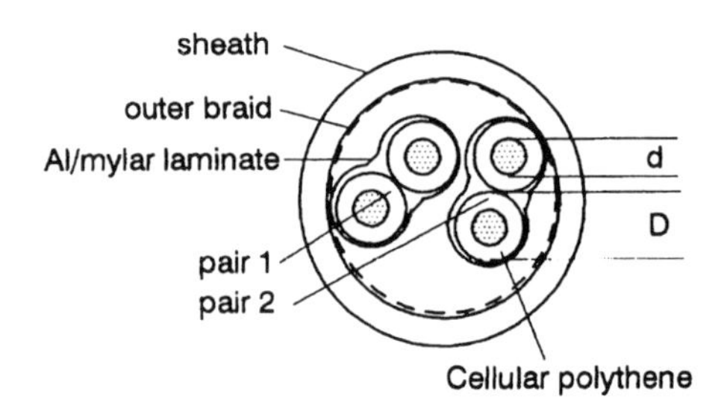

Figure 8.21
Local area network cable

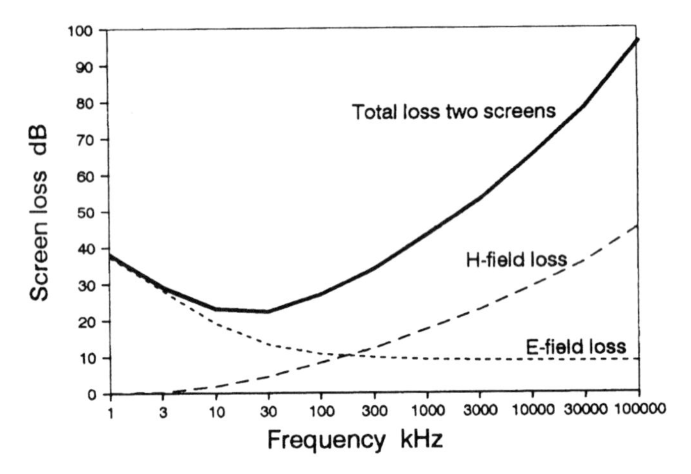

Figure 8.22
Losses due to 9 μm aluminium screen at 1.5 mm radius

screens. The expected mutual inductance for unscreened pairs is calculated by Pearson's formula, Equation (8.19) and including the associated capacitance from Equation (8.22), but taking the quality factor for 0.63 mm conductors as 5.2 and (r/a) as 0.25. This is calculated for an elementary length of 0.2 m as 0.54 nH.

The NEXT accumulated to 100 m according to Equation (8.26) is shown in Figure 8.23 before and after applying the calculated screening effect. The NEXT with screens is above 100 dB at all frequencies in the band up to 100 MHz. To measure this on a network analyser it was necessary to reduce the measuring bandwidth to 1 Hz and increase the sweep rate to about two minutes. In this way the noise level in the analyser was kept below −115 dB in the band from 60 kHz to 6 MHz. The measured values of NEXT with screen are shown as crosses in Figure 8.23 and show a satisfactory agreement in both level and shape, in this low-noise band. The drop-off at low and high frequency is due to the inherent noise level of the analyser. Note that the typical measurement of screened NEXT at say 1 kHz bandwidth and 5 s sweep, is merely a measurement of the analyser noise. Similar results were obtained for ELFEXT with screens, which calculated above 105 dB for this 100 m length, but the measurements were complicated by earth loop effects.

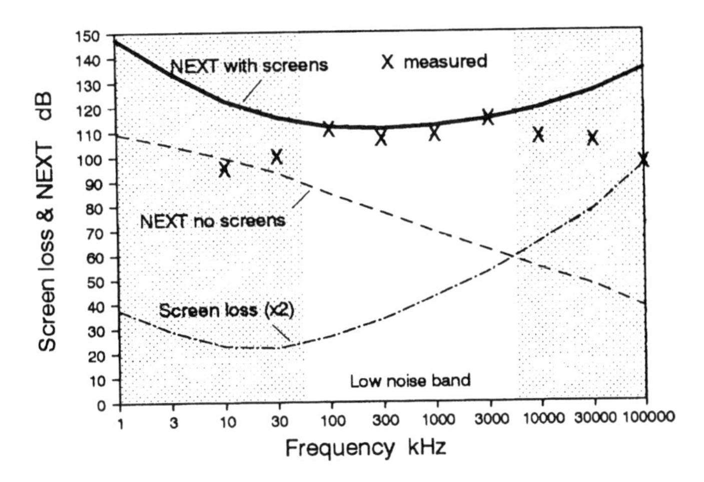

Figure 8.23
Screened pair NEXT on 100 m length

Analysis of PCM Cable Crosstalk Measurements

A sketch of the cross-section of an 80-pair PCM cable is shown in Figure 8.24. It consists of four 20-pair units with the 'go' units 1 and 2, separated from the 'return' units 3 and 4, by a transverse screen. The screen is a 70 μm aluminium foil laminated on both sides with polythene. The cable is screened overall by an aluminium polythene laminate sheath. The 0.63 mm conductors are insulated to 1.45 mm with solid polythene to give a mean pair capacitance of 41 nF/km, a 1 MHz impedance of 120 Ω and attenuation at 1 MHz of less than 12 dB/km. The pairs each have a different lay length with a mean value of 90 mm and a mean difference of 4 mm. Routinely, measurements of ELFEXT are made at 1 MHz between all combinations of pairs within each unit and NEXT at 1 MHz between all combinations of pairs across the transverse screen.

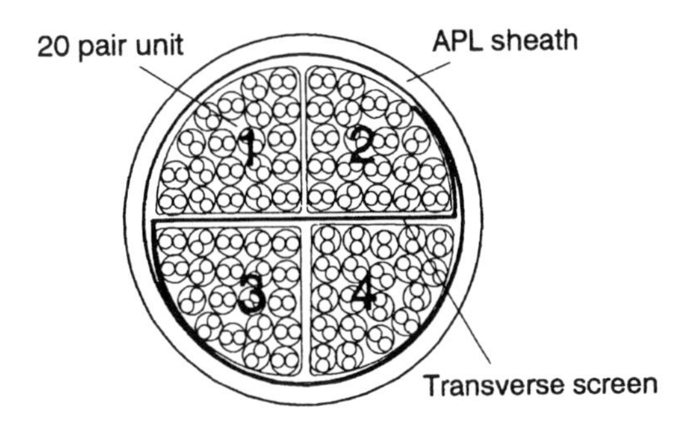

Figure 8.24
PCM cable with longitudinal screen

To assess the effect of the screen, compare the average NEXT between adjacent unscreened units, e.g. 1-to-2, with the average NEXT between units across the screen e.g. 1-to-3. This requires the derivation of the elementary length crosstalk, A_e of Equations (8.26) and (8.28), from the ELFEXT measurements using Equation (8.28). This is then used to calculate the average NEXT within unit using Equation (8.26) and the average NEXT between adjacent unscreened units, 1-to-2, by using the spacing term a in Equation (8.19).

NEXT for long lengths is $A_N = A_e + 10 \log 4\alpha$ from Equation (8.26) with A_e and α in dB/m. Similarly ELFEXT is $A_F = A_e - 10 \log L$ from Equation (8.28) with L in m. Typically ELFEXT within units has a mean value of 66.5 dB on 1000 m lengths, so that

$$A_e = \{\text{ELFEXT}\} + 10 \log L = 66.5 + 30 = 96.5\,\text{dB for a 1 m elementary length}$$

Now derive the NEXT within unit from this elementary crosstalk using a measured pair attenuation of 11.5 dB/km (or 0.0115 dB/m)

$$\text{NEXT within unit} = A_N = A_e + 10 \log 4\alpha$$

$$= 96.5 + 10 \log(4 \times 0.0115) = 96.5 - 13.4 = 83.1\,\text{dB}$$

Examination of the cross-section allows the assumption to be made that this mean value of NEXT within units corresponds to a mean pair spacing of 'one-between' (i.e. the spacing $a = 2 \times$ pair diameter). Similarly for NEXT between units assume that the average pair spacing increases to 3-pair diameters. From Equation (8.19) the ratio of the average mutual impedances between and within units will be $M_{\text{bu}}/M_{\text{wu}} = (2/3)^{3/2} = 0.55$, which is equivalent to a 5.3 dB increase in NEXT attenuation, hence

$$\text{NEXT between adjacent unscreened units} = 83.1 + 5.3 = 88.4\,\text{dB}$$

The average NEXT measured between adjacent screened units is typically 112 dB, thus the effect of the transverse screen is to add $112 - 88.4 = 23.6$ dB to the crosstalk attenuation.

Since the mean pair combinations considered are already electrostatically screened by the intervening pairs, this screen loss is due only to H-field losses by reflection and attenuation in the aluminium. By calculation using Equations (8.35) and (8.45):

Attenuation in screen

$$\alpha = \sqrt{\frac{\omega\mu\sigma}{2}} = \sqrt{\frac{2\pi \times 10^6 \times 4\pi \times 10^{-7} \times 3.5 \times 10^7}{2}} = 11.75\,\text{N/mm} \equiv 102\,\text{dB/mm}$$

For 70 μm thickness the attenuation is 7.1 dB.

Radial impedance for the H-field in air is $Z_{\rho H} = j\omega\mu\rho = j \times 7.9\rho\,\text{m}\Omega$ (ρ in mm)

Radial impedance for the H-field in aluminium is

$$Z_\rho = \sqrt{\frac{j\omega\mu}{\sigma}} = \sqrt{\frac{j \cdot 2\pi 10^6 \times 4\pi \times 10^{-7}}{3.5 \times 10^7}} = \sqrt{j} \times 0.474\,\text{m}\Omega$$

Thus $K_H = Z_{\rho 1}/Z_{\rho 2} = 7.9\rho/0.474 = \sqrt{j} \times 16.7\rho$

Reflection loss $A_r = 20\log\left|\frac{(K+1)^2}{4K}\right| = 20\log\frac{(16.7\rho+1)^2}{66.7\rho}$

In order that this reflection loss should equal that derived from the crosstalk measurements, namely 23.6 − 7.1 = 16.5 dB, the wavefront radius must equal 1.49 mm. This is approximately the radius of a single pair, implying that the wavefront is composed of wavelets re-radiated from the pairs adjacent to the screen (analagous to Huygen's wavefront principle in optics).

Now use Pearson's formula, Equation (8.19) with the quality factor $k = 5.2$ appropriate to 0.63 mm conductors, to derive A_e using the same spacing assumptions. After stranding with a 600 mm lay, the maximum pair lay length is 107 mm and the minimum is 46 mm. The geometric mean lay is then 70 mm, and the geometric mean lay ratio relative to this is $(107/70)^{1/\sqrt{10}} = 1.14$. (Since it is ratios that are being considered, the geometric means are appropriate). Using Pearson's formula for '1-between' spacing

$$|\bar{M}| = k\left(\frac{r}{a}\right)^{3/2} \cdot \frac{1}{\{5 \times 10^{-6} + (1/p_1)\sqrt{1-(p_1/p_2)^2}\}}$$

$$= 5.2 \times (0.125)^{3/2} \times 144.4 = 33.2\,\text{nH/km}$$

$$= 33.2 \times 0.5(0.001 + \sqrt{0.001}) = 0.541\,\text{nH/m}$$

Thus from Equation (8.7)

$$A_e = 20\log\left(\frac{2Z_0}{\omega M}\right) = 20\log\left(\frac{2 \times 120}{2\pi \times 10^6 \times 0.541 \times 10^{-9}}\right) = 97.0\,\text{dB}$$

This is in good agreement with the value of $A_e = 96.5$ dB calculated from the mean measured ELFEXT.

Statistical Considerations

Thus the above analysis of a rather complex cable shows that Pearson's formula and Schelkunoff's theory may be used successfully to design for the *mean value* of crosstalk attenuations. However to meet a specification requiring a minimum value of crosstalk attenuation requires a knowledge of the statistical spread of crosstalk to be expected. Figure 8.25 shows the distribution of the measured ELFEXT within unit. This approximates a Gaussian curve calculated from the measured mean and standard deviation but is skewed significantly towards the lower attenuations.

In Figure 8.26 is shown the distribution of the moduli of the inductive couplings calculated from the crosstalk attenuations. The shape of this curve is typical of the distribution of the moduli of two independent variables having approximately the same standard deviation and described mathematically by

$$y = \frac{dF}{dz} = \frac{N}{\sigma^2} \cdot z \cdot \varepsilon^{-(z^2/\sigma^2)} \tag{8.46}$$

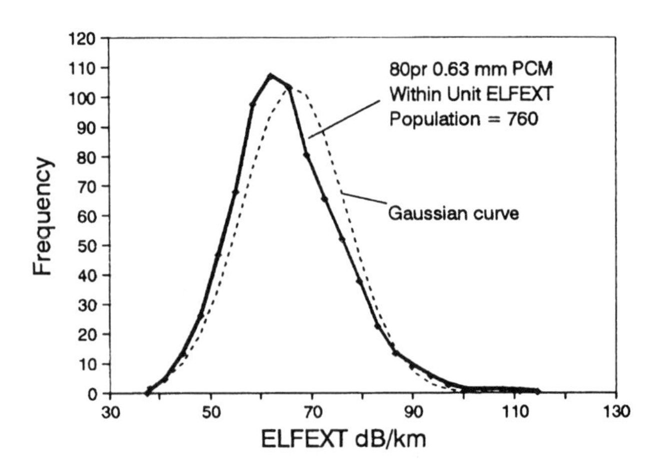

Figure 8.25
Distribution of ELFEXT within unit of PCM cable

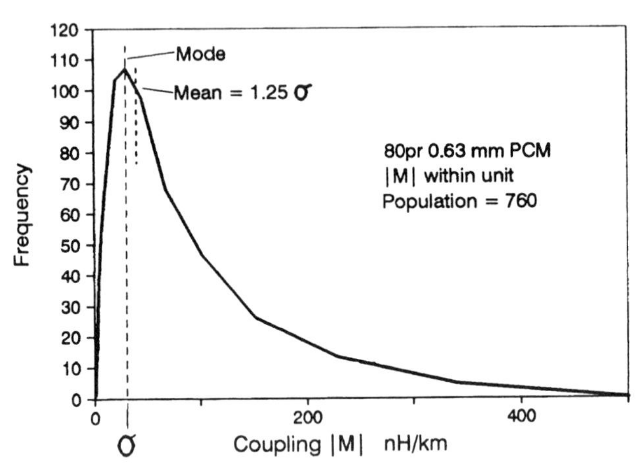

Figure 8.26
Distribution of inductive coupling moduli

where z is the variate, N the size of the population, F the statistical frequency and σ the standard deviation. It may be shown that in this case that the mode of the distribution occurs at $z = \sigma$ and the mean value of $z = 1.25\sigma$. For this population the mode and standard deviation are at 30 nH/km and the mean is 37.5 nH/km (cf. the calculated mean modulus of 33.2 nH/km).

By calculating the crosstalk attenuations at $|\bar{M}| \pm \sigma$ the apparent standard deviation of the crosstalk attenuation can be derived

$$\sigma = M/1.25 = 0.8M$$

$$M \pm \sigma = M(1 \pm 0.8)$$

$$A_e - \sigma_C = 20\log\left|\frac{2Z}{\omega M}\right| - 20\log(1.8) \quad \text{and} \quad A_e + \sigma_C = 20\log\left|\frac{2Z}{\omega M}\right| - 20\log(0.2)$$

whence

$$\sigma = \frac{20\log(1.8) - 20\log(0.2)}{2} = 9.54\,\text{dB}$$

This shows that the apparent standard deviation of the crosstalk attenuation is independent of the mean value of coupling or crosstalk and depends only on the shape of the distribution of Figure 8.26. This varies a little depending on the ratio of the individual variate standard deviations. In practice the standard deviation calculated from the measured crosstalk varies from 8.3 dB to 10.5 dB.

The specification for this PCM cable calls for a 1000 m length at 1 MHz to have

ELFEXT within unit: mean > 58 dB and std dev. = 8 dB

NEXT across screen: mean > 92 dB and std dev. = 8 dB

The system requirement for NEXT is given by

$$(\text{Mean} - \sigma) = \text{section loss} + (\text{Signal/Noise} + 6) + 10\log(n)$$

$$= 35 + 32 + 10\log(40) = 83\,\text{dB} \tag{8.47}$$

where (Signal/Noise + 6) is the system S/N plus a factor of safety of 6 dB and n is the number of systems on the cable. Thus the specification is really for (mean $-\,\sigma$) and a larger standard deviation can be accomodated by a larger mean crosstalk attenuation. Notice that the effect of the number of simultaneous signals on the other pairs is to degrade the crosstalk by the square root of n.

COAXIAL PAIR CROSSTALK

Mechanism of Coaxial Crosstalk

Since the return signal current completely surrounds the the sent current in a coaxial pair, there is no electromagnetic field outside the outer conductor. The mechanism causing crosstalk between coaxial conductors must therefore be different from that causing crosstalk between balanced pairs. The source of the crosstalk is the current flowing in the outer conductor producing a longitudinal voltage on its outer surface.

If the outer conductors of the coaxials are in electrical contact throughout their length, then the voltage on the outer conductor of the first coaxial is directly transferred. This was the situation considered on p. 149 under 'Systematic accumulation of crosstalk'. If the coaxials are insulated from each other, the longitudinal voltage causes a current to flow in the third circuit formed between two coaxials, which in turn generates a longitudinal voltage on the outer surface of the second coaxial. This voltage then causes a crosstalk current to flow in the second coaxial.

As the frequency of the signal increases, the skin-effect causes the current in the first coaxial outer conductor to be concentrated increasingly on its inner surface, reducing the voltage generated on the outer surface. Similarly the current flowing in the third

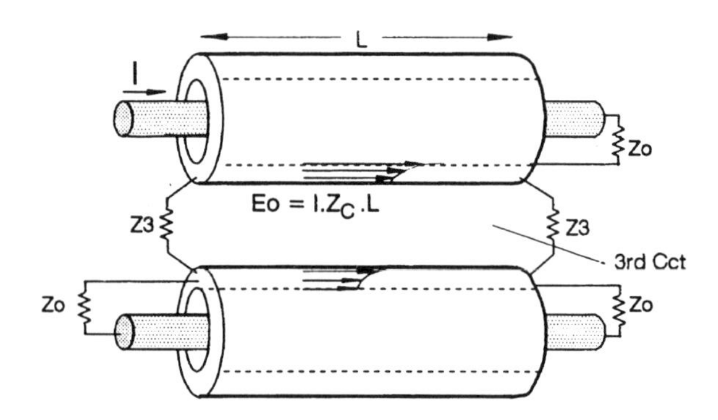

Figure 8.27
Coaxial pair crosstalk mechanism

circuit concentrates on the outer surfaces of the coaxial outer conductors reducing the crosstalk voltage on the inner surface of the second coaxial outer conductor. Thus the crosstalk between coaxials diminishes as the frequency increases in contrast to the situation between balanced pairs. This is illustrated in Figure 8.27

Coupling Impedance

The coupling between the two coaxial circuits is characterized by the coupling impedance, Z_c, which is defined on a short length of coaxial by the ratio of the voltage measured on the outer surface of the coaxial E_0, to the current I, sent down the inner conductor

$$Z_c = \frac{E_0}{I} \tag{8.48}$$

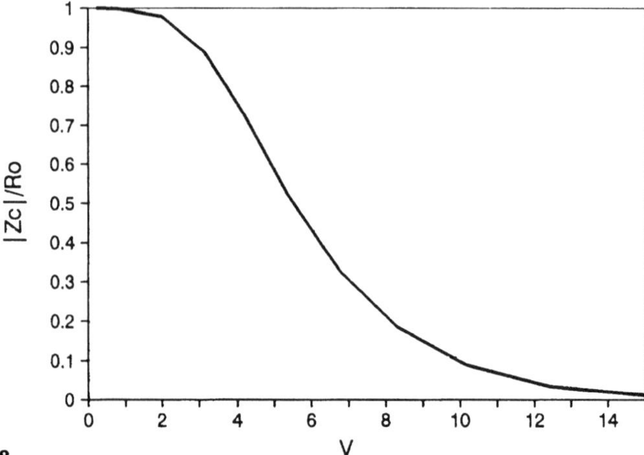

Figure 8.28
Coupling impedance ratio of tubular screen versus parameter 'v'

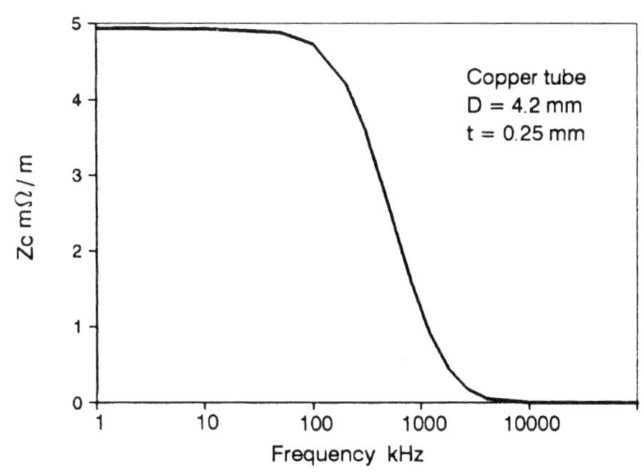

Figure 8.29
Coupling impedance of copper tube versus frequency

For a coaxial outer conductor in the form of a thin metal tube, the coupling impedance can be calculated using an asymptotic approximation to the Bessel function solution. This results in the formula

$$\frac{|Z_c|}{R_0} = \frac{v}{\sqrt{\cosh v - \cos v}} \tag{8.49}$$

where $v = t\sqrt{2\omega\mu\sigma}$, $\omega = 2\pi \times$ frequency, μ and σ are the permeability and conductivity of the metal tube, t is its thickness and R_0 is its d.c. resistance in Ω/m. This expression is plotted against v in Figure 8.28 and against frequency (for a copper tube of diameter 4.2 mm and thickness 0.25 mm) in Figure 8.29. It will be seen that the coupling impedance decreases very rapidly as the skin effect is established in the outer conductor.

Calculation of Crosstalk

The calculation of the crosstalk between two coaxials via the third circuit formed by their outer conductors is complicated, but basically follows the same steps used in the derivation of Equations (8.30) and (8.31) for coaxials in contact. It involves two integrations, the first to establish the current in the third circuit and the second to establish the voltages generated in the second coaxial circuit by this current. It results in the following for identical coaxials, where Z_0 and $\gamma = \alpha + j\beta$ refer to the coaxial circuits and Z_3 and $\gamma_3 = \alpha_3 + j\beta_3$ refer to the balanced third circuit.

For near-end crosstalk between the coaxial pairs

$$\text{NEXT} = 20\log\left|\frac{4Z_0Z_3}{Z_c^2}\right| - 20\log|S_n| \tag{8.50}$$

$$S_n = \frac{(1 - \varepsilon^{-2\gamma l})}{\gamma(\gamma_3 + \gamma)} \cdot \left[1 + \frac{2\gamma}{\gamma_3 - \gamma} \cdot \frac{\varepsilon^{-(\gamma_3+\gamma)l}}{(1 - \varepsilon^{-2\gamma l})}\right] \tag{8.51}$$

For $\gamma_3 \gg \gamma$ the term in the square brackets approaches unity, leaving

$$\text{NEXT} = 20\log\left|\frac{4Z_0Z_3}{Z_c^2}\right| - 20\log\left|\frac{1}{\gamma}\cdot\frac{1-\varepsilon^{-2\gamma l}}{\gamma_3}\right| \tag{8.52}$$

and for large values of l,

$$\text{NEXT} = 20\log\left|\frac{4Z_0\gamma\cdot Z_3\gamma_3}{Z_c^2}\right| \tag{8.53}$$

But $Z\gamma = R + j\omega L$, for both circuits, hence since $\omega L > R$,

$$\text{NEXT} = 20\log\left|\frac{4\omega^2 L_0 L_3}{Z_c^2}\right| \tag{8.54}$$

Thus the NEXT attenuation approaches a limiting value as the length increases and this limiting value also increases as the square of the frequency. When $\gamma_3 \rightarrow \gamma$ the term in square brackets becomes dominant and much greater than unity, severely degrading the NEXT attenuation. To avoid this γ_3 should be $\geq 3\gamma$.

Similarly for the FEXT attenuation

$$\text{FEXT} = 20\log\left|\frac{4Z_0Z_3}{Z_c^2}\right| - 20\log|S_f| + \alpha l \tag{8.55}$$

where

$$S_f = \frac{2\gamma_3 l}{\gamma_3^2-\gamma^2} - \frac{1-\varepsilon^{-(\gamma_3+\gamma)}}{(\gamma_3+\gamma)^2} - \frac{1-\varepsilon^{-(\gamma_3-\gamma)l}}{(\gamma_3-\gamma)^2} \tag{8.56}$$

For $\gamma_3 > \gamma$, $S_f = 2\cdot l/\gamma_3$ hence

$$\text{FEXT} = 20\log\left|\frac{2Z_0Z_3\gamma_3}{Z_c^2\cdot l}\right| + \alpha l \tag{8.57}$$

$$= 20\log\left|\frac{2Z_0\cdot\omega L_3}{Z_c^2\cdot l}\right| + \alpha l \tag{8.58}$$

The FESN (or ELFEXT) is given by the first term only, and this decreases as the length increases but increases as the frequency increases.

Comparing Equations (8.31) and (8.29) with Equations (8.57) and (8.53), it will be seen that

$$Z_k = \frac{Z_c^2}{Z_3\gamma_3} = \frac{Z_c^2}{(R_3 + j\omega L_3)} \tag{8.59}$$

So the effect of creating a third circuit in the crosstalk path is to reduce the effective coupling impedance proportionally to the third circuit series impedance, thus increasing the crosstalk attenuations.

Applying this theory to a practical situation, if the third circuit parameters are calculated for two coaxials having a plain copper tape outer conductor insulated with paper lappings, it will be found that $\gamma_3 \rightarrow \gamma$. To avoid this dangerous situation the outer conductors are lapped with two steel tapes each. This brings $\gamma_3 \geq 4\gamma$, and is the usual construction for multi-coaxial cables. Measured values of NEXT and FESN are

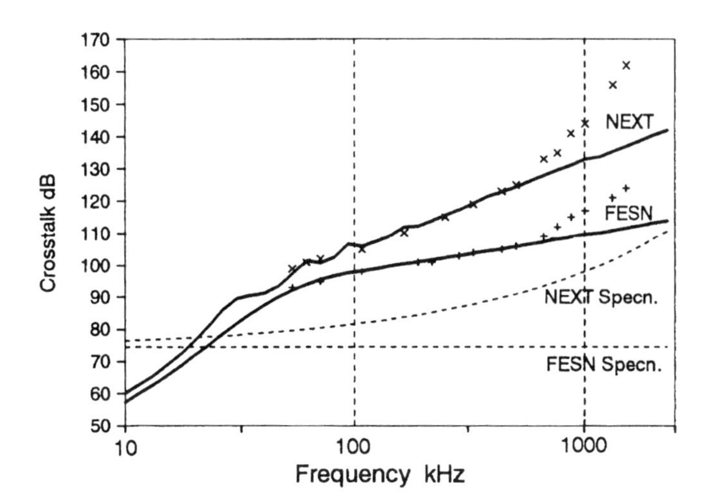

Figure 8.30
Repeater section crosstalk for coaxial type 163 (reproduced by permission of BICC Cables Ltd)

shown by the crosses in Figure 8.30, for adjacent type 163 coaxial pairs in a 4-coaxial cable, for a 3.75 km repeater section. Notice first of all that the measured values of both crosstalks are well above the specification levels (also shown) above 60 kHz, which is the lowest frequency for the system.

From Equation (8.54) the NEXT would be expected to climb as the square of the frequency increase. Although it starts this way it soon departs from this law, climbing more slowly up to about 500 kHz where it starts to increase more rapidly again. Similar behaviour is shown by the FESN attenuation. Using Equations (8.53) and (8.57) with calculated values for the third circuit parameters, the effective coupling impedances can be calculated from the measured crosstalk attenuations. These are shown in Figure 8.31.

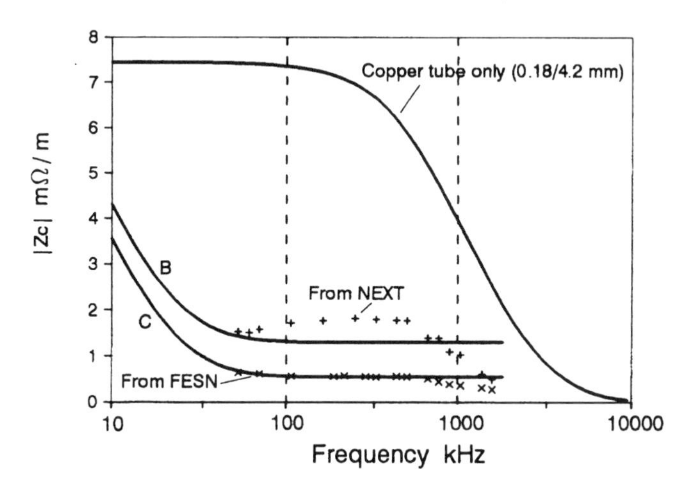

Figure 8.31
Coupling impedance for coaxial type 163

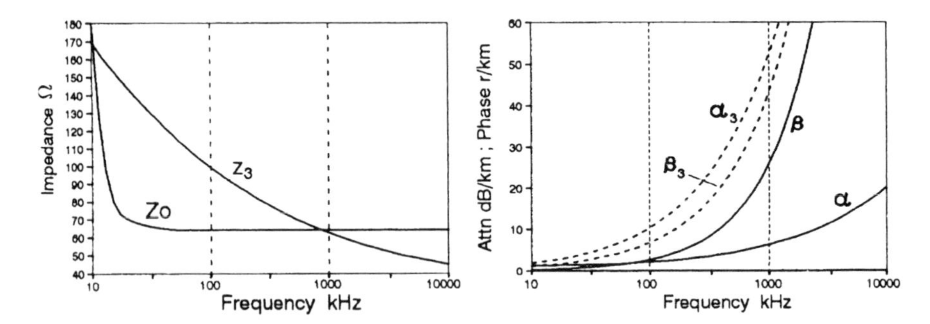

Figure 8.32
Impedance and third circuit impedance for coaxial 163 cable

The values of coupling impedance calculated from the crosstalk measurements are shown by the crosses. They tend to level out between 60 kHz and 500 kHz, and then continue to decrease.

The coupling impedance calculated for a plain copper tube similar to the outer conductor of the coaxials is shown for reference. The inclusion of the steel tapes reduces the coupling impedance considerably. There are three effects which could explain this behaviour

(a) The calculations made of the third circuit parameters and the coupling impedance assumed that the outer conductor with steel tapes could be represented by a metal tube having the same dimensions and conductivity as the copper foil, but with a permeability of 200. The behaviour of the laminated structure actually used could well be different.

(b) It is noted in the literature that copper outer conductors made from longitudinally formed foil can distort during manufacture to give a slight opening of the foil and this was advanced as a reason for the directly measured coupling impedance levelling off at about 0.3 mΩ/m.

(c) The measurements of crosstalk were made without terminating the third circuits at the ends of the repeater section. Although the attenuation of the third circuits is high, promoting self-termination, this could be a disturbing influence and could affect the NEXT and FESN attenuation measurements differently.

The curves B and C on the graph are calculated from Equation (8.49) using the outer conductor dimensions and conductivity but making the permeability 200 and adding a limiting value of 1.3 or 0.55 mΩ/m respectively. These values of coupling impedance were also used to calculate the NEXT and FESN curves in Figure 8.31 and give good agreement up to 500 kHz. The parameters used in the calculations above are shown in Figure 8.32.

9

Power Line Interference

When a conductor is in an electromagnetic field, voltages are induced in it which at best cause noise interference and at worst can be so high as to be dangerous. Both electric and magnetic fields produce these voltages but as it was seen in Chapter 8 even a thin metallic screen surrounding the conductor will protect the conductor from the electric induction due to the almost complete reflection of the E-field. The H-field loss however is much less, either by reflection or attenuation, depending on the frequency of the field. At low frequencies such as power-line frequencies, the H-field loss is very small, even for a magnetic screen. It is at these frequencies moreover that high fields can be present from, for example, nearby electric traction circuits. Other means of protection have therefore to be used in these circumstances [22].

MAGNETICALLY INDUCED VOLTAGES AND CONDUCTIVE SCREENING

In Figure 9.1 a current I amps flows in a line parallel to the conductor for L km. Both lines are connected to earth at their ends. The mutual impedance between the lines is $Z_{12}\,\Omega$/km and the induced voltage in line 2 is

$$e = -I \cdot Z_{12}\,\text{V/km} \tag{9.1}$$

The negative sign indicates that the e.m.f. is in the opposite direction to the inducing current. This e.m.f. would be balanced about the centre point of of the exposure L, but when one end of the disturbed line is earthed, the full e.m.f., eL, appears to earth at the other end of the line. In a typical traction situation Z_{12} could be 1 mH/km and I could be 1500 A at 50 Hz. Thus the e.m.f. on a 25 km exposure would be

$$eL = 1500 \times 2\pi \times 50 \times 0.001 \times 25 = 11\,775\,\text{V}$$

Such a longitudinal voltage would be dangerous to both personnel and equipment.

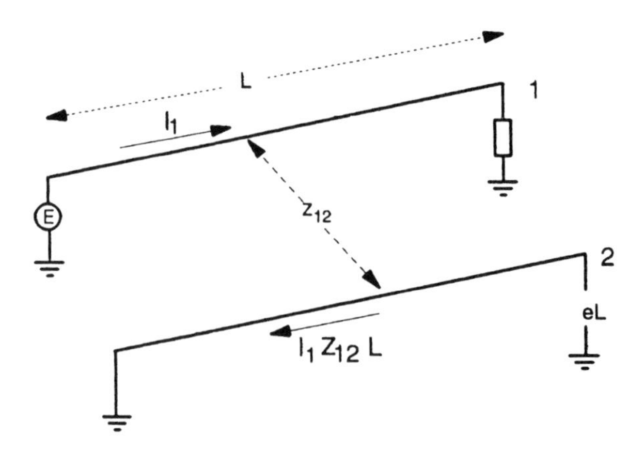

Figure 9.1
Induced e.m.f. in parallel line

Inductances of Conductors with Earth Return

These were given in Chapter 6 Equations (6.48) and (6.49). Since with power line induction the conductors are usually within a few metres of the physical earth surface, and at power line frequencies the effective earth depth H is of the order of several hundred metres, these formulae can be simplified by neglecting terms in y/H to give

$$H = 330\sqrt{\frac{\rho}{f}}\ \text{m} \tag{9.2}$$

$$L_{SE} = 0.2\ln\left(\frac{4H}{D}\right)\ \text{mH/km} \tag{9.3}$$

$$M = 0.2\ln\left(\frac{2H}{S}\right)\ \text{mH/kM} \tag{9.4}$$

where ρ is the earth resistivity in $\Omega \cdot \text{m}$, f is frequency in Hz, D is the conductor diameter in m and S the line separation in m. The expressions are graphed in Figures 9.2 and 9.3 for a range of earth resistivities.

Electromagnetic Screening

If an additional conductor, S, is introduced parallel to the disturbing conductor 1, and to the disturbed conductor 2, as in Figure 9.4, and if S is earthed at each end with negligible resistance, then 1 induces an e.m.f. into both S and 2. The e.m.f. induced in S causes a current I_S to flow, and this in turn induces an e.m.f. to flow in 2. This e.m.f. is oppositely directed to the e.m.f. induced directly in 2 from 1. Hence the effect of S is to reduce the resultant e.m.f. It therefore has a screening effect.

The screening factor of S is defined by

$$k = e/E \tag{9.5}$$

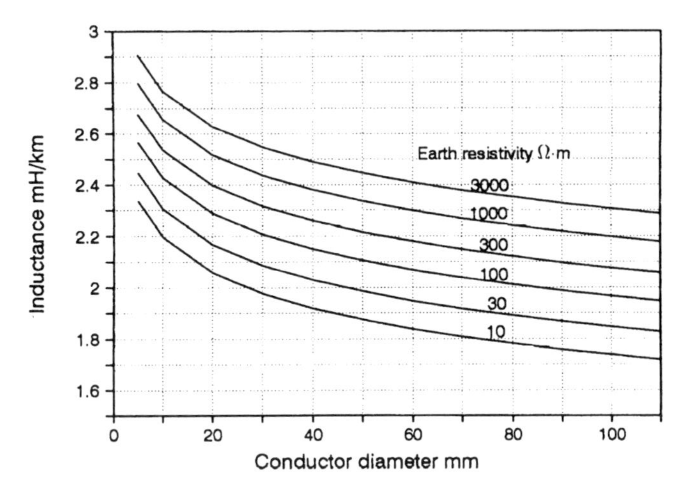

Figure 9.2
Inductance of line with earth return

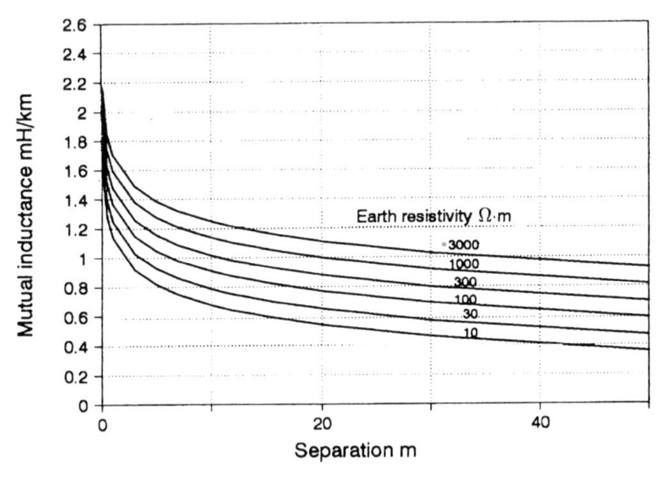

Figure 9.3
Mutual inductance between parallel lines with earth return

For circuit S with earth return

$$Z_{SE}I_S \cdot L - Z_{1S}I_1 \cdot L = 0$$

For circuit 2 with earth return

$$Z_{12}I_1 \cdot L - Z_{2S}I_S \cdot L = eL$$

where Z_{SE} is the self-impedance per unit length of S with earth return and Z_{2S} is the mutual impedance with earth return between 2 and S. Whence

$$e = \frac{Z_{12}Z_{SE} - Z_{1S}Z_{2S}}{Z_{SE}} \cdot I_1$$

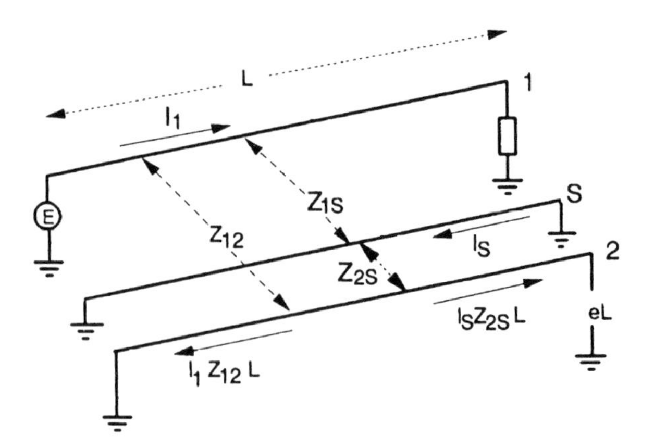

Figure 9.4
Effect of screening conductor

and

$$k = \frac{e}{E} = 1 - \frac{Z_{1S}Z_{2S}}{Z_{SE}Z_{12}} \tag{9.6}$$

If the screening conductor is much closer to 1 than 2, then $Z_{2S} \approx Z_{12}$ and

$$k = 1 - \frac{Z_{1S}}{Z_{SE}} \tag{9.7}$$

If the screening conductor is much closer to 2 than 1, then $Z_{1S} \approx Z_{12}$ and

$$k = 1 - \frac{Z_{2S}}{Z_{SE}} \tag{9.8}$$

In this case

$$k = \frac{Z_{SE} - Z_{2S}}{Z_{SE}} = \frac{R_{SE} - R_{2S} + j\omega(L_{SE} - M_{2S})}{Z_{SE}} \tag{9.9}$$

where

$$Z_{SE} = R_{SE} + j\omega L_{SE} \quad \text{and} \quad Z_{2S} = R_{2S} + j\omega M_{2S}, \quad \text{with } \omega = 2\pi f$$

Screening by Cable Sheath

In this case S is very near to 2, so that Equation (9.9) holds. In addition the mutual inductance between S and 2 is so very nearly equal to the self-inductance of S that $\omega(L_{SE} - M_{2S})$ can be neglected. Also any a.c. losses in the sheath–earth circuit are due to the flux which is common to both S and 2 so that $R_{SE} - R_{2S} \to R_{dc}$, the d.c. resistance of the sheath. Hence

$$k = \frac{R_{dc}}{Z_{SE}} = \frac{R_{dc}}{R_{SE} + j\omega L_{SE}} \tag{9.10}$$

The screening factor can thus be reduced (i.e. the screening can be improved) by reducing the d.c. resistance of the sheath and/or by increasing the impedance of the sheath–earth circuit. This may be done by using low resistivity sheathing material and also by adding inductance to the sheath–earth circuit (so long as the mutual inductance between S and 2 is increased by a like amount). This is done by enclosing both S and 2 within the same steel tape lappings. Note, if S does not surround 2, steel tape lapping of S is ineffective since the self-inductance is increased more than the mutual inductance.

As an example, consider a cable with a lead sheath of internal diameter 38 mm and a lead thickness of 2.5 mm. The d.c. resistance of the lead sheath will be 0.75 Ω/km and the inductance of the sheath–earth circuit will be of the order of 2 mH/km (Figure 9.2). Hence $k = 0.77$ at 50 Hz. Replacing the lead sheath with an aluminium sheath, the d.c. resistance falls to 0.1 Ω/km and the screening factor becomes 0.16. If the inductance is now increased by lapping with two steel tapes, the added inductance is about 10 mH/km and the added resistance due to iron losses is typically 1.0 Ω/km. For a lead sheath the screening factor will be about 0.18 at 50 Hz and for an aluminium sheath about 0.025 at 50 Hz. The induced longitudinal voltage calculated from Equation (9.1), reduces in proportion to k.

Calculation of Z_{SE}

For low frequencies where there is no skin-effect, the impedance of the sheath earth circuit is

$$Z_{SE} = R_{dc} + j\omega\left(0.2\ln\frac{4H}{D} + L_S\right) \times 10^{-3}\ \Omega/\text{km} \qquad (9.11)$$

H is the effective earth plane depth and depends on the earth resistivity (Equation (9.2)) and L_S is the added inductance, in mH/km, due to steel tape lapping. The earth resistivity depends on the type of ground (and its moisture content) down to several hundred metres. An indication of its value is given in Table 9.1. Urban land and suburban land have low effective resistivities owing to the many buried sevices using metallic components or carrying water.

Table 9.1
Typical earth resistivities

Earth resistivities	Ω · m
Urban land	3
Suburban land	10
Agricultural soil, loam	30
Moist sand	100
Moist gravel	300
Dry sand	1000
Stony ground	3000
Gneiss, granite	10 000

In order to reduce the screening factor, the d.c. resistance of the sheath must be made as low as possible. A cheaper alternative to using a swaged or extruded aluminium sheath is to use a layer of aluminimium or copper wires (as in wire armouring) beneath an APL sheath (to give corrosion protection).

The added inductance due to steel tape lapping may be calculated as follows

$$L_S = \frac{\text{Flux through steel tapes}}{I} = \frac{B \times \text{cross-sectional area}}{I}$$

$$= \frac{\mu \cdot I}{2\pi r} \cdot \frac{mn \cdot t}{I} = \frac{4\pi 10^{-7} \cdot mn \cdot t}{2\pi r}\ \text{H/m}$$

$$L_S = 0.4 \cdot \frac{mn \cdot t}{D} (\mu_R - j\mu_I)\ \text{mH/km} \tag{9.12}$$

n is the number of tapes, t = thickness of tapes and D = mean diameter of tape lappings. $m = (1 - g/100)$ where g = percentage gap of the lappings. The relative permeability of the steel tapes is $\mu_r = \mu_R + j\mu_I$ and is complex due to losses in the steel tapes. When multiplied by $j\omega$ in Equation (9.11) the imaginary part produces R_S, a real (i.e. resistive) addition to R_{SE}. With perfect earthing of the sheath at both ends, R_{SE} is equal to the d.c. resistance of the sheath and

$$k = \frac{R_{dc}}{R_{dc} + R_S + j\omega(L_{SE} + L_S)} \tag{9.13}$$

$$|k_0| = \frac{R_{dc}}{\sqrt{(R_{dc} + R_S)^2 + \omega^2(L_{SE} + L_S)^2}} \tag{9.14}$$

This is known as the intrinsic screening factor of the sheath. The permeabilities of typical mild steel tapes, after application, are shown in Figure 9.5. These permeabilities

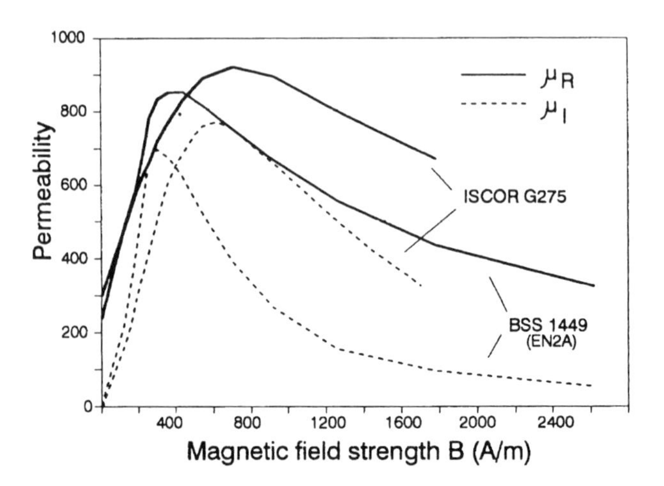

Figure 9.5
Magnetic permeabilities of typical steel tapes

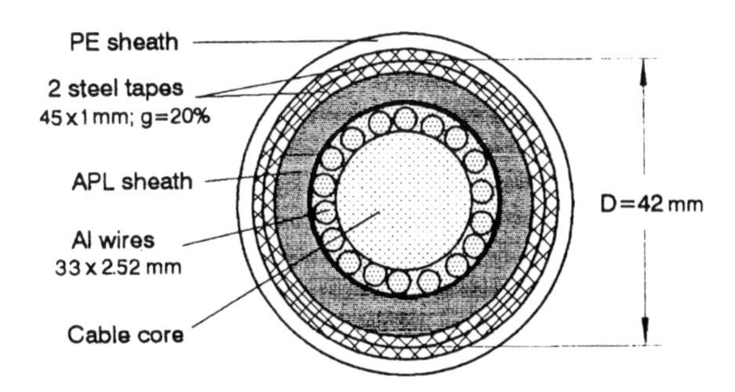

Figure 9.6
Electromagnetically screened cable

vary by about 10% depending on material batch. Since they vary with the magnetic field strength so will the intrinsic screening factor. Thus $|k_0|$ can be calculated for a range of B, knowing the dimensional details of the cable.

The induced longitudinal voltage in the sheath can be calculated from B as follows

$$I = B \cdot \pi \cdot D \text{ A}$$

$$E = I \cdot Z_{SE} = I \cdot R_{dc}/|k_0|$$

Thus

$$E = \frac{B \cdot \pi \cdot R_{dc}}{1000 \cdot |k_0|} \text{ V/km} \tag{9.15}$$

where R_{dc} is in Ω/km and D is in mm.

For the typical railway lineside cable shown in Figure 9.6, the intrinsic screening factor, calculated and measured, is shown in Figure 9.7. The measurement of screening factor is dealt with in Chapter 13.

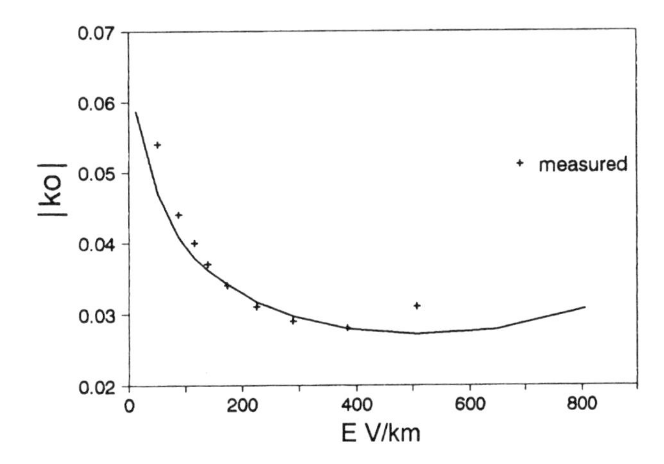

Figure 9.7
Intrinsic screening factor of screened

EFFECT OF FINITE EARTHING RESISTANCES

If the connections to earth at the ends of the cable have a finite resistance, R_E each, these resistances add to Z_{SE} and R_{dc} and Equation (9.13) becomes

$$k = \frac{R_{dc} + 2R_E}{(R_{dc} + 2R_{SE}) + j\omega(L_{SE} + L_S)} \tag{9.16}$$

Physically, the finite earth resistances reduce the induced screening current and increase (i.e. degrade) the screening factor.

Effect of Distributed Earths

If the earthing of the screening conductor is uniformly distributed along its length, thus forming a leakance G S/km, and if the screening conductor is infinitely long (or the screen–earth circuit is terminated in its characteristic impedance) the screen current at a point x from the origin (taken as one end of the exposure) is given by

$$\begin{aligned} 0 < x < l \qquad & I_S(x) = \frac{E}{2Z_{SE}}(2 - \varepsilon^{-\gamma x} - \varepsilon^{-\gamma(l-x)}) \\ x > l \qquad & I_S(x) = \frac{E}{2Z_{SE}}((1 - \varepsilon^{-\gamma l}) \cdot \varepsilon^{-\gamma(x-l)}) \\ x < 0 \qquad & I_S(x) = \frac{E}{2Z_{SE}}((1 - \varepsilon^{-\gamma l}) \cdot \varepsilon^{\gamma x}) \end{aligned} \tag{9.17}$$

where E = the field strength at the screening conductor and γ = the propagation constant of the screen–earth circuit

$$\begin{aligned} \gamma = \sqrt{(R + j\omega L)(G + j\omega C)} &= \sqrt{Z_{SE} \cdot G} \quad \text{since } G \gg \omega C \\ &= \sqrt{\frac{R_{dc} \cdot G}{k_0}} \quad \text{from Equation (9.13)} \end{aligned} \tag{9.18}$$

Correspondingly, the potential difference from the screening conductor to earth at x is given by

$$\begin{aligned} 0 < x < l \qquad & V_{SE}(x) = \frac{E}{2\gamma} \cdot (\varepsilon^{-\gamma(l-x)} - \varepsilon^{-\gamma x}) \\ x > l \qquad & V_{SE}(x) = \frac{E}{2\gamma} \cdot ((1 - \varepsilon^{-\gamma l})\varepsilon^{-\gamma(x-l)}) \\ x < 0 \qquad & V_{SE}(x) = \frac{E}{2\gamma} \cdot ((1 - \varepsilon^{-\gamma l})\varepsilon^{-\gamma x}) \end{aligned} \tag{9.19}$$

The effect of Equations (9.17) is that inside a long exposure, in the portions remote from the ends, the lines of current flow in the screening conductor and the earth are parallel. Consequently there is no interchange of current between them and the screening effect in this portion is independent of G, thus the screening factor has the

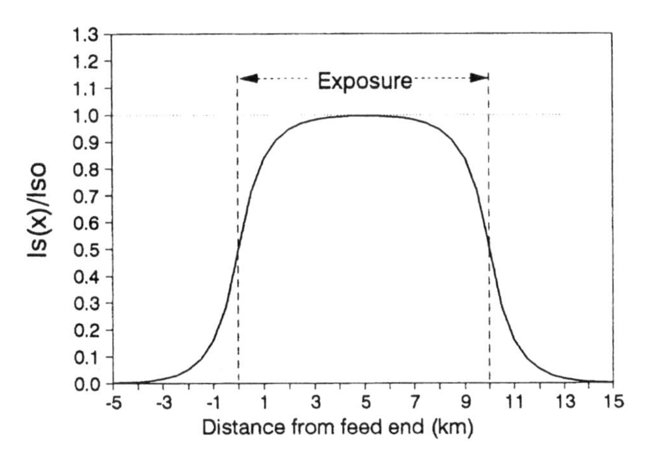

Figure 9.8
Current in sheath for cable with distributed leakance

value k_0. As the ends of the exposure are approached however, current flows between the screening conductor and earth, the current in the screening conductor becomes progressively smaller and its screening effect diminishes, typically to half its value.

The distribution of current and voltage for the cable of Figure 9.6 installed with a uniform leakance to earth of 0.25 S/km at a field strength of $E = 290$ V/km are shown in Figures 9.8 and 9.9. The exposure is taken as 10 km with the cable being longer than the exposure at both ends. Under these conditions at 50 Hz

Intrinsic screening factor	$\lvert k_0 \rvert = 0.03$	
Sheath d.c. resistance	$R_{dc} = 0.152$	Ω/km
Sheath a.c. resistance	$R_{dc} + R_S = 0.189$	Ω/km
Sheath–earth inductance	$L_S + L_{SE} = 16.38$	mH/km
Sheath–earth impedance	$\lvert Z_{SE} \rvert = 5.15$	Ω/km
For $G = 0.25$ S/km	$\lvert \gamma \rvert = 1.134$	neper/km

To find the effective screening factor in this case of non-uniform screening current, define it as

$$\text{For } l_1 > l \qquad k_e = \frac{El - \bar{I}_S \cdot Z_{2S} \cdot l_1}{El} = 1 - \frac{\bar{I}_S \cdot Z_{2S} \cdot l_1}{El} \tag{9.20}$$

where $l_1 > l$ is the length of the screening conductor and $\bar{I}_S$ is the mean screening current. For perfect earthing the screen current is uniform at a value of $I_{SO} = E/Z_{SE}$ and there is no current in the screening conductor outside the exposure. Hence the intrinsic screening factor is

$$k_0 = 1 - \frac{I_{SO} \cdot Z_{2S} \cdot l}{El} \tag{9.21}$$

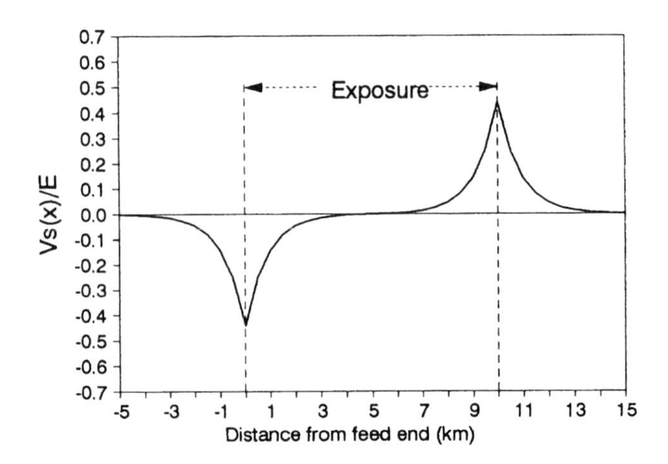

Figure 9.9
Voltage from sheath to earth for cable with distributed leakance

whence

$$k_e - k_0 = (1 - k_0) \cdot \left(\frac{1 - \bar{I}_S \cdot l_1}{I_{SO} \cdot l} \right) \tag{9.22}$$

The difference $k_e - k_0$ is called the earthing penalty.

When the screening conductor coincides in length and position with the exposure, $l_1 = l$, and by integration of Equation (9.17)

$$\bar{I}_S = I_{SO} \left[1 - \left(\frac{1 - \varepsilon^{-\gamma l}}{\gamma l} \right) \right] \tag{9.23}$$

whence from Equation (9.22)

$$k_e - k_0 = (1 - k_0) \left(\frac{1 - \varepsilon^{-\gamma l}}{\gamma l} \right) \tag{9.24}$$

$$\text{For } \gamma l > 3 \qquad k_e \approx k_0 + \frac{(1 - k_0)}{\gamma l} \qquad \text{and for } k_0 < 0,1 \qquad k_e \approx k_0 + \frac{1}{\gamma l} \tag{9.25}$$

For the example above

$$k_e = 0.03 + \frac{1 - 0.03}{1.134 \times 10} = 0.03 + 0.086 = 0.116$$

Discrete Periodic Earths

In practice it is essential to protect the sheath and armour from corrosion by a plastic oversheath. This insulates the screen from earth and earthing can only be provided at discrete points in the exposure. If these occur at regular intervals and the earthing resistance is $R_0\,\Omega$ at intervals of D km, and there are n earthing points in the exposure, then the effective value of G in Equation (9.18) becomes

$$G' = \frac{1}{DR_0}\left(1 + 4.35\sqrt{\frac{D}{nR_0}}\right)\ \text{S/km} \tag{9.26}$$

This correction operates to increase γ and to reduce the earthing penalty. Using $4\,\Omega$ earths every 1 km, $G' = 0.42$ S/km and $\gamma = 1.47$ in the example. Thus the earthing penalty reduces to 0.066 and $k = 0.096$. If the $4\,\Omega$ earths are connected every 300 m, $G' = 1015$ S/km, $\gamma = 2.29$, the earthing penalty reduces to 0.042 and the effective screening factor is 0.072. This example shows the importance of earthing and the maintainance of the earths in good condition, in order to achieve good screening from interfering power frequency fields.

Screening by External Conductors

If the screening conductor is external to the disturbed cable but relatively close to it, Equation (9.9) gives

$$k = \frac{R_{SE} - R_{2S} + j\omega(L_{SE} - M_{2S})}{Z_{SE}} = \frac{R_{dc} + j\omega(L_{SE} - M_{2S})}{Z_{SE}}$$

$$= k_0 + \frac{j\omega(L_{SE} - M_{2S})}{Z_{SE}} \tag{9.27}$$

where k_0 is the intrinsic screening factor for the screening conductor surrounding the cable. The term $|k| - |k_0|$ is called the separation penalty. When the screening conductor is a solid cylinder of diameter D m and the separation is S m, then from Equations (9.3) and (9.4)

$$L_{SE} - M_{2S} = 0.2\ln\left(\frac{2S}{D}\right)\ \text{mH/km} \tag{9.28}$$

For an aluminium conductor having the same d.c. resistance as the sheath in the example cable, $D = 27$ mm. Taking $L_{SE} = 2$ mH/km, at 50 Hz, the separation penalties are as given in Table 9.2.

Surrounding the screening conductor with steel tapes increases the self-inductance much more than the mutual inductance and considerably increases the separation penalty. Therefore a steel tape armoured cable provides little screening effect for adjacent cables. In addition the earthing considerations above apply equally to external screening conductors.

Table 9.2
Separation penalties

Separation (m)	0	0.25	0.5	1.0	2.0
Separation penalty	0	0.18	0.23	0.20	0.31
Screening factor	0.24	0.42	0.46	0.51	0.55

Rail Screening Factor

The resistivity of the steel rails is relatively high (about 26 times that of copper) but the large cross-sectional area gives the rails a low d.c. resistance of about 0.045 Ω/km each. However the high magnetic permeability increases their self-inductance much more than the mutual inductance to the disturbed cable and hence the separation penalty, from Equation (9.27), is very large.

The rails carry 50% to 70% of the return current (the remainder taking a long path deep into the earth) and this is in the opposite direction to the catenary current feeding the locomotive. Thus the rail current tends to cancel the field due to the catenary current providing a significant screening effect. The screening factor of the rails depends on their number and their position relative to the disturbed cable and computation is complicated. Values ranging from 0.25 to 0.6 have been measured in practice. For system design purposes the following values are usually assumed

Single track	0.6
Double track	0.5
Four tracks	0.3

Rail joints must be well bonded electrically.

Overall Screening Factor

The overall screening factor applying to a system is the product of the individual screening factors i.e.

$$k = k_1 \times k_2 \times k_3 \times \ldots.$$

TRACTION CURRENT CONSIDERATIONS

Running Conditions

A typical 25 kV single-phase locomotive can draw a starting current of 600 A. In order to avoid wheel-slip this current is built up over about one or two minutes, using tap-changing or thyristor control. In modern locomotives this is controlled automatically. Thus at maximum current the locomotive uses about 15 MW of power equivalent to about 11 200 h.p.

On heavy freight and ore traffic routes it is not uncommon for the train to be double- or triple-headed (i.e. having two or three locomotives). Thus the starting current can rise as high as 1800 A after a few minutes. This is held during the acceleration period and then reduced as the desired train velocity is reached. Thus the worst induced voltages occur for relatively short times during the train accelaration from rest.

Short-Circuit Conditions

Under fault conditions or when accidents occur, short-circuiting of the catenary can occur and very high currents can flow until the breakers operate (usually within three cycles or 0.6 s). The magnitude of the short-circuit current depends on the distance, x, from the feed point

$$I_{sc} = \frac{E_f}{A + Bx} \tag{9.29}$$

where E_f is the feed voltage and A and B are constants. A is identifiable with the feed impedance (sometimes deliberately augmented by leakage inductance in the feed transformer) and B is the catenary–rail/earth circuit impedance in Ω/km. Thus for a 20 mm diameter hard-drawn copper catenary having 1.3 mH/km inductance to rail/earth

$$B = 0.06 + j\omega \times 0.0013\,\Omega/\text{km}$$

$$|B| = 0.41\,\Omega/\text{km at } 50\,\text{Hz}$$

It is usual to to provide power feeding to 50 km sections of track from the centre point, thus limiting the feed distance (and the inducing exposure) to 25 km. If it is required that a triple-headed train can draw its full 1800 A at a distance of 25 km without bringing out the breakers, then the short-circuit current at 25 km must at least be equal to this full-load current. As $Bx = 10.3\,\Omega$ then $A = 3.6\,\Omega$.

However as a triple-headed train drawing 1800 A represents an impedance of 13.9 Ω, only half the feed voltage is available at 25 km and hence only one quarter of the power can be developed. Thus the value of A must be a compromise between current available at maximum feed distance and allowable short-circuit current. For A as low as 0.62 Ω the short-circuit current at 25 km would be 2300 A but the available running voltage would be 14 kV and the power developed only 31%. Thus a triple-headed train at 25 km would develop only 93% of the power of a single locomotive at zero feed distance.

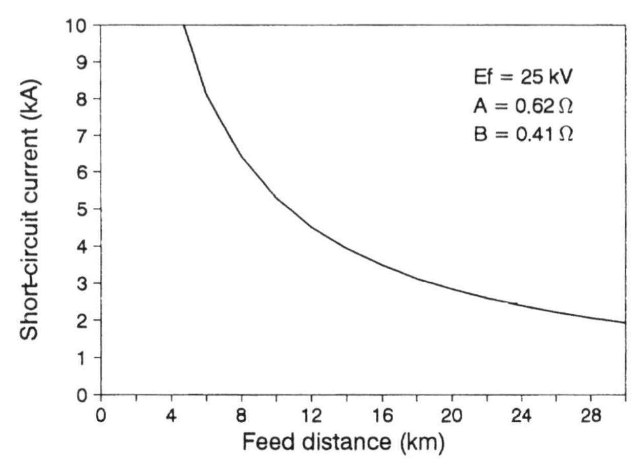

Figure 9.10
Typical short-circuit current

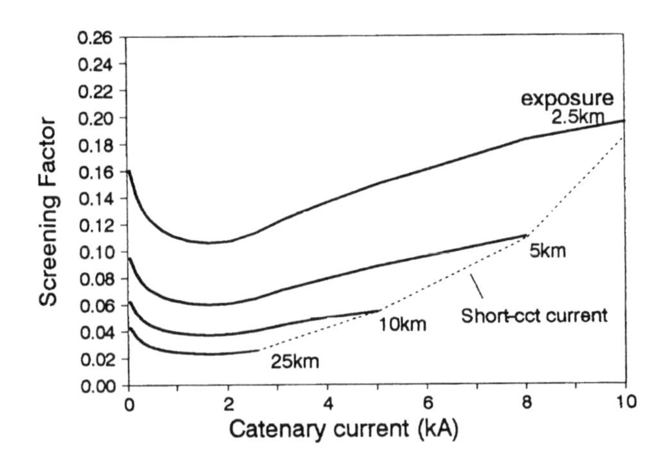

Figure 9.11
Screening factor versus catenary current and exposure

It is usual for the traction designers to provide information on short-circuit currents vs. feed distance for the system. For illustration, Figure 9.10 shows this information for $A = 0.62\,\Omega$ and $B = 0.41\,\Omega/\text{km}$. The cable screening factor will vary with exposure (because the earthing penalty decreases as the exposure increases) and with the traction current drawn (because the permeability of the steel tape armour changes with field strength). This is shown in Figure 9.11 for the example cable.

The induced longitudinal voltage on the cable will also vary with the same factors, and will be reduced by any additional screening. This is shown in Figure 9.12 for a double-track railroad with a rail screening factor of 0.5. This voltage will be balanced about the centre-point of the exposure, as shown in Figure 9.9, and so the voltage to local earth will be, at worst, half that shown on the ordinate axis of Figure 9.12. The CCITT requirements for induced voltage are that for non-isolated circuits it should not exceed 60 V during running conditions. This would just be achieved in the example for

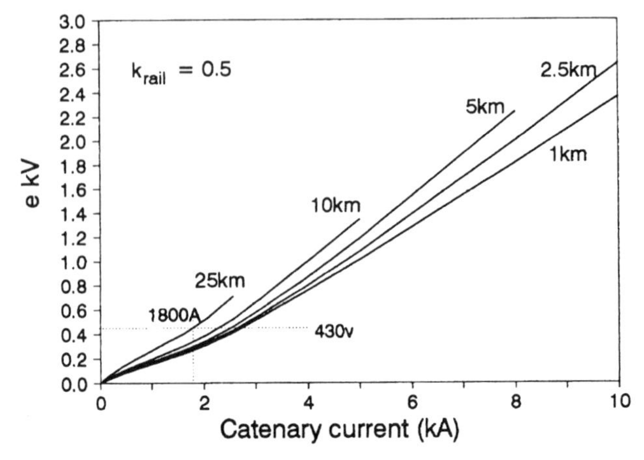

Figure 9.12
Induced voltages versus catenary current and exposure

single-heading out to 10 km. If the circuits are isolated by transformers the induced voltage may be allowed to rise to 430 V. This would be achieved even for triple-heading out to 25 km.

For short-circuit conditions, the voltage between the conductors and screen must not exceed the breakdown voltage of the insulation (1 kV for paper and 20 kV for Pe). For paper insulated cable there will be breakdown of the insulation to screen during short-circuit conditions at the ends of the exposure, for exposures less than 10 km. However paper insulation is self-healing for short duration breakdown. For plastic insulation between the conductors and screen, the voltage withstand is greater than 10 kV and it is unlikely to breakdown even at the shortest exposures.

INDUCED NOISE VOLTAGES

Associated with the fundamental frequency of the traction current there will be harmonics from the feed lines and also extra harmonics generated by the locomotive control circuits. These harmonics will be present in the same proportion in the induced longitudinal e.m.f. on the disturbed circuits. Because of small unbalances between the conductors of the pairs, these will cause transverse voltages on the pairs which are superimposed on the signal as unwanted noise.

Psophometric Voltage

The perceived effect of these noise voltages on the listener depends on the frequency characteristics of the receiver and of the listener's ear response. The measured noise is therefore weighted according to the frequency of the noise in accordance with an agreed 'psophometric weighting' characteristic. The CCITT psophometric weighting curve is shown in Figure 9.13. Similar curves are specified by AT & T for USA areas of influence. When measuring the transverse noise e.m.f., a high-impedance electronic

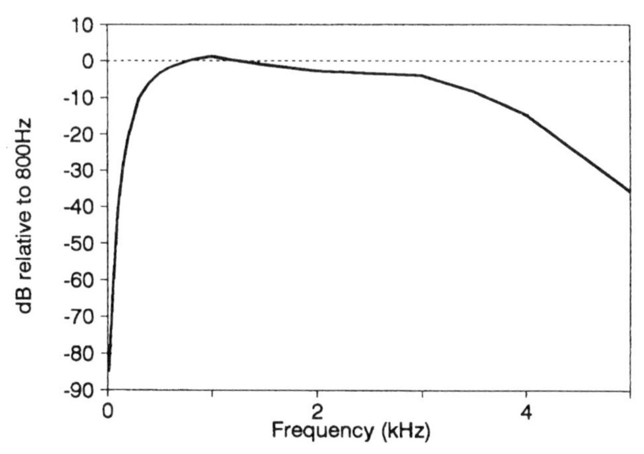

Figure 9.13
Psophometric weighting curve (no reply from ITU)

voltmeter fitted with a network to give the required frequency shaping is used and this is called a 'psophometer'. The CCITT requirement calls for a maximum psophometric e.m.f. of 1 mV when measured at the terminals of the subscriber's telephone with the instrument in the receive condition.

Form Factor

To make an accurate assessment of the disturbing effect it is necessary to to evaluate the longitudinal induced voltage at each harmonic frequency, apply the appropriate weighting factor and find the r.m.s. summation

$$E_{ps} = \sqrt{\sum (\omega M_f \cdot p_f \cdot k_f \cdot I_f)^2}\ \text{V/km} \tag{9.30}$$

where M_f, p_f, k_f, and I_f are respectively the mutual inductance, weighting factor, screening factor and harmonic current at the harmonic frequency f Hz. A more convenient approximate procedure is to apply the weighting factors relative to 800 Hz to the components of the disturbing current and, by r.m.s. summation of these, determine the 'equivalent 800 Hz disturbing current'. The ratio of this to the fundamental disturbing current is the 'Form factor', F. Then

$$E_{ps} = \omega M_f \cdot k_f \cdot F \cdot I_c\ \text{V/km} \tag{9.31}$$

where I_c is the fundamantal disturbing current and $f = 800$ Hz. The form factor varies with the type of locomotive control, and with the fundamental frequency. Typical values for F are

16.7 Hz	tap-changing/rectifier	0.005–0.01
50 Hz	tap-changing/rectifier	0.01–0.02
50 Hz	Thyrister control	0.025–0.05

Sensitivity Factor

The transverse psophometric e.m.f. is generated from the longitudinal psophometric e.m.f. by unbalances to earth of the pairs and of the equipment connected to them. This is accomodated by defining a Sensitivity factor, S, which for representative cases takes the values

Cables	0.0001–0.001
Open wires	0.001–0.05
Equipment	0.001–0.1

The total sensitivity factor is given by the r.m.s. sum of the cable and equipment values. For heavy traffic conditions, balanced isolating transformers will probably be used to limit the effects of longitudinal voltage, in which case the equipment unbalance can be neglected. The transverse psophometric e.m.f. is given by

$$e_{ps} = \omega M_f \cdot k_f \cdot F \cdot I_c \cdot S \, \text{V/km} \tag{9.32}$$

Mutual Inductance at 800 Hz

For a given earth resistivity and spacing, the 800 Hz mutual inductance becomes, from Equations (9.2) and (9.4)

$$M_{50} - M_{800} = 0.2 \ln\left(\frac{2H_{50}}{S}\right) - 0.2 \ln\left(\frac{2H_{800}}{S}\right) = 0.2 \ln\left(\sqrt{\frac{800}{50}}\right) = 0.28 \, \text{mH/km}$$

Thus the mutual inductance decreases by 0.28 mH/km from 50 Hz to 800 Hz irrespective of earth resistivity or spacing.

Screening Factors at 800 Hz

From Equation (9.14) the intrinsic screening factor of a steel tape armoured cable will decrease almost inversely proportional to frequency. From Equations (9.18) and (9.25) the earthing penalty will decrease but proportionately to the square root of frequency. Thus in the example for 4 Ω earths at 1 km spacing, the effective screening factor for a 10 km exposure at 800 Hz will be

$$0.03 \times 50/800 + 0.066 \times \sqrt{50/800} = 0.0019 + 0.0165 = 0.018 \text{ (cf. 0.06 at 50 Hz)}$$

For a 25 km exposure, the effective screening factor at 800 Hz becomes 0.0045 owing to the earthing penalty reducing with increase of exposure.

For an external screening conductor, although its intrinsic screening factor will change approximately as the inverse of frequency, the separation penalty Equation (9.27) will not change significantly with frequency. Thus for a 2 m separation the 800 Hz screening factor will be

$$0.24 \times 50/800 + 0.31 = 0.015 + 0.31 = 0.32 \text{ (cf. 0.55 at 50 Hz)}$$

Similarly the rail screening factors will decrease by approximately 70% at 800 Hz compared with 50 Hz.

Numerical Example of Noise Calculation

For triple-heading on a 25 km exposure, using the numerical data above:

Catenary cable spacing of 5 m, at 800Hz	$M = 0.72$ mH/km
Good quality cable	$S = 0.003$
Effective screening factor of cable at 800 Hz	$k_e = 0.0045$

Double track rail screening factor at 800 Hz	$k_r = 0.34$
At 25 km, triple heading draws, say	$I_c = 1400$ A
Thyristor control, optimistically	$F = 0.03$

Then, from Equation (9.32)

$$e_{ps} \cdot l = l \cdot S \cdot \omega \cdot M_{800} \cdot (k_e \cdot k_r)_{800} \cdot I_c \cdot F \text{ V}$$

$$= 25 \times 0.0003 \times 2\pi \cdot 800 \times 0.72 \times 10^{-3} \times 0.0015 \times 1400 \times 0.03 = 1.71 \text{ mV}$$

If the locomotives drew their full 1800 A this would rise to 2.2 mV. Due to the fact that the earthing penalty increases as the exposure decreases, almost exactly compensating for length, the psophometric e.m.f. stays constant at 2.2 mV for 1800 A traction current for all exposures less than 30 km. Since the CCITT specifies the noise e.m.f. with a terminated condition, this calculated noise will be halved to 1.1 mV which nearly meets the requirement.

In the absence of other measures the earthing must be improved. For 1 Ω earths at 1 km spacing the terminated noise e.m.f. drops to 0.45 mV at all exposures. Alternatively since 1 Ω earths may be difficult to achieve, 2 Ω earths at 250 m spacing will reduce the terminated noise to 0.6 mV. Reduction in earth spacing is therefore not as effective as reduction in earth resistance.

TRANSFORMERS

Booster Transformers

If a return conductor is installed parallel and close to the catenary traction circuit, and if transformers are arranged in both circuits as shown in Figure 9.14, a current equal and opposite to the catenary current is forced to flow in the return conductor. This effectively sucks the return current out of the earth and reduces the area of the inducing loop. The effective screening factor of this arrangement is 0.025 for disturbed cables >8 m from the track. For lineside cables the screening factor is larger depending on cable and return conductor placement and varies between 0.025 and 0.1. However if the locomotive is 'in section', i.e. between the booster and earth strap, there will be a significant length of rail between adjacent transformers carrying return current. In this section the screening factor rises to 0.15.

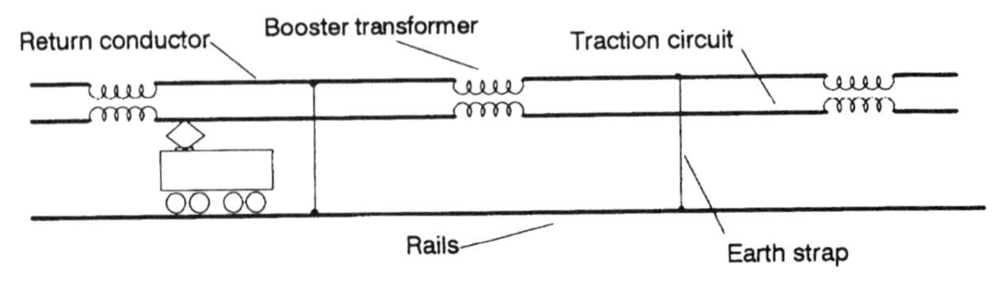

Figure 9.14
Booster transformers with return conductor (Not received)

Booster transformers when used are usually spaced about 3 km apart and give good protection at the fundamental power frequency (especially for non-lineside cables e.g. PTT cables). Although in theory the effect of booster transformers should be independent of frequency and should therefore be equally effective against noise induction, in practice the leakage reactance of practical transformers severely affects the audio frequency screening factor. Screening factors at audio frequencies for lineside cables are about 0.4 to 0.8. The use of booster transformers using the rails as the return conductor requires insulating joints in the rails and the screening factor is not so good, varying from 0.15 to 0.4 depending on frequency and transformer spacing. Booster transformers increase the impedance and losses of the traction circuit, necessitate stronger supports and with the requirement for a separate return conductor generally increase the costs of installation.

Sectioning and Neutralizing Transformers

Although sectioning the lineside cables with transformer coupling will considerably reduce the induced longitudinal voltages, it will not help to reduce the noise e.m.f. which is transmitted throught the transformers with the required signal. A. Rosen [23] extended the notion of lump-loaded lines to the problem of increasing the cable screening factor of small lineside cables. Essentially the cable is wound around a laminated transformer core at regular intervals. This increases the inductance of the screen, increasing Z_{SE} in Equation (9.10), without adding inductance to the signal circuits. It is not known whether this has ever been used in practice.

EARTHING

The importance of earthing has been demonstrated above. The difficulty of providing low resistance earths depends very much on the earth resisivity and the nature of the ground. The three basic ways of providing earths are

- Deep-driven vertical rod earths
- Buried metal plates
- Long horizontal trench earths

The approximate formulae for the resistances given below were developed by H. B. Dwight [17] and are quoted from C. F. Boyce [18], correcting for printing errors and making minor changes.

Vertical Rod Earths

The resistance of a single deep-driven earth rod of length L m and diameter d m is given by

$$R = \frac{\rho}{2\pi L}\left(\ln\frac{8L}{d} - 1\right)\,\Omega \qquad (9.33)$$

where ρ is the earth resistivity in $\Omega \cdot m$. The resistance is not very sensitive to the rod diameter which is therefore chosen on the basis of strength and cost. If several rods are used in parallel, they should be spaced apart by approximately their depth. The reduction in resistance for n rods compared with a single rod is then

$$R_n = \frac{R}{\{n^{0.76}\}} \tag{9.34}$$

For closer spacing the index of n approaches 0.5 i.e. a square-root law.

Plate Earths

For a rectangular plate of sides L m $\times$ W m, at a depth of S m from the top edge

$$R = \frac{\rho}{2\pi}\left\{\frac{3.1(1+0.037/W)-1}{\sqrt{WL}} + \frac{1}{4S}\right\}\Omega \tag{9.35}$$

For a square plate $W = L$. For multiple plate earths spaced at least 2 m apart, the reduction in resistance follows the same law as for rod earths.

Trench Earths

The resistance of a long straight wire of length L m and diameter d m buried to a depth of S m, and for a connection made at one end, is given by

$$R = \frac{\rho}{2\pi L}\left(\ln\frac{4L}{d} + \ln\frac{L}{S} - 2 + \frac{S}{L} - \frac{S^2}{L^2} + \frac{S^4}{2L^4}\right)\Omega \tag{9.36}$$

Again the resistance is not sensitive to the diameter, nor to the depth of the wire, but the wire resistance does limit the earth resistance for long earth runs. For multiple trench earths at equiangular spacing the resistance follows approximately the same law as for multiple rod earths.

Comparison of Earthing Methods

The earth resistances calculated from the above formulae are shown in Figure 9.15 for an earth resistivity of $100\,\Omega \cdot m$. For other earth resistivities multiply by $\rho/100$. In the graph the rod diameter is 0.015 m, the plate earth is taken as a square, and the trench earth wire diameter is 0.005 m. $S = 0.5$ m in the case of plate and trench earths. For comparable dimension L, the rod and trench earths have approximately the same resistance, the difference being solely due to diameter differences. For the plate earth the resistance is approximately one-half of this, as expected intuitively.

It is unlikely that for rod or plate earths that L can exceed 5 m even in the best of soils. Also the use of more than four rods or two plates is unproductive. Hence for this earth resistivity the achievement of an earth resistance of less than $4\,\Omega$ is difficult by

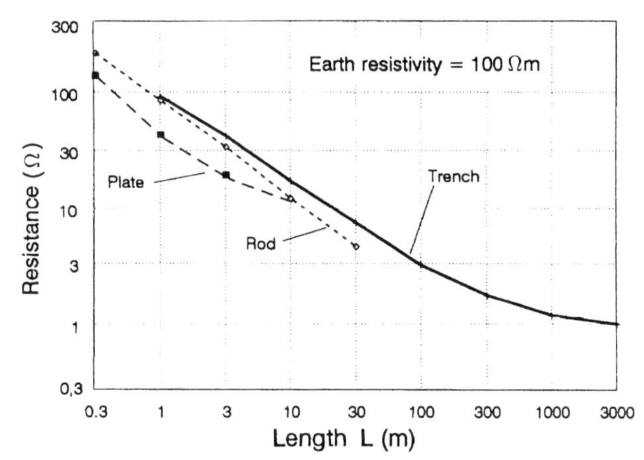

Figure 9.15
Comparison of earthing methods versus linear dimension

these methods. Plate earths of these dimensions are costly and difficult to install. Rod earths are difficult to drive in rocky ground and find their best application in low resistivity areas ($<50\,\Omega\cdot$m).

Using a trench earth connected at one end, a length of 100 m would give a resistance of 3 Ω and 200 m connected at the mid-point would give 1.5 Ω. Notice that for very long trench earths the resistance is limited by the wire resistance (0.088 Ω). Despite the cost of trenching this is the only feasible method of achieving low resistance earth in moderate-to-high earth resistivities. The use of mole ploughing is recommended. If the cable itself is directly buried, the earth wire can be incorporated in the same trench. Using two earth wires in place of one improves the security of the earth.

Measurement of Earth Resistivity [18]

Consider two point electrodes touching the surface of the earth with a current I entering the ground at one electrode and the second electrode being sufficiently remote that its presence may be neglected. The current flows radially into the earth from the first electrode. The area of the hemispherical surface centred on the electrode is $2\pi S^2$ and the radial current density is $4I/(2\pi S^2)$. If the earth resistivity is ρ, the radial electric intensity in the ground at a distance S is

$$E_S = \frac{I\rho}{2\pi S^2}$$

The potential at a distance S is the integral of the electric force between S and an infinitely remote point

$$P = \int_S^{\infty} E_S \cdot dS - \frac{I\rho}{2\pi S}$$

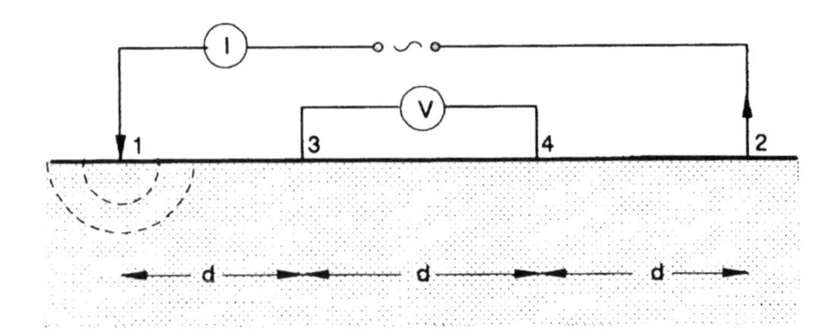

Figure 9.16
Measurement of earth resistivity

and the ratio of potential to current is the resistance

$$R = \frac{\rho}{2\pi S} \tag{9.37}$$

If the current enters the ground at electrode 1 and leaves at electrode 2, the mutual resistance to a third electrode 3 is the difference between R_{13} and R_{23}

$$R_{13} - R_{23} = \frac{\rho}{2\pi}\left(\frac{1}{S_{13}} - \frac{1}{S_{23}}\right) \tag{9.38}$$

For the measurement of earth resistivity, four equi-spaced electrodes in a straight line are used, as in Figure 9.16. Current is passed between electrodes 1 and 2, and the voltage between electrodes 3 and 4 is measured. The measured mutual resistance is the ratio of this voltage to the current.

$$R = \frac{V}{I} = R_{13} - R_{23} - R_{14} + R_{24} = \frac{\rho}{2\pi}\left(\frac{1}{S_{13}} - \frac{1}{S_{23}} - \frac{1}{S_{14}} + \frac{1}{S_{24}}\right) = \frac{\rho}{2\pi}\left(\frac{2}{d} - \frac{1}{d}\right)$$

whence

$$\rho = 2\pi \cdot d \cdot R \tag{9.39}$$

Alternating current is used to avoid polarization effects at the electrodes. A null method of measuring the voltage avoids error due to the resistance in the potential electrodes. The electrodes should be 1 m deep in high resistivity ground. The depth of penetration of the current is approximately equal to the separation distance d.

Measurement of Earth Resistance

To measure the resistance of an earth electrode it is sufficient to measure the ratio of V to I for current flowing into the earth electrode and leaving via a remote electrode at a distance approximately equal to the distance between earthing points. The same precautions as above should be observed.

10

Stranded Conductors and Braids

In order to improve the flexibility of small cables it is common to use stranded conductors and in the case of screens and coaxial outer conductors, to use braids of fine wire. Since wire drawing of fine wires is an expensive process and both these forms of construction use a lot of wire in terms of length, the requirement for flexibility must be carefully justified. Also, braiding must be one of the slowest of cable making operations.

STRANDED CONDUCTORS

Strictly speaking, stranded conductors should comply with the 'perfect' lay-up arrangements discussed in Chapter 4 and need to be accurately stranded together. However in many cases it is sufficient to bunch the wires together in one twisting operation particularly if the wires are accurately placed at the start of the operation and run into the twisting machine through a diamond die of the correct diameter.

Typical wire arrangements with their overall diameters d as a multiple q of the wire diameter d_w are shown in Table 10.1. The act of stranding the conductors together increases the diameter of the stranded wire since the cross-section of the layer wires

Table 10.1
Wire stranding configurations

No. of strands	$q = d/d_w$
3	2.16
$1 + 6 = 7$	2
$3 + 9 = 12$	4.16
$1 + 6 + 12 = 19$	5
$3 + 9 + 15 = 27$	6.16
$1 + 6 + 12 + 18 = 37$	7

Table 10.2
Increase in overall diameter due to stranding

p	5	10	15	20	30
Increase in q %	4.3	2.3	1.0	0.7	0.3

taken at right angles to the axis of the strand becomes elliptical, the more so as the lay length is reduced. They thus require a larger diameter pitch circle to fit together into a layer. If the lay length P, is expressed as a multiple p, of the overall diameter,

$$P = p \times d = p \times q \cdot d_w \tag{10.1}$$

the percentage increase in overall diameter is as given in Table 10.2. It is good practice therefore to keep the lay factor $p \geq 15$ to limit the increase in q to $<1\%$. That is the lay length should be more than 15 times the overall diameter of the stranded conductor. For the same diameter, a stranded conductor will have a higher d.c. resistance and a higher internal inductance than a solid wire of the same material.

Effective Diameter of Stranded Conductor

Because of the varying profile of the stranded conductor, it is to be expected that the effective diameter for electrical purposes will be slightly less than the overall diameter. Meyers [25] has calculated the effective diameter for the number of strands in the outer layer as

$$d_e = k_1 d \tag{10.2}$$

This is given in Table 10.3, together with the overall diameter factor, q, and the surface stress factor (also due to Meyers) referred to later.

The effective diameter is used in place of the solid conductor diameter to calculate the capacitance, external inductance and characteristic impedance of the line using stranded conductors. For example the h.f. characteristic impedance of a coaxial line using a stranded inner conductor becomes

$$\begin{aligned} Z_\infty &= \frac{60}{\sqrt{K}} \ln \frac{D}{d_e} = \frac{60}{\sqrt{K}} \ln \frac{D}{k_1 d} \\ &= \frac{60}{\sqrt{K}} \ln \frac{D}{d} - \frac{60}{\sqrt{K}} \ln k_1 \ \Omega \end{aligned} \tag{10.3}$$

Table 10.3
Effective diameter and stress factor of stranded conductors

Total no. of strands	1	3	7	12	19	27	37
Overall diameter factor, q	1	2.16	3.00	4.16	5.00	6.16	7.00
Effective diameter factor, k_1	1	0.871	0.939	0.957	0.970	0.976	0.980
Surface stress factor, k_2	1	1.459	1.408	1.403	1.397	1.396	1.395

Thus the effect of using a stranded conductor in place of a solid cylindrical wire for the inner conductor is to increase the impedance, since k_1 is fractional, by a small amount which is independent of the conductor diameter ratio. In the case of a 7-strand inner conductor for a coaxial with solid polythene dielectric the impedance increase would be 2.50 Ω.

Internal Inductance of Stranded Conductor

The twisting of the strands causes the current to flow in a helical path. The conductor therefore behaves like a solenoid which increases its internal inductance. The inductance of a solenoid of cross-sectional area A and having n turns/m is

$$L = 4\pi n^2 A \times 10^{-7}\ \text{H/m} \tag{10.4}$$

Since from Equation (10.1) the lay length $P = pd$ the number of turns/m is $1/P$ and $A = \pi(k_1 d)^2/4$, the added inductance becomes

$$L = 0.1\,\frac{\pi^2 k_1^2}{p^2}\ \text{mH/km} \tag{10.5}$$

For a 7-strand conductor with $p = 15$, $L = 3.9$ mH/km, which is about 1.5% of the external inductance of a typical coaxial pair. The effect on the impedance is an increase of 0.75% or about 0.5 Ω.

Resistance of Stranded Conductors

Stranding the conductors increases their length relative to the axis. This should be taken into account when calculating the d.c. resistance of the strand. The increase in length is the take-up factor K, Equation (3.1)

$$K = \sqrt{1 + \left(\frac{\pi d_p}{P}\right)^2}$$

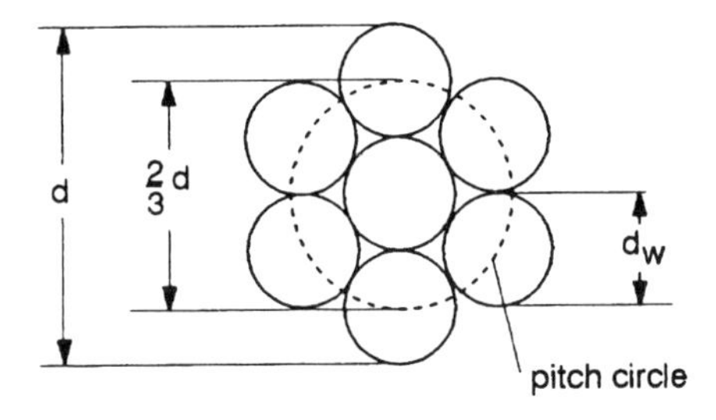

Figure 10.1
7-strand conductor

d_p is the pitch circle diameter of the stranded wires. In the case of a 7-strand conductor the pitch circle diameter is $\frac{2}{3}d$ and the lay $P = p \cdot d$, so that

$$K = \sqrt{1 + \left(\frac{2\pi}{3p}\right)^2} \tag{10.6}$$

The total resistance R, is that due to the axial wire, r, in parallel with 6 strands of resistance, rK each, i.e.

$$R = r \cdot \left(1 + \frac{6}{\sqrt{1 + (2\pi/3p)^2}}\right)^{-1} \tag{10.7}$$

Thus for a 7-strand conductor with a lay factor $p = 15$, $R/r = 0.145$. Without stranding the d.c. resistance of the bunch would be one-seventh of that of a single wire, namely 0.143. Similar calculations may be made for other strand numbers and arrangements and the effect of stranding is found to be an increase over the unstranded case of about 2%.

The high-frequency resistance of stranded conductors is complicated by skin and proximity effects. The current distibution is confined to the outer layer of strands in the manner shown in Figure 10.2 with the current concentration and field density being greatest at the centre of the shaded arcs. There is also an effect due to the longer helical path about the effective diameter, d_E. It is usual to lump all these effects together into a factor of 1.25, multiplying the h.f. resistance of a cylindrical conductor of the same overall diameter, d. This factor has been confirmed experimentally.

Effect of Stranding on Voltage Rating of Coaxial Cable

Two effects are evident from Figure 10.2, namely the smaller effective diameter of the stranded inner conductor and the the higher field concentration at the centre-points of the arcs of current flow. Thus the stress produced by the application of a voltage V is allowed for by the diameter factor k_1 and the surface stress factor k_2 given in Table 10.3

$$E = \frac{2k_2 V}{d} \cdot \ln\left(\frac{D}{k_1 d}\right) \text{ V/m} \tag{10.8}$$

where D and d are in metres.

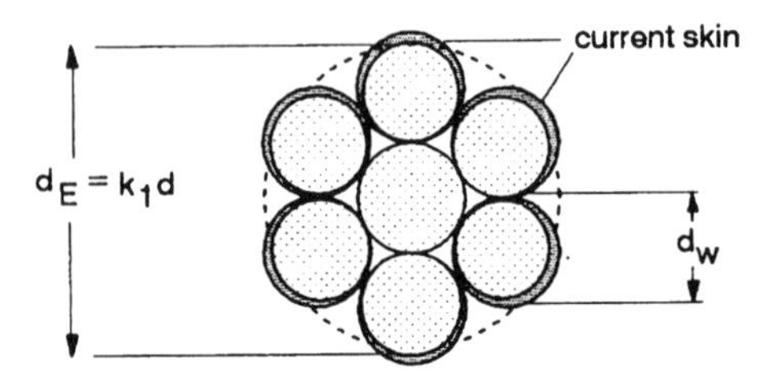

Figure 10.2
High-frequency current distribution in stranded conductor

BRAIDED SCREENS AND CONDUCTORS

Braids are made by weaving in maypole fashion a number of bobbins (on spindles) around the cylindrical cable or insulated wire to be screened. Each bobbin carries a number of parallel wires (or ends) previously wound on to it without twisting. There are two sets of bobbins which counter-rotate around the cable axis, weaving in and out of each other. This produces a 'twill weave' in which each spindle of wires crosses over two other spindles and is then crossed by two further spindles. The same effect is produced in the faster machines by having two counter-rotating carriages of bobbins and moving the spindles of wires in and out by guides. Figure 10.3 is a sketch of a braid.

If one wire is unwound for one lay length, P, it will make an angle θ with the braid axis and

$$\tan\theta = \pi D/P \tag{10.9}$$

The length of an individual wire relative to the axial length is the take-up factor, K

$$K = \sqrt{1 + \left(\frac{\pi D}{P}\right)^2} = \frac{1}{\cos\theta} \tag{10.10}$$

An American term used in connection with braids is 'picks/in' which is defined as the number of spindle crossings per inch along the surface of the braid parallel to the axis. If the number of spindles is m and the lay length is expressed in inches, the number of picks/in is

$$p_K = m/(2P) \tag{10.11}$$

Since there are two counter-rotating helices of wires, for complete coverage each set must cover half the surface. From Equation (3.9) for tape lapping at 50% coverage

$$\sin\theta = \frac{W}{2P} \tag{10.12}$$

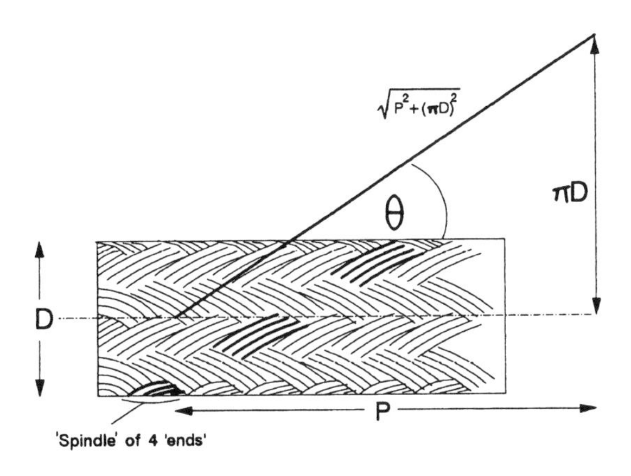

Figure 10.3
Braid geometry

If W is compared with nmd_w the effective tape width, the 'filling factor' of the braid becomes

$$K_F = \frac{nmd_w}{2P \sin\theta} = \frac{nmd_w}{2\pi D}\sqrt{1 + \left(\frac{\pi D}{P}\right)^2} \quad (10.13)$$

where m and n are the number of spindles and wires respectively and d_w is the individual wire diameter. It is virtually impossible to get $K_F = 1$ without serious problems of riding wires. The best braids achieve $K_F = 0.95$. For less stringent screening requirements the filling factor should lie between 0.3 and 0.7 but for good screening with adequate flexibility K_F should be ≥ 0.7.

To improve the screening effect with a smaller filling factor, a thin metal foil is sometimes applied longitudinally under the braid in the same manufacturing process. A metal foil laminated with Mylar is generally preferred since unsupported foils tend to crack under repeated flexing.

Conduction in Braided Conductors

In the coaxial case, the current flowing in the inner conductor tries to induce an equal and opposite longitudinal current in the braid. This results in equal currents flowing in the left- and right-hand spindles, the resultant of these being an equivalent longitudinal flow. As both wires are at the same potential at a crossing point there will be no current transfer between the wires. Thus tarnishing of the wires with age does not affect the conduction or screening of braided coaxial pairs.

In the case of the twin cable, the currents induced in the braid vary around the circumference. At points in the screen perpendicular to the plane of the twin the induced currents are zero; at points on the screen in the plane of the twin they are at a maximum. In order to maintain this equilibrium there must be current transfer between the right-handed and left-handed braid wires at intermediate points around the circumference. If the transfer of current is restricted by tarnishing (or has deliberately been prevented by using enamelled wires) it is impossible for a resultant longitudinal current to flow: the braid ceases to act as a screen and has little effect on the transmission parameters of the twin. However, flexing of the braid can clean the contacts and restore the screening. Thus with tarnished braid wires the transmission and screening parameters will be unstable. The longitudinal application of a laminated metal foil tape under the braid can remove this problem and improve the braid performance at high frequencies.

Resistance of Wire Braid at Lower Frequencies

When the frequency is low enough for the skin depth to be large compared with the braiding wire diameter, its resistance may be compared with that of a solid copper tube of thickness equal to twice the braid wire diameter. Compared with such a tube the resistance of the braid will be larger because the cross-section of the wires is less than that of the tube and also the helical path of the wires is greater than the axial length of the tube. Hence the ratio of braid to tube resistance will be

$$K_B = \frac{\pi D \cdot 2d_w}{mn \cdot (\pi/4)d_w^2} \cdot \sqrt{1 + \left(\frac{\pi D}{P}\right)^2}$$

$$= \frac{8D \cdot K}{mnd_w} = \frac{8D \cdot K}{K_F \cdot 2P \sin\theta} = \frac{4K^2}{\pi K_F} \qquad (10.14)$$

Thus it will be seen that at the lower frequencies the resistance is proportional to the square of the take-up factor K, and inversely proportional to the filling factor K_F.

Resistance of Wire Braid at High Radio Frequencies

At frequencies high enough to make the current skin small compared with the braid wire diameter, e.g. at or above about 100 MHz, additional losses arise owing to circumferential circulating currents flowing in the braid wires on the inside of the braid. Mildner [26] advanced a semi-empirical formula for the braiding factor at high frequencies

$$K_B = \frac{K^2}{2K_F} + \frac{\pi K^2 K_F}{4(1+a)^2} \qquad (10.15)$$

with

$$a = \sqrt{\frac{\rho_0}{\mu f}} \times \frac{K^4}{2\pi d_W (K^2 - 1)} \qquad (10.16)$$

where ρ_0 is the resistivity of the braid wires in $\Omega \cdot$ m, μ is the permeability $= 4\pi \times 10^{-7}$ H/m and f is the frequency in Hz. d_W is in metres. (Note. Both Mildner and Blackband use a lay factor $K_1 = K^2$)

BRAIDED COAXIAL ATTENUATION

The attenuation of a coaxial with solid conductors is given as in Equation (7.51), by

$$\alpha = \sqrt{\frac{\varepsilon\omega}{2}} \frac{1}{\ln D/d} \left(\frac{\sqrt{\rho_i}}{d} + \frac{\sqrt{\rho_0}}{D}\right) + \frac{1}{2}\sqrt{\mu_0 \varepsilon}\, \omega \tan\delta \text{ N/m} \qquad (10.17)$$

$$= \frac{4.578 \times 10^{-3} \sqrt{k \cdot f}}{\ln D/d} \left(\frac{\sqrt{\rho_i}}{d} + \frac{\sqrt{\rho_0}}{D}\right) + 9.096 \times 10^{-6} \sqrt{k} \cdot f \cdot \tan\delta \text{ dB/100 m} \qquad (10.18)$$

where f is the frequency in Hz and d, D are in metres. k is the effective permittivity and δ is the loss factor of the dielectric. ρ_i and ρ_0 are the resistivities in $\Omega \cdot$ m of the inner and outer conductors $\varepsilon = (k/36\pi) \times 10^{-9}$ F/m.

For a coaxial with a stranded inner conductor ($K_S = 1.25$) and a braided outer conductor, this becomes

$$\alpha = \frac{4.578 \times 10^{-3}\sqrt{k \cdot f}}{\ln D/k_1 d}\left(\frac{K_S\sqrt{\rho_i}}{d} + \frac{K_B\sqrt{\rho_0}}{S}\right)$$
$$+ 9.096 \times 10^{-6}\sqrt{k} \cdot f \cdot \tan\delta \,\text{dB}/100\,\text{m} \qquad (10.19)$$

The parameter a of Equation (10.16) is insensitive to the value of K and is less than 0.1 for frequencies above 50 MHz for all practical lay factor values, as shown in Figure 10.4. Also the function $1/(1+a)^2$ is sensibly equal to 0.95 above 200 MHz. Equation (10.15) may thus be rewritten as

$$\frac{K_B K_F}{K^2} + \frac{1}{2} + \frac{0.95\pi}{4} K_F^2 \qquad (10.20)$$

This enables a check on the theory, since if $K_B K_F/K^2$ is plotted against K_F^2 for a range of measured values of K, K_F and K_B, the result should be a straight line with slope $0.95\pi/4$ and an intercept on the y-axis of 0.5.

At the request of Blackband, this test was carried out by Hinchliffe et al. [27], using a precision slotted line to measure the attenuation of a range of twelve stranded and braided coaxials to an estimated accuracy of 1%. Using Equation (10.19) in which the second term represents the dielectric attenuation and the first term the total conductor attenuation, the latter was derived from the measured attenuation by subtracting the calculated value of the second term (but using measured values of permittivity and loss factor). K_B was then calculated from the total conductor attenuation using the measured dimensions of the coaxials and taking $K_S = 1.25$.

Two basic sizes of coaxial were used. The first size, of nominal 50 Ω impedance, used a stranded inner conductor of 7/0.029 in (7/0.737 mm) insulated with polyethylene to a nominal diameter of 0.285 in (7.239 mm) with a range of spindles, ends and lays of 0.0076 in (0.193 mm) plain copper wire strands. The second size of coaxial, of nominal

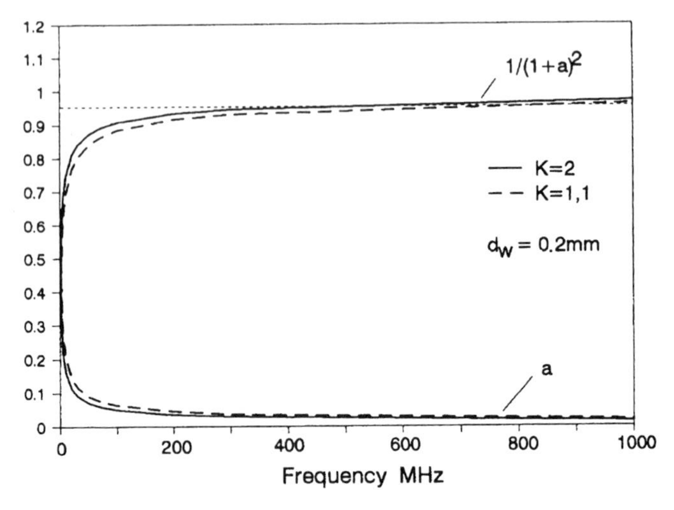

Figure 10.4
Variation of parameter a with frequency

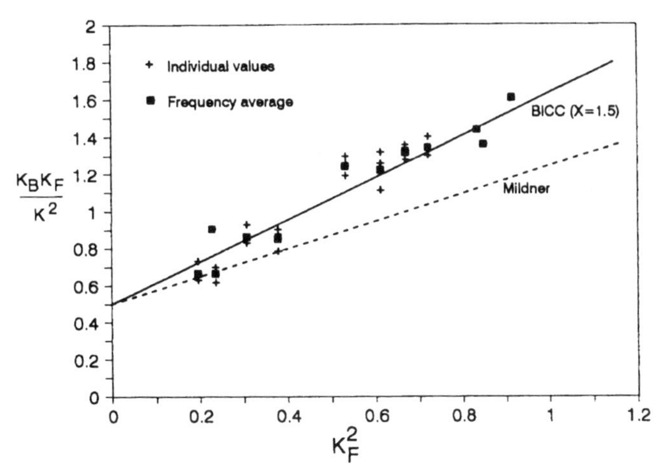

Figure 10.5
Test of Equation (10.20) (Reproduced by permission of BICC Cables Ltd)

75 Ω impedance, used 7/0.0076 in stranded copper inner insulated with polyethylene to a nominal diameter of 0.128 in (3.25 mm) again with a range of plain copper wire (0.0076 in diameter) braids.

When Equation (10.20) is plotted as in Figure 10.5, it can be seen that the y-axis intercept is satisfactorily equal to 0.5 as required by Mildner's theory, thus confirming the first term of Equations (10.15) and (10.20) but that the slope of the best-fit line is 1.5 times that predicted by the theory. BICC proposed that a factor X be included in the second term of Equation (10.20) resulting in

$$\frac{K_BK_F}{K^2} = \frac{1}{2} + \frac{X\pi}{4(1+a)^2} K_F^2 \tag{10.21}$$

It was suggested that this was due to the induced current in the braid wire being not circular as assumed by Mildner, but elliptical as shown in Figure 10.6. X would then be the ratio of the circumference of the ellipse to that of Mildner's circle, increasing the resistive loss in the same ratio. Thus

$$X = \frac{c\{1 + 1/(\sin\theta)\}}{2} \tag{10.22}$$

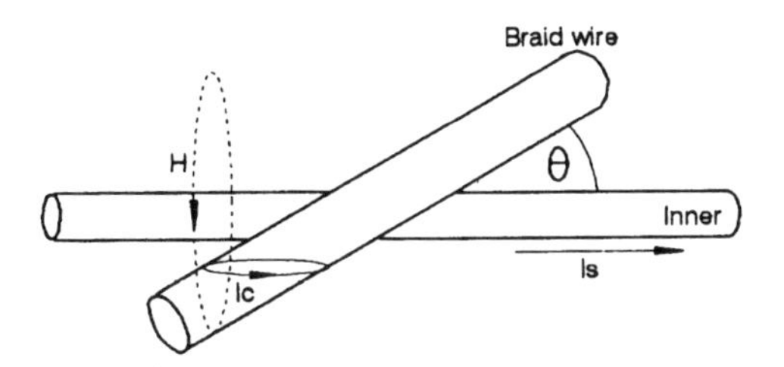

Figure 10.6
Circulating current in braid wire (Reproduced by permission of BICC Cables Ltd)

where c for practical braid angles is approx $= 1.002$. The values of X calculated for the above coaxials have a mean value of 1.46 as required by Figure 10.5.

Figure 10.7 compares the values of K_B calculated by Mildner's theory with those calculated by the extension to the theory by BICC. It will be seen that the extended theory correlates better with the measured values. The mean value of Mildner's values is 20% too low whereas the mean of the BICC values is only 3% too low. Since the braid contributes only 30% to the total attenuation this represents a 1% accuracy for the total attenuation.

Also shown in Figure 10.7 are measurements made on the coaxials at 1.62 MHz. The measurements were made on an admittance bridge giving an accuracy of about 2%. At this frequency the factor a is no longer small compared with unity and the factor X is of less importance. The reasonable agreement of the measured and calculated values is confirmation of the factor a, in Mildner's theory. Given the difficulty of making both sets of measurements at that time, the agreement with the extended theory over the whole frequency range is very satisfactory.

If a metal foil or laminate is applied under the braid, as suggested above to improve the screening factor, this also acts as the outer conductor and the a.c. resistance, attenuation and impedance are calculated in accordance with Chapters 6 and 7. The braid has no effect above about 10 MHz according to measurements made on such a construction using a modern network analyser.

If Equation (10.19) is rewritten to emphasise the different contributions to the attenuation of inner, outer and dielectric, we have, in dB/m, for f in MHz and d, D in metres

$$\alpha_t = \alpha_i + \alpha_0 + \alpha_\delta = A\sqrt{f}\left(\frac{K_S}{d} + \frac{K_B}{D}\right) + B \cdot f \cdot \tan\delta$$

$$\alpha_R = \alpha_t - \alpha_\delta = A\sqrt{f}\left(\frac{K_S}{d} + \frac{K_B}{D}\right)$$

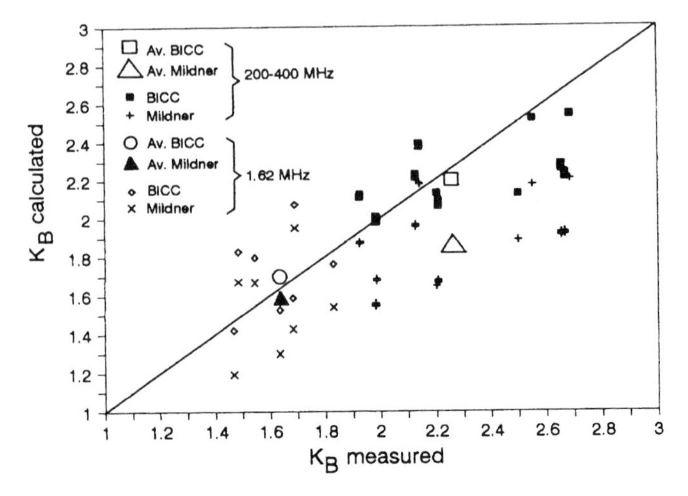

Figure 10.7
Braid factor K_B correlation with theory (Reproduced by permission of BICC Cables Ltd)

where

$$A = \frac{0.458\sqrt{k\rho}}{\ln\dfrac{D}{k_1 d}} \quad \text{and} \quad B = 0.091\sqrt{k}$$

Above about 1 MHz the a.c. resistance of the coaxial is given by the h.f. approximation Equation (7.41)

$$R = 2\alpha_R \cdot Z_R = 2A \cdot Z_R\sqrt{f}\left(\frac{K_S}{d} + \frac{K_B}{D}\right)$$

and also

$$Z_R = \frac{60}{\sqrt{k}} \ln \frac{D}{k_1 d}$$

thus

$$A \cdot Z_R = 60 \times 0.0458\sqrt{\rho}$$

Separating the resistance of the outer braid conductor we have

$$R_0 = (2 \times 60 \times 0.0458)\sqrt{\rho}\,\frac{K_B}{D}\sqrt{f}\ \Omega/\mathrm{m}$$

K_B is the braid factor derived from the extended theory above. However this approximation will decrease to zero at zero frequency and so we must add the d.c. resistance derived from Equation (10.14). So finally we arrive at the full expression for the braid resistance

$$R_0 = 5.5\sqrt{\rho}\,\frac{K_B}{D}\sqrt{f} + \frac{\rho}{\pi D \cdot 2d_W} \cdot \frac{4K^2}{\pi K_F}$$

$$= 5.5\sqrt{\rho}\,\frac{K_B}{D}\sqrt{f} + \frac{2\rho K^2}{\pi^2 D d_W K_F}\ \Omega/\mathrm{m} \tag{10.23}$$

This can only be an approximate solution since it ignores the transition region between about 30 kHz and 1 MHz.

BRAIDED CONDUCTORS AS SCREENS

In order to describe the behaviour of braids as screens it is necessary to look first at the screening effect of thin solid cylindrical conductors. In Chapter 8 on coaxial crosstalk the coupling impedance was defined in Equation (8.48) as $Z_c = E_0/I$. In screening theory it is usual to work in terms of the 'transfer impedance' Z_T which is identical to Z_c in all respects. Figure 10.8 indicates how the current density in the outer conductor diminishes towards the outside as the skin-effect develops with frequency causing Z_T to diminish with frequency. Also, as in Equation (8.49) the transfer impedance is given by the asymptotic approximation to the Bessel solution as

$$\frac{|Z_T|}{R_{dc}} = \frac{v}{\sqrt{\cosh v - \cos v}} \tag{10.24}$$

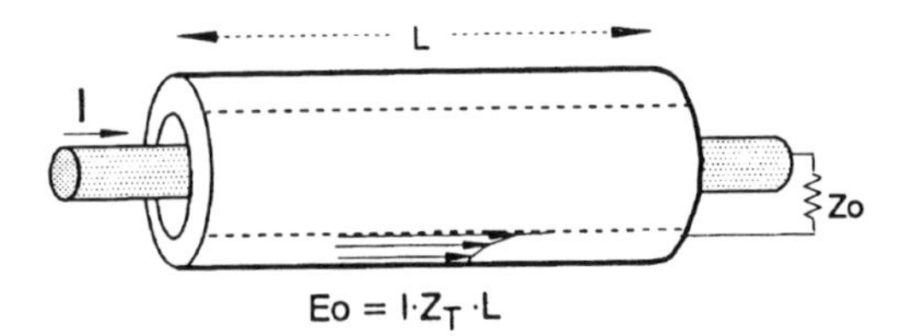

Figure 10.8
Coaxial transfer impedance mechanism

where

$$v = t\sqrt{\frac{2\omega\mu}{\rho}}$$

The tube thickness is t m, $\omega = 2\pi f_{Hz}$, $\mu = 4\pi \times 10^{-7}$ and ρ is the resistivity in $\Omega \cdot$ m. The transfer impedance for various copper tube thicknesses is plotted in Figure 10.9. It will be seen that at low frequencies the transfer impedance is equal to the d.c. resistance of the tubes, and that as the skin-effect is established with rising frequency, it diminishes to very low values.

In contrast, the measured tranfer impedance of a single braid outer conductor, having approximately the same d.c. resistance as a 50 µm copper foil, remains constant up to about 1 MHz and then starts to rise, becoming proportional to the square root of frequency. This is owing to the current, rather than being increasingly confined to the inside of the braid as the frequency increases, being constantly brought to the ouside by the weave of the braid. Consequently the voltage developed on the surface increases as the braid resistance increases with frequency. The screening of the braid at higher frequencies can be improved by including a foil under the braid. The effect can be

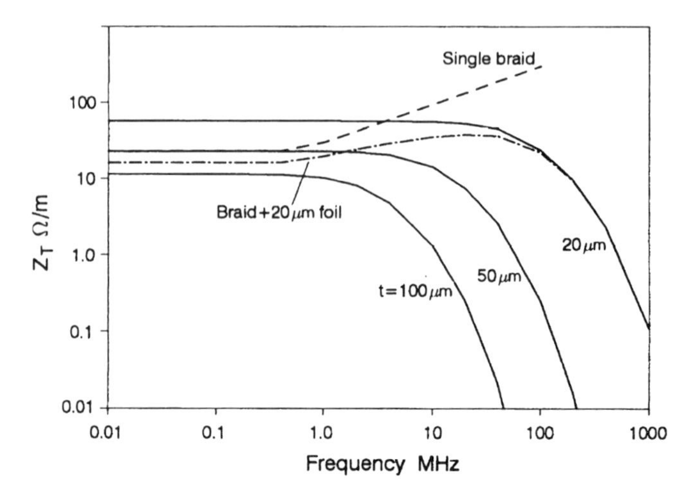

Figure 10.9
Transfer impedance of thin copper tubes

estimated by considering the the respective transfer impedances to be in parallel. Thus the overall transfer impedance of the single braid with a 20 μm copper foil beneath it would be as shown in Figure 10.9.

Screening Efficiency (SE) and Screening Factor (SF)

The above terms are sometimes used to describe the screening of the outer conductor of coaxials. If a current is induced in the outside of the outer conductor, either from an external field or from the triaxial arrangement of Figure 10.10, the screening efficiency can be defined in terms of the current transfer ratio, and the screening factor in terms of the voltage transfer ratio. Starting from the relations

$$I_S = \frac{E}{Z_1} \quad \text{and} \quad I_R = \frac{e}{2Z_0} = \frac{I_S Z_T l}{2Z_0} = \frac{E Z_T l}{2Z_0 Z_1}$$

The screening efficiency is given by

$$SE(dB) = 20\log\left|\frac{I_S}{I_R}\right| = 20\log\left|\frac{2Z_0}{Z_T \cdot l}\right| \tag{10.25}$$

whence

$$20\log|Z_T| = 20\log\left|\frac{2Z_0}{l}\right| - SE(dB) \tag{10.26}$$

The screening factor is given by

$$SF(dB) = 20\log\left|\frac{E}{e}\right| = 20\log\left|\frac{Z_1}{Z_T \cdot l}\right| \tag{10.27}$$

whence

$$20\log|Z_T| = 20\log\left|\frac{Z_1}{l}\right| - SF(dB) \tag{10.28}$$

Equating (10.26) and (10.28) and rearranging gives the relationship between *SE* and *SF*:

$$SE(dB) = SF(dB) + 20\log\left|\frac{Z_0}{Z_1}\right| \tag{10.29}$$

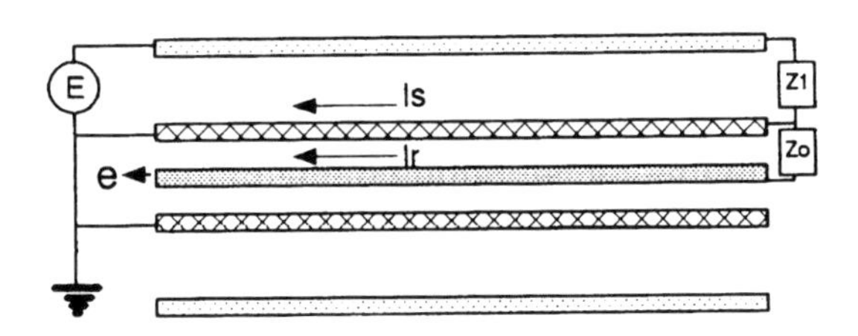

Figure 10.10
Triaxial test arrangement for screening

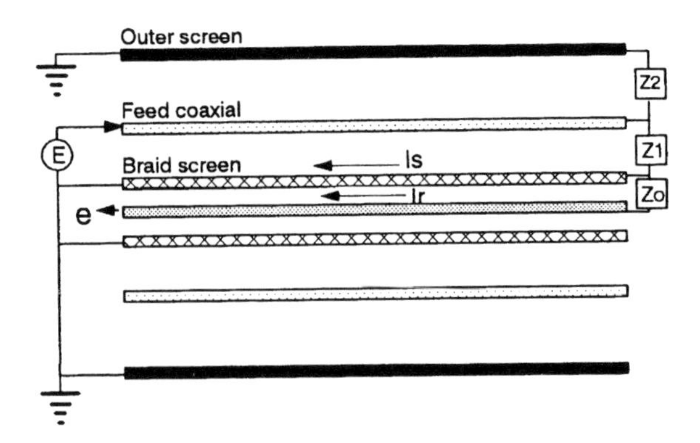

Figure 10.11
Quadaxial arrangement for screening

As a definition of a parameter relating to the screen, *SE* is to be preferred to *SF* since the latter depends on the impedance of the feeding circuit while the former only concerns the coaxial itself. However the two are easily converted if Z_1 is known and *SF* is easier to measure than *SE*.

Screen Measurements

Either of these screen parameters can be measured by the triaxial arrangement but the fact that the outer conductor is 'hot' is a disadvantage. The quadaxial arrangement of Figure 10.11 avoids this but with a loss of sensitivity since the feed current divides between Z_1 and Z_2 in the ratio $F = Z_1/(Z_1 + Z_2)$. Equations (10.25) and (10.26) are unchanged but in Equations (10.27), (10.28) and (10.29), Z_1 becomes Z_1/F. Z_1 and Z_2 must be chosen for constructional practicality and sensitivity of measurement. Coaxial crosstalk can be computed from Z_T using Equation (8.51) *et seq.* from Chapter 8.

Published Results of Screening Measurements

Published results of transfer impedance and screening efficiency sometimes appear to suffer from measurement difficulties, particularly those using the triaxial arrangement. However useful information can be gleaned from them. In particular some results by Knowles and Olsen [28] are worth mentioning. The measurements were made using the quadaxial arrangement and all the braids had a good fill factor of about 0.83. They indicate that there is an optimum braid angle (relative to the axis) of about 30° which improves the SE, relative to an angle of 60°, by about 20 dB. Also the use of multiple braids indicates an improvement in SE of 20 dB per braid over a wide frequency band (1–100 MHz).

11

Installation

Some aspects of installation methods obviously affect the cable design. These are generally associated with the cable mass, tensile strength and protection. In this chapter various installation topics and their appropriate design requirements are considered.

DIRECT BURIAL

For this requirement two methods are generally employed. In the first, an open trench is dug, either manually or with a mechanical digger. The depth of burial depends on the probability of disturbance by future land works, but is usually 0.5 to 1.2 m. In order to protect the cable from penetration damage to the sheath by stones, a 150 mm layer of sifted soil is required to be spread on the bottom of the trench before installing the cable. To avoid abrasion damage to the sheath, the cable is first laid alongside the trench on roller supports and then lifted into the trench. The backfill, up to 150 mm above the cable, should also be of sifted soil. For jointing, a wider excavation is generally required for the jointer to work. This should be supported by timbers against inadvertent collapse.

Mole Ploughing

An alternative way of direct burial which can be economical in the correct circumstances, is mole ploughing of the cable. This method is illustrated in Figure 11.1. A tractor carrying the drum of cable draws a plough blade through the ground. The cable is led off the drum and down through an oversize tube attached to the plough blade, in such a way that it is laid in the ground at slightly less than the depth of the plough. The plough blade forces aside any stones or loose rock that might damage the cable and the only tension on the cable is that required to pull it off the drum and through the feed pipe. However any tension that is in the cable is immediately locked in by the backfilling.

This works well in ground containing only loose stones or rock. Where shallow layers of rock exist it is possible to carry out a preliminary ploughing with a ripping blade to break up the rock layers. The plough connection to the tractor requires

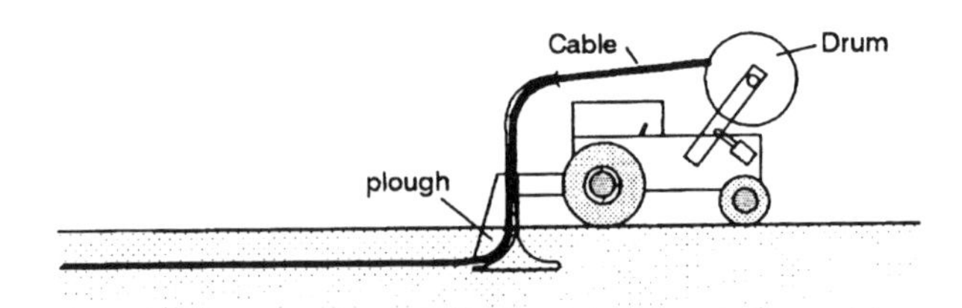

Figure 11.1
Mole ploughing of cable

degrees of freedom in the roll and yaw axes. Even so a major sideways slippage of the tractor rear wheels can result in a serious kinking of the cable. This means that the route must be capable of supporting the tractor and be free of soft spots into which the tractor wheels could slip.

Rodents and Termites

In areas where rodents such as gophers are present, directly buried cables may be gnawed and the sheath penetrated. This generally only happens when the cable diameter is small, so that the rodents' teeth, with the mouth wide open, can get a purchase on the sheath. It has been estimated from laboratory experiments that a cable with a diameter of about 50 mm or more is safe from gnawing.

For smaller diameter cables several protective measures have been used. The simplest is to oversheath the cable with a thin layer of nylon which is more resistant than polythene to gnawing. Otherwise various forms of armour can be used, such as lapping with thin steel, copper or bronze tapes. If this can be part of the sheath design, for example the STALPETH sheath (see Chapter 5), then no extra cost is involved. The use of copper or bronze however, is to be preferred, since the armour will be subject to corrosion after attack. For APL sheaths it is possible to increase the aluminium laminate thickness to 0.3 mm to obtain reasonable protection from gnawing.

If the cable crosses a termite run, or is installed near to wood which has become infested with termites, it is possible that it will be attacked by the termites. However termite damage is usually potential rather than actual as the number of proven cases of termite damage to cables is low. Termites have been shown to be able to penetrate lead, polythene, PVC, neoprene and rubber sheaths.

True ants have also been responsible for damage to small diameter cables. Ants generally forage above ground but in loose sandy soils they can penetrate to considerable depths. Again ants can penetrate most plastic sheaths and the tearing of an 0.08 mm aluminium laminate by a species of large ant has been reported, although this material, especially in larger thicknesses, is generally considered to be immune to attack.

The use of insecticides such as γDDT, Chlordane, Aldrin and Dieldrin in the back-fill at fairly high concentrations, or the incorporation of Aldrin and Dieldrin in the cable sheathing plastic at 0.25% concentration have been shown to be effective for a few years. However the handling of these materials, or sheaths incorporating them, is very hazardous in manufacture and installation. Their use has also been banned in many countries as being too dangerous enviromentally. The recommended protective measures are now the same as those for rodents (i.e. metallic armour or nylon oversheaths).

The hardness of ant and termite mandibles is usually about 3 on Mohs' scale. However the incorporation of harder abrasive particles in the sheaths of cables has not been shown to be effective in trials.

In the laboratory, termite 'runs' are often defined by petroleum jelly barriers. It was thought that the PJ contaminated the mandibles of the termites. However it has been noticed that polythene sheathed cables with the interstices filled with PJ (for water penetration protection) are also shunned by termites as if they can 'smell' the PJ through the sheath. Similarly cable sheaths wrapped with grease-impregnated tape (Denso or Densopol) have been shown to reduce termite attack in field trials.

DUCT INSTALLATION

In heavily populated urban and suburban areas, and where expansion of the distribution network is ongoing, it is often more economic and more sightly, to install vitreous clay ducts (in 6-, 8- and 12-bore formations) or multiple collections of pitch-fibre ducts, for subsequent drawing in of cables as required. These are generally of 80 mm bore diameter and one bore can often accomodate two or three small diameter cables or one cable of up to 70 mm diameter. Although the installation of ducts is quite expensive, with good planning it only has to be done once in a given area, resulting in less inconvience to road users. Ducts are generally installed about 0.8 m below the surface.

From the exchange, in European practice, large pairage (up to 4800 pair) small conductor gauge (0.5 mm to 0.32 mm) cables are led through ducts to roadside cabinets, whence smaller pairage (up to 100 pair/0.9 mm to 0.4 mm) cables are led through ducts to pillar-mounted distribution terminal boards or distribution points (often pole-mounted). This allows for considerable flexibility for subscriber connections. For high-capacity carrier or PCM cables, installation in ducts laid between exchanges provides the necessary security from damage by road-works, etc.

Duct routes are provided with surface access by manholes or surface boxes at regular intervals, depending on the cable lengths that can be manufactured and handled. In the days of lead sheathed cables the lengths were often limited by weight to 176 yards (160 m), but with the smaller mass of polythene sheathed cables, lengths of from 300 m to 1000 m can be handled. Joints are accomodated in these access chambers on wall-mounted bearers.

The ducts are rodded with a brush, to clean out any accumulated dirt or soil, and a draw rope of wire or hemp is used to draw in the cable. The tension required for this depends on the mass of the cable (and hence the length to be drawn in) and the coefficient of friction between the cable and the duct. The latter, denoted by μ, does not have a well-defined value, but for purposes of estimation it is generally taken as 0.6 for static friction which is fairly high and 20% less for sliding friction. Thus the drawing tension in Newtons for a straight-line pull is

$$T = \mu \cdot g \cdot M \cdot L \text{ N} \tag{11.1}$$

where M is the cable mass in kg/m, L is the actual length in the duct in m and g is the acceleration due to gravity = 9.81 m/s^2. Ducts cannot always be in straight lines and bends increase the tension by a factor

$$\frac{T_0}{T_i} = \varepsilon^{\mu\theta} \tag{11.2}$$

where θ is the angle in radians through which the duct turns.

In the days of lead sheathed cables, the cable was often greased to reduce the friction. With polythene sheathed cables this tends to increses the susceptibility to environmental stress cracking (Chapter 5) and a paste of kaolin in water must be used if required. As an illustration, a 500 pr/0.5 mm APL sheathed cable with a mass of 2.7 kg/m pulled through a straight duct with a coefficient of friction of 0.6 would have a linearly increasing tension reaching about 7.3 kN after drawing in 500 m of cable. If there were a 30° bend in the duct 100 m from the entry, the pulling-in tension would increase by a factor of 1.37 through the bend as shown in Figure 11.2 and the total tension after a 500 m pull would be about 8 kN. However a further 30° bend at 390 m would again increase the tension beyond this point by a factor of 1.37 resulting in the total tension after 500 m rising to 10.5 kN (about 1000 kgf or 1 tonnef). The maximum tension in the cable occurs at the pulling end. Immediately after removal of the pulling force, the tension in the cable relaxes to that shown by the dotted line, so that the maximum tension left in the cable is approximately half that due to the maximum tension in a straight pull. With ground vibration due to traffic, even this disappears eventually.

In order that the applied tension is evenly spread among the conductors and sheath, the cable end is 'spiked' before fitting the pulling sleeve (an open cylindrical steel wire braid about 0.75m long). The pulling force is applied uniformly via a driven winch, which should be provided with a dynamometer to indicate the tension.

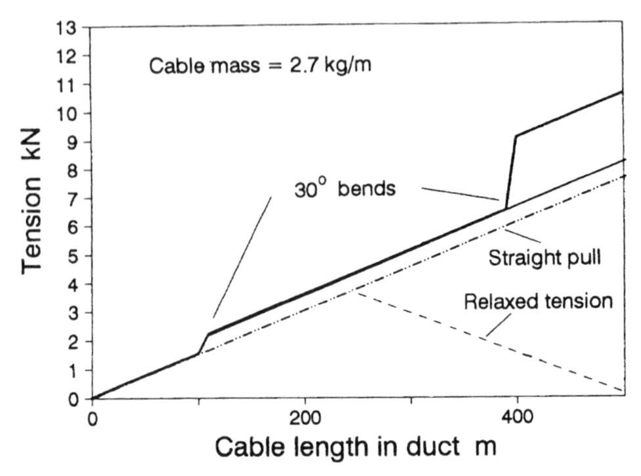

Figure 11.2
Cable pulling tension during installation in duct

The fractional stretch in the cable can be estimated from the formula

$$\frac{dL}{L} = \frac{T}{\{(EA)_1 + (EA)_2\}} \tag{11.3}$$

where the subscripts 1 and 2 refer to the conductors and sheath respectively, E is the tensile modulus and A the total cross-sectional area of conductors or sheath in appropriate units. In the case of the illustration used, the maximum force of 10.5 kN would produce a cable stretch of 0.05%.

When drawing-in, snatches should be avoided since these could apply much higher tensions to the cable causing permanent stretch of the copper conductors. More fragile materials such as aluminium conductors or quartz fibres could be broken in these circumstances.

When the duct route is on a slope to the horizontal of angle θ, traffic vibration can cause the cable to slip downhill. This puts the cable into tension. If the cable is anchored at the uphill end the maximum tension will be at the anchor end and will be

$$T = M \cdot g \cdot L \cdot \sin\theta \, \mathrm{N} \tag{11.4}$$

where L is the length of cable on the slope. To limit this tension on a long slope the cable must be anchored on the downhill side of each of the joints. For the cable used in the illustration, 300 m of cable at a 10° slope would build up to a tension of about 1.4 kN at the anchor.

AERIAL INSTALLATION

The cheapest form of cable laying is probably to lay it straight on the ground. But here it is liable to damage and theft. The next cheapest therefore would be to suspend it from poles or buildings. For distribution cables this is the method used in many countries. Even in countries where most cables are buried, the final run of smaller cables to subscribers is often aerially supported. Also the connection of the subscriber's premises to the nearest pole distribution point is by means of single-pair drop-wire. In order to limit the sag of the cable between aerial supports it is tensioned to an extent which its construction can safely withstand. In most cases this requires a steel suspension member.

The suspension member may be a separate stranded steel cable, to which the cable is later attached by leather/plastic saddles at intervals, or the cable may be lashed to the suspension by a continuous smaller diameter steel wire or finally the suspension strand and the cable may be sheathed together using a 'figure-of-8' sheath, in this case it is known as an Aerial Self-Supporting (ASS) cable. The methods are illustrated in Figure 11.3.

A uniform flexible strand when hanging freely between end-supports hangs in a shape called a catenary curve. In order to calculate tensions and sags in aerial cables, use is made of the fact that for sags which are small compared with the length, the catenary curve approximates a parabolic curve. The general rule of thumb for aerial cables is that the sag should be approximately one-sixtieth of the length.

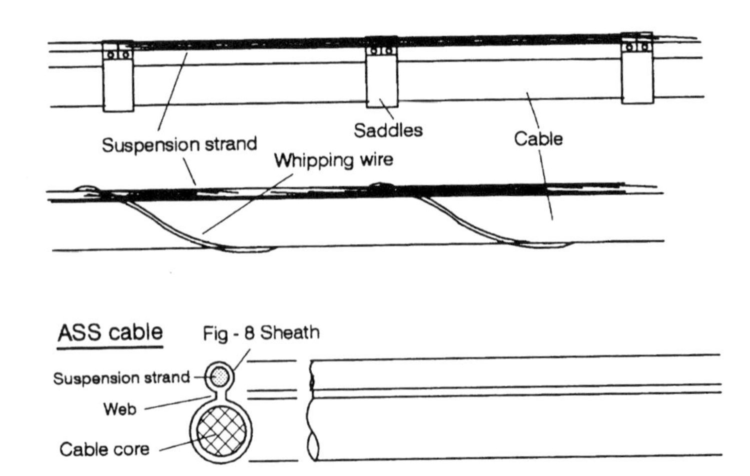

Figure 11.3
Aerial suspension of cables

Using the parabolic approximation the tension is related to the sag by

$$T = \frac{M \cdot g \cdot L^2}{8 \cdot s} \text{ N} \tag{11.5}$$

where M is the mass of cable plus suspension in kg/m, $g = 9.81\,\text{m/s}^2$, L is the span between the supports in m and s is the sag in m. The variation of tension with span is thus a hyperbola as shown in Figure 11.4 with the tension rising to very high values at small sags. For the recommended sag of one-sixtieth of the span the tension varies linearly with the span as shown in Figure 11.5

$$T = \frac{M \cdot g \cdot L^2}{8 \cdot L/60} = 7.5M \cdot g \cdot L \text{ N}$$

In both figures the tension is proportional to the total mass per metre of the cable plus suspension strand.

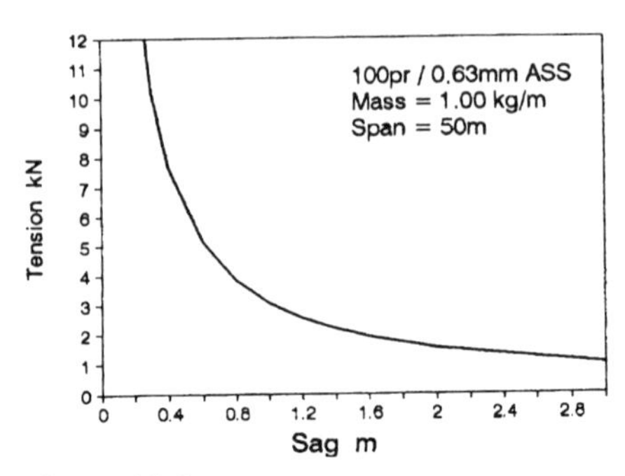

Figure 11.4
Tension variation with cable sag

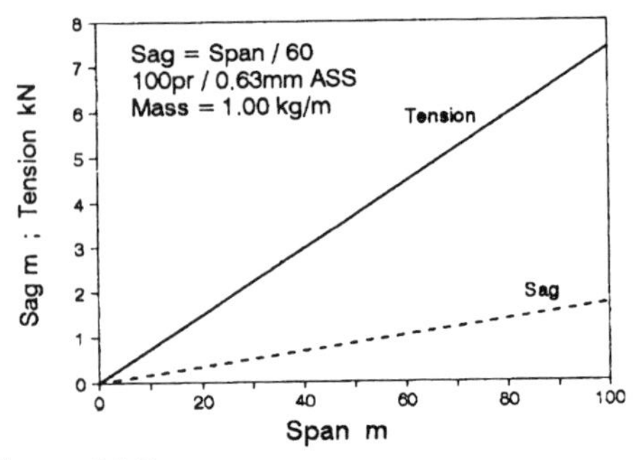

Figure 11.5
Tension variation with span length

Table 11.1
Typical suspension strands

Construction no./diam. mm	Diam. mm	Min. breaking load (kN) Grade 1150	Grade 1300	Approx. mass kg/1000 m
7/0.56	1.7	1.98	2.24	14
7/0.71	2.1	3.19	3.60	28
7/0.85	2.6	4.57	5.15	31
7/0.90	2.7	5.12	5.80	35
7/1.00	3.0	6.32	7.15	43
7/1.25	3.8	9.88	11.15	67
7/1.40	4.2	12.35	14.00	84
7/1.60	4.8	16.20	18.30	110
7/1.80	5.4	20.50	23.20	140
7/2.00	6.0	25.30	38.60	170
7/2.36	7.1	35.20	39.80	240

Suspension Strand

In most cases this is a stranded high-tensile steel cable. The standard galvanized steel wire strands of BSS 183:1972 are Grade 1150 (with a tensile modulus $E = 190\,\text{kN/mm}^2$) and Grade 1300 (with $E = 220\,\text{kN/mm}^2$). The grade number is the UTS in N/mm^2. The table gives the diameter, min. breaking load and mass per 1000 m of these grades for a range of sizes.

Variation of Tension with Ambient Conditions

The values of tension and sag calculated above apply only for the conditions of installation with no wind or ice loading, and at a constant temperature. If the temperature rises after installation the suspension strand will expand, increasing the sag and reducing the tension in the suspension strand. If the temperature drops the suspension strand will contract and the sag reduces, thus increasing the tension. Also winds and ice loading increase the total mass, again increasing the tension.

These increased tensions need to be calculated for the expected conditions after installation in order that the suspension strand does not exceed its breaking load (with a suitable factor of safety – FOS). This is done using the following equations

$$\left(\frac{M_1 P}{T_1}\right)^2 - T_1 = \left(\frac{M \cdot P}{T}\right)^2 - T - EA \cdot \alpha \cdot \Delta\theta \qquad (11.6)$$

giving

$$T_1^3 + Q \cdot T_1^2 - (M_1 P)^2 = 0 \quad \text{to be solved for } T_1 \qquad (11.7)$$

where

$$P = g \cdot L\sqrt{\frac{EA}{24}} \quad \text{and} \quad Q = \left(\frac{M \cdot P}{T}\right)^2 - T - EA \cdot \alpha \cdot \Delta\theta \tag{11.8}$$

E = tensile modulus of strand in N/m^2
A = cross-sectional area of strand in m^2
M = mass of suspension + cable (no wind or ice loading)
M_1 = mass of suspension + cable with wind and ice loading
T = installed tension (no wind or ice loading, normal temperature)
T_1 = worst tension with wind and ice loading and lowest temperature
α = temperature coefficient of expansion of suspension (per °C)
$\Delta\theta$ = drop in temperature expected below installed temperature (°C)

To calculate for wind pressure W in kg/m

$$W = 4.72 \times 10^{-6} \cdot D \cdot v^2 \tag{11.9}$$

where

D = cable diameter in mm $\qquad v$ = wind velocity in k.p.h.

A maximum wind pressure of 700 Pa (N/m^2) is often specified. This corresponds to a wind velocity of about 95 k.p.h. for round cable, and about 120 k.p.h. for a 'flat' ASS cable at right angles to the wind.

To calculate for ice loading at a radial thickness of ice of i mm, the mass of the ice is

$$m = \pi D \cdot i \cdot 0.915 \text{ kg/m} \tag{11.10}$$

Since m and M are in the same vertical direction and W is at right angles to this direction

$$M_1 = \sqrt{(M + m)^2 + W^2} \text{ kg/m} \quad \text{and} \quad \beta = \tan^{-1}\{W/(\mu + m)\} \text{ rad} \tag{11.11}$$

where β is the resultant angle to the vertical.

Procedure

Starting from the installed conditions with the sag equal to one-sixtieth of the span, estimate the installation tension. With an FOS of about 5, choose the appropriate suspension strand and calculate M and T. Then apply the worst conditions expected, calculate P, Q, M_1 and solve Equation (11.7) for T_1 and check that the specified minimum FOS (usually 2) is met. Calculate the new sag and the angle from Equations (11.4) and (11.11) and check that they are acceptable.

$$\text{FOS} = \text{Tension/UTS of suspension}$$

Equation (11.7) is a cubic, having in general the possibility of three roots. The required root is the highest positive root. A plot of Equation (11.7) is shown in Figure 11.6 for a 100 pr/0.63 mm ASS cable installed at 25 °C on a 50 m span with a sag of 0.83 m and then subjected to a temperature drop of 30 °C, a wind of 120 k.p.h. horizontally at right angles to the cable and an ice loading of 3 mm radial thickness. The root is the value of T_1 at which the value of Equation (11.7) is zero.

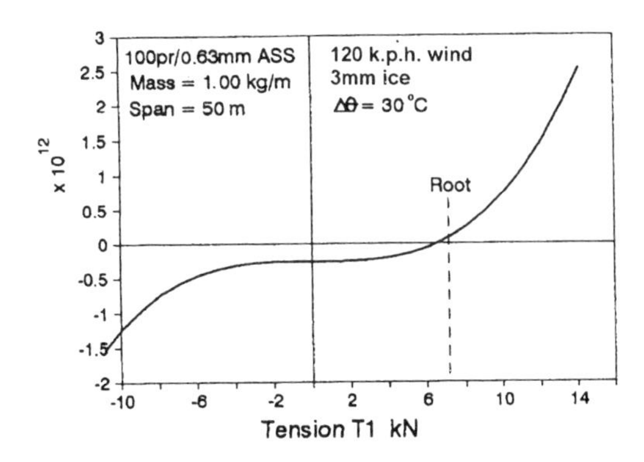

Figure 11.6
Plot of Equation (7.7)

The installation conditions are

Sag = 0.83 m
Suspension strand = 7/1.60 mm
$D = 40.4$ mm
FOS = 4.38

Tension = 3695 N
(Grade 1150, $E = 190$ kN/mm^2)
Mass = 1.00 kg/m

After applying the worst expected ambient conditions, solving Equation (11.7) gives

$M_1 = 3.07$ kg/m
FOS = 2.21

$T_1 = 7340$ N
Sag = 1.28 m at 63° to vertical.

Equation (11.7) is readily solved by trial-and-error interpolation on a spreadsheet, after plotting it roughly to estimate the root. The effect of varying the parameters of temperature drop, ice loading and wind velocity on the sag is shown in Figure 11.7. The worst FOS achieved is 2.1.

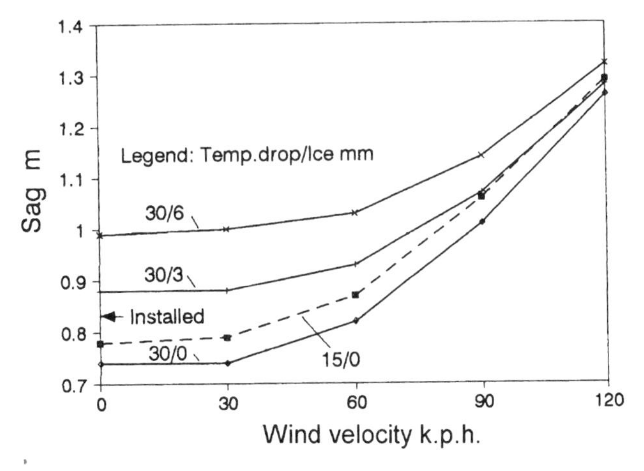

Figure 11.7
Variation of cable sag with ambient conditions after installation

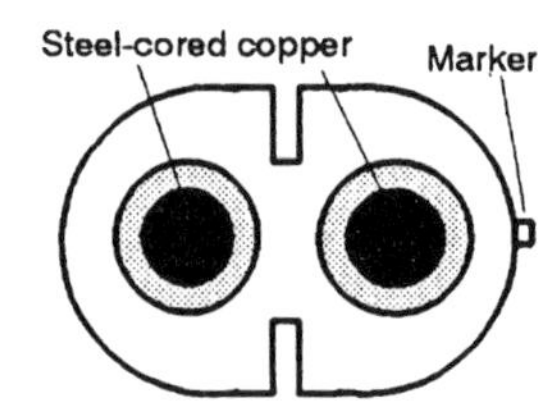

Figure 11.8
Subscriber drop-wire

Installation

The cable is installed by pulling it through temporary suspension loops at each pole, anchoring one end and tensioning it to the calculated installation sag using a ratchet mechanism, then anchoring the pulling end. The temporary loops are then replaced with permanent clamps which grip the suspension strand and the figure-of-8 web.

Galloping

When the aerial cable is not perfectly circular, the wind will create vortices around the cable which will produce a varying aerodynamic lift. This will cause the cable to vibrate and 'gallop', especially at harmonics of the span. To counteract this it is necessary to twist the cable through about six twists per span around the suspension strand, before tensioning.

Drop-wire

An important component of a telephone system is the connection from the distribution cable to the subscriber. For domestic subscribers this is nearly always an aerial self-supporting single pair from a distribution pole to the premises, and is known as a drop-wire. A lightweight construction is desirable and a common form, known as a double-D construction is sketched in Figure 11.8. For strength, the conductors may be of hard-drawn copper, cadmium copper or copper-clad steel. Of these, copper-clad steel is preferred with 40% conductivity of an equal-diameter copper wire. For 0.7 mm diameter the UTS is 380 N per conductor, and the mass of the wire is 3.12 g/m. With an effective tensile modulus of 172 kN/mm^2 the dropwire may be used in spans of 30 m (initial sag = 0.5 m at a tension of 15 N) and will withstand a 30 °C drop in temperature and 120 k.p.h. wind with an FOS of 1.95.

WATER PENETRATION

If the sheath of a buried cable is damaged or a joint closure develops a leak in the presence of groundwater, the entry of water into the cable can cause extensive damage before remedial measures can be taken. The effect on paper insulated cable is to soak the paper and cause a severe drop in the insulation resistance and an increase in the loss angle. Wet paper tends to swell and produce a blockage which limits the spread of water along the cable to a few metres, so that only a short length, about 30 m, of cable may need replacing (although in ducted cable this may not be possible).

In plastic insulated cable considerable penetration of water into the cable occurs. This may not produce any immediate drastic effects, but with time the water can travel many hundreds of metres into the cable and in due course disclose its presence through the inevitable pinhole faults in the insulation and deterioration of transmission quality. Once water has gained access to a cable length in this way, drying out, even using sophisticated techniques like flushing with acetone, is seldom successful in the long term and cable replacement is the only certain remedy. Several methods have been developed to prevent water penetration and/or to give early warning of sheath or joint failures.

Pneumatic Protection

The use of pneumatic pressure for testing cable sheath and joint integrity after installation has long been practiced. The extension of this to a permanently pressurized cable route was used as long ago as 1843, but was brought to a higher degree of sophistication in the 1950s. Typically the cable is pressurized to 400 mbar and pressure sensitive contactors are fitted in the joints, connected across an alarm pair, so that a short circuit across the pair occurs if the pressure falls to 300 mbar. This not only gives an alarm but the resistance to the contactor gives a rough location of the fault. This is illustrated in Figure 11.9 which also shows the pressure profiles with time after a fault occurs. Contactor c3 will indicate a fault within about 5 hours of occurence.

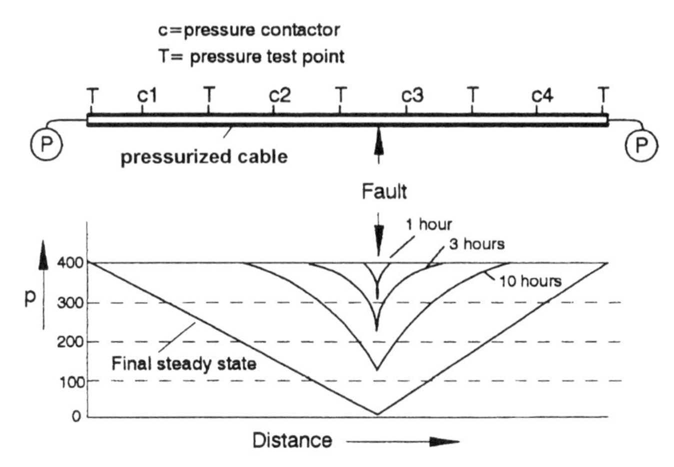

Figure 11.9
Transient pressures after fault occurrence in pressurized cable

Cable Pneumatic Parameters

In order to have a theoretical basis for system evaluation it is necessary to have some idea of the pneumatic characteristics of cables. Since cables are primarily designed for electrical characteristics these pneumatic parameters are not generally well controlled, nor is the accuracy of measurement very high (about 5% at best). Although the pressure drop down a cable with a constant mass flow of gas is actually parabolic with length, it is close enough to linear to enable planning calculations to be made, analogously to the electrical current flow through a resistance under the influence of applied voltage. Under most circumstances the flow of gas in a cable can be taken as streamlined flow. Thus the following parameters can be established

- Pressure difference (voltage) in mbar (1 bar $= 10^5$ N/m^2)
- Flow (current) in g/h (1 g/h = 0.78 l/h at NTP)
- Pneumatic Resistance in mbar/(g/h)/km {(mbar · h)/(g · km)}
- Pneumatic Capacitance in g/mbar/km {g/(mbar · km)}
- Pneumatic Leakance in g/h/mbar/km {g/(mbar · h · km)}

Pneumatic Resistance to Gas Flow

Figure 11.10 shows the 'best fit' lines of pneumatic resistance for 0.63 mm and 0.9 mm polythene insulated pair cables measured by 5 manufacturers on five sizes of plastic sheathed cables, published by US IPCEA in 1965. The spread of measurements about these lines was a factor of 2 for 60% confidence limits. The measurements have been converted from Imperial units.

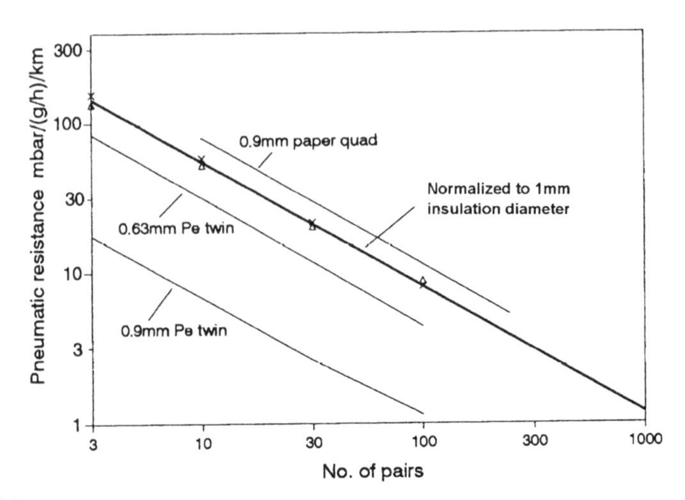

Figure 11.10
Pneumatic resistance of cables

Since the pneumatic resistance depends on the size and number of interstices in the cable, and since the size of the interstices depends on the insulation diameter of the conductors, these lines can be normalized to 1 mm insulation diameter using an inverse fourth power law of diameter (Poiseuille's law), as shown. From these results the following empirical law is derived

$$R = \frac{366}{N^{0.83} D^4} \text{ mbar/(g/h)/km} \tag{11.12}$$

where N is the no. of pairs and D is the insulation diameter in mm. This sensitivity of the resistance to the interstice diameter explains the great variability of the measurements.

Also shown in Figure 11.10 is the pneumatic resistance of 0.9 mm paper quad cable. This has the same slope with pair count, but although the insulation diameter is comparable, the resistance is about 15 times higher than the comparable polythene twin cable. A factor of about two is explainable by the fact that quad cable has only half the number of interstices for the same pair count, the remainder, a factor of about 7.5 is probably due to the interstices being less well defined owing to crushing of the insulation. Thus a paper insulated twin cable would have a pneumatic resistance about 7.5 times that given by Equation (11.12).

Pneumatic Capacitance

This expresses the capability of the cable to store gas under pressure. It is a parameter in the determination of transient performance of the cable to sudden faults. It is obviously related to the free air space under the cable sheath. Using Equation (4.4) for a plastic insulated unit twin cable, this free air space is given approximately by

$$\pi N D^2 \left(\frac{1.9^2}{4} - \frac{1}{2} \right) = 1.26 \cdot N D^2$$

Hence the pneumatic capacitance, at 1.282 g/l for air at NTP, is given by

$$C = 1.62 \times 10^{-3} \cdot N D^2 \text{ g/mbar/km} \tag{11.13}$$

where the insulation diameter D is again in mm. For paper tape insulated cable the free air space will be about 2.5 times this owing to the extra air space within the insulation (it may be estimated by taking the difference in area between the inside diameter of the sheath and the bare conductors).

Sheath Leakance

For a lead sheathed cable or for an APL/polythene sheath, the permeability of the sheath to gas is very low. For a plain plastic sheath, gas permeation is significant. It is expressed as a parameter G in units of g/h per mbar pressure difference per km. From Equation (5.1) we have

$$G = \frac{P \cdot \pi D_S}{T} \text{ g/h/mbar/km} \tag{11.14}$$

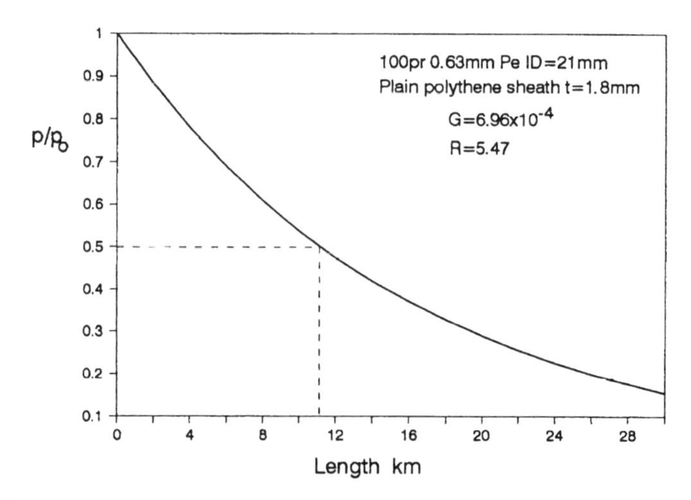

Figure 11.11
Pressure variation with length in cable with permeable sheath

where D_S is the sheath diameter and T its thickness and P in the table on page 62 is multiplied by 0.132 to correct for the units of pressure, mass and length. Thus for air (one O_2 to four N_2) permeating through low-density polythene, P becomes 1.9×10^{-5} g/h/mbar/km and for high-density polythene P becomes 0.3×10^{-5} g/h/mbar/km.

In a similar way to the electrical characteristics, it is possible to define the parameters

$$Z_0 = \sqrt{\frac{R}{G}}\,\text{mbar/g/h} \quad \text{and} \quad \alpha = \sqrt{RG}\,\text{ratio/km} \tag{11.15}$$

Thus the fall-off of pressure p with length inside a permeable sheath is given by

$$p = p_0 \varepsilon^{-\alpha L} \tag{11.16}$$

where p_0 is the input pressure into a semi-infinite length. The decay is exponential as shown in Figure 11.11. The flow of gas into the cable required to keep the pressure p_0 at the input constant is given by

$$f = \frac{p_0}{Z_0}\,\text{g/h} \tag{11.17}$$

It will be seen from Figure 11.11 that the pressure in the cable falls to half in about 11 km for a 100/0.63 mm Pe cable ($\alpha = 0.062$/km). The required flow rate to maintain 400 mbar at the input ($Z_0 = 88.4$) is 4.5 g/h equivalent to 3.5 l/h at NTP.

Pressurized Cables

The time required to pressurize a cable, sealed at the far end, or the time required to discharge the pressure from a pressurized cable through an open end, is governed by the time constant $R \times C$ in the same way as charging or discharging an electrical capacitance through a resistance.

$$\text{for charging} \quad p = p_0(1 - \varepsilon^{-t/RC}) \qquad \text{for discharging} \quad p = p_0 \cdot \varepsilon^{-t/RC} \tag{11.18}$$

In one time constant the pressure will rise to within 37% of the applied pressure, or will fall to 37% of the starting pressure. In $t = 4RC$ these percentages become 2%.

For a 1 km length of 100 pr/0.63 mm polythene insulated cable pressurized to 400 mbar the pressure will fall when one end of the cable is opened, to 147 mbar in $5.46 \times 0.196 = 1.07$ h and to 8 mbar in 4.3 h. Note that these times are proportional to the square of the cable length in km.

If the same cable has a permeable polythene sheath and is sealed at both ends with an internal pressure of 400 mbar, this pressure will fall with a time constant of C/G which is not length dependent. Thus for a low-density plain polythene sheath of i.d. = 21 mm and thickness 1.8 mm, $G = 0.0007$ g/h/mbar/km and the pressure will fall to 147 mbar in 0.196/0.0007 = 280 h or 12 days. This was an unpleasant surprise in the early days of plastic sheathing. For a 1000 pr/0.63 mm cable with i.d. = 66 mm and a sheath thickness of 3 mm, the 12 days becomes 1.7 years.

Fault Leakance

Although the irregular shape of an actual fault prevents the calculation of its leakance, some idea of the flow can be gained from the calculated value for a circular hole in the sheath of the same order of size. Bernoulli's theorem gives

$$F = 44.3 D_h^2 \sqrt{\Delta p}\ \text{g/h} \tag{11.19}$$

where Δp is the pressure difference in mbar and D_h is the hole diameter in mm. Thus F is proportional to the square of the hole diameter and to the square root of the pressure difference. At 400 mbar pressure difference a 1 mm diameter hole would thus show a leakage of 44.3 g/hr (34.6 l/h at NTP).

Protective Air Flow for Circular Hole

The nature of a fault can vary from complete severance of the cable to a small hole. In the first case the pneumatic system can only limit the extent of water damage but the fault will be easily detected and found by electrical tests. In the latter case the pneumatic flow may be able to prevent water ingress. For air to exit through a circular hole in a sheath under water, there must be a minimum pressure difference which is equal to the excess pressure in a bubble of diameter equal to the hole diameter. This is given by

$$\Delta p = \frac{4T}{100 D_h}\ \text{mbar} \tag{11.20}$$

where T is the surface tension of the water in dyne/cm and d is the hole diameter in mm. For pure water at 5 °C, T is about 75 dyne/cm, but for contaminated water it is lower. The pressure difference thus varies from 3 mbar for a 10 mm diameter hole to 30 mbar for a 0.1 mm hole. Thus for a cable under 1 m head of water (98 mbar) the internal pressure in the cable must exceed $(98 + \Delta p)$ mbar for air to bubble out. It is tempting to substitute Equation (11.20) into (11.19) to find the protective flow of air, but this is erroneous since the flow is no longer streamlined and Equation (11.19) will not hold. CCITT recommendations [29] indicate that empirically, a protective airflow of about

$$F_p = D_h^2 \text{ g/hr} \tag{11.21}$$

is required for a circular hole of D_h mm diameter, to prevent water entering a water-immersed cable.

Pneumatic Scavenge Systems

In Chapter 5, the permeation of water through plastic sheaths was considered. This builds up humidity in the cable which under temperature variation can result in water condensation. To counteract this it is possible to send a low flow rate of dry air (about 35 to 180 g/h) through the cable thus scavenging the humid air to the far end of the cable and out to the atmosphere through a controlled leak. In the event of sheath damage the sudden increase in flow rate can also sound an alarm. This method has been used successfully both for plastic sheathed trunk cables and for distribution cable systems.

Petroleum Jelly Filling

An alternative method of preventing water penetration down cables was developed by BICC in the UK in the 1950s. This consisted in blocking all the cable interstices throughout the length with petroleum jelly (the so-called 'fully filled' cable. The name 'fully filled' was used to contrast this method to that of including discrete 0.5 m water blocks at 10 m intervals throughout the cable length). The effect of this on the electrical parameters was to increase the electrical capacitance. This could be overcome either by increasing the insulation diameter of the conductors (thus increasing the overall cable diameter), or by using cellular polythene insulation at the same diameter (which reduced the voltage withstand by about half). This method of protection not only prevented water penetration from sheath or joint damage, but also protected the cable from water permeation through the sheath, which could therefore be of plain polythene.

A whole new area of cable technology grew from this development. The PJ had to be carefully formulated to avoid stress cracking of the insulation and sheath. It had to be non-dermatitic. It had to be able to withstand moderately high temperatures without dripping while still being able to be applied in a semi-molten state to the cable during manufacture. The test of the effectiveness of water blocking was set to the withstand of a 1 m head of water without water penetration of more than 1 m down the cable. Early cables had the PJ applied to the closing dies, layer by layer on the strander. Later developments were able to pressure-apply molten PJ through the cross-section of the stranded cable in one operation, in tandem with the sheathing extruder. Bell

Laboratories, coming late to this development, improved the filling material by blending PJ with low-melting-point polythene. This gave a gum-arabic sort of texture which was much easier for the jointers to handle and had very few drip problems.

Absorbent Powder

Another development in the 1970s applied a water absorbent powder to the cable cross-section, either by electrostatic attraction, or by using string or tape carriers, during stranding. This had a much smaller effect on cable capacitance when dry. In the event of water penetration, the powder absorbed the water and rapidly swelled up to provide an effective water-block of about 0.5 m length. This block could withstand a considerable head of water (several metres).

JOINTING OF CABLES

Conductor Jointing

The common practice of stripping the insulation from the wires, twisting them together (soldering the tip of the twist in the case of trunk and junction cables) and then insulating the joint with a small paper or polythene sleeve has now been largely superseded by the use of mechanical connectors. These come in a number of forms, one of the earliest being the American 'B-connector'. This consists of an inner tinned phosphor-bronze piece with punched tangs to penetrate the insulation and contact the conductors, surrounded by a steel piece which maintains pressure on the inner piece after crimping. The whole unit is insulated with a polythene sleeve, closed at one end. The unit may be filled with silicone grease for water protection.

In use, the insulated conductors to be jointed are slipped together into the connector which is then crimped down with a special tool. Automatic jointing machines using these connectors are available for large pairage cables. The conductor joints are staggered along the joint to reduce build-up of the joint diameter, and the completed joint is hand-lapped with paper or plastic tape to consolidate it before sheath closure.

Sheath Closures

The displacement of lead or lead alloy sheaths by polythene brought many economic and operational advantages, but also brought a major problem in the form of a need to produce an adequate sheath closure. The requirements for a successful sheath closure are

- Good mechanical strength against movement or tension
- Wide temperature range – e.g. – 10 °C to 40 °C
- Internal pressure withstand – pneumatic protection

- Resistance to water permeation
- Low level of locked-in stress to avoid ESCR
- Simple to apply in field
- Reopenable when necessary
- Applicable to multiple joints – i.e. large to several small cables

Jointing Techniques

Practical jointing techniques fall into three categories all of which have been proved sucessful in the field.

Mechanical joints

These rely on compression onto the cable and jointing sleeve, using rubber bungs or sealing rings. This type of closure is applicable to both polythene and PVC sheathed cables. Since the compression can vary with temperature there should be plenty of reserve compression in the design of the joint piece parts.

Fusion methods

These are principally used with polythene sheaths. In the BICCweld system, a perspex mould is clamped around the sleeve and a polythene diameter-adaptor, and molten

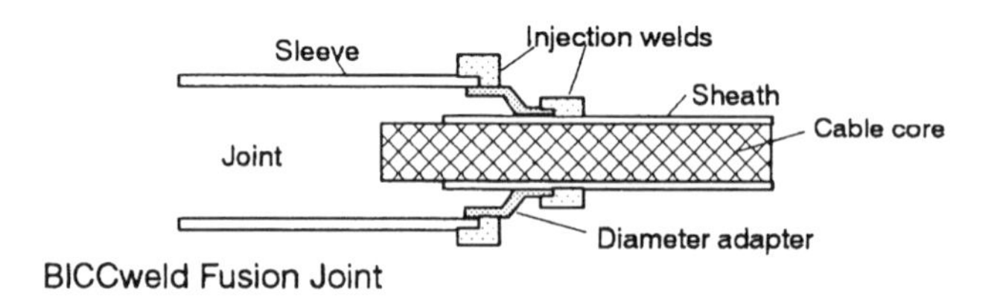

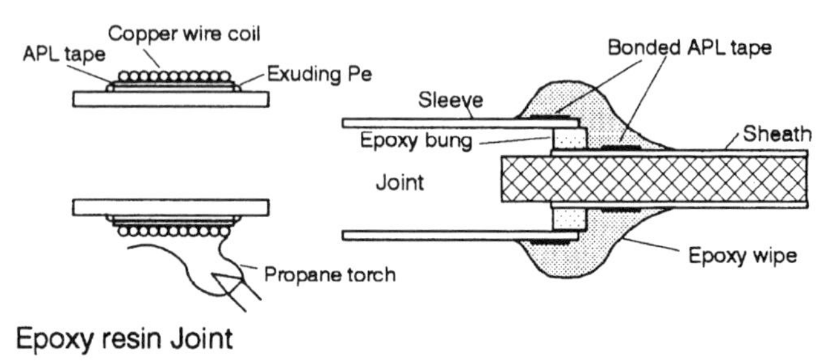

Figure 11.12
Polythene sheath joint closures

polythene from a hand-held ram extruder is forced into the mould and out through a bleed hole at the opposite side until the black sheathing material is seen to flow through (thus cleaning the sleeve and adaptor polythene surfaces of oxides). The extruder is then removed and both the feed and bleed holes are fitted with screw rams to keep the cooling polythene under pressure until it sets. Later versions used simple strip metal/ masking tape moulds made up on site. The method met all the requirements listed above but it was used mainly on trunk and junction cables, where it is by far the most reliable of sheath closures. This is illustrated in Figure 11.12.

Adhesive methods

Various adhesive methods are available, ranging from adhesive tapes to epoxy putties. Since polythene is a non-polar material adhesion is very difficult to achieve. The polythene must therefore first be degreased and then oxidized using a propane flame. Alternatively an intermediate material must be used, such as a wrap of APL tape which is bonded to the sheath by heating under pressure. The method of bonding is shown in Figure 11.12. The APL foil is wrapped, polythene inside, around the degreased sheath and wrapped with a single layer coil of 0.9 mm copper wire which is drawn tight. The coil is heated gently with a propane flame until polythene exudes from the edge of the APL tape. Heating is then discontinued and when cool enough to handle the copper wire is removed.

Some epoxy resin formulations will adhere to the aluminium foil directly, others with less free resin require a primer of nitrile rubber solution to be applied first. A moist 'moleskin' is used to wipe the putty into a bulbous shape as shown, and it is then allowed to cure for 1 to 3 hours before applying test pressure.

Adhesive tape joints generally require two layers of a polythene-based self-amalgamating tape, followed by about ten layers of tightly applied high-tensile PVC tape to apply pressure.

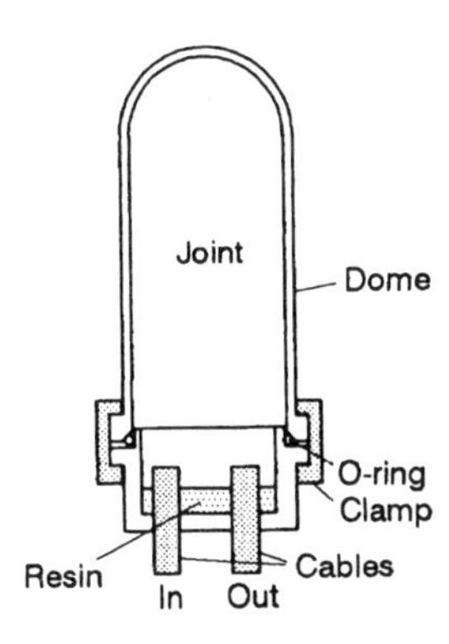

Figure 11.13
Dome joint closure

Dome joint

The dome joint illustrated in Figure 11.13 is a mixture of a mechanical closure and an adhesive resin seal to the cable sheaths. This is a popular method for distribution cables and also for optical fibre cables. The pieces of the dome joint are moulded from black polythene. This has the advantage of equalizing the temperature coefficients of expansion.

12

Lightning

PHYSICS OF LIGHTNING

One more hazard that cables, both aerial and buried, are subject to, is the possibility of damage by lightning storms. To examine this danger some knowledge of the physics of lightning storms is desirable, although owing to the difficulty of investigation much remains to be learnt. A reasonable visualization of the processes, derived from the literature [30, 31] is illustrated in Figure 12.1.

The Development of Thunderclouds

The earth carries an electric charge of −500 000 C. In fine weather conditions this is neutralized by a space charge of positive and negative ions formed by cosmic ray bombardment, with the positive ions congregating near the earth in the lower atmosphere producing a downwardly directed electric field strength of about 0.15 kV/m at the surface and a mean potential relative to earth of +300 kV in the lower atmosphere.

In moist air conditions, owing to the vertical lapse of air temperature of about 10 °C/km, cloud formation can develop owing to the condensation of water onto dust particles about the 0 °C level. In still conditions with little vertical momentum of the air, the cloud can grow into a dense cumulus cloud. As the cloud develops it is polarized in the electric field and the field strength at the earth surface increases fairly rapidly to as much as 15 kV/m. When the water particles coalesce to a sufficient size they precipitate into drizzle or light rain and the cloud and the field can dissipate slowly.

If the moisture in the air exceeds about 5 g of water vapour/kg of dry air, and there is significant vertical momentum due to uplift by an approaching cold front or by thermal convection from the ground, the cloud formations extend to considerable altitudes and are known as cumulo-nimbus clouds. Solar heating of about 1 kW/m^2 can raise the temperature of a 1 m layer of air near the surface by 1 °C/s at midday in the summer months. As a result of this heating from below, the density of the surface air is reduced and the atmosphere can become unstable for vertical motions. Strong updraughts develop in polygonal cells accompanied by downdraughts around the cell periphery.

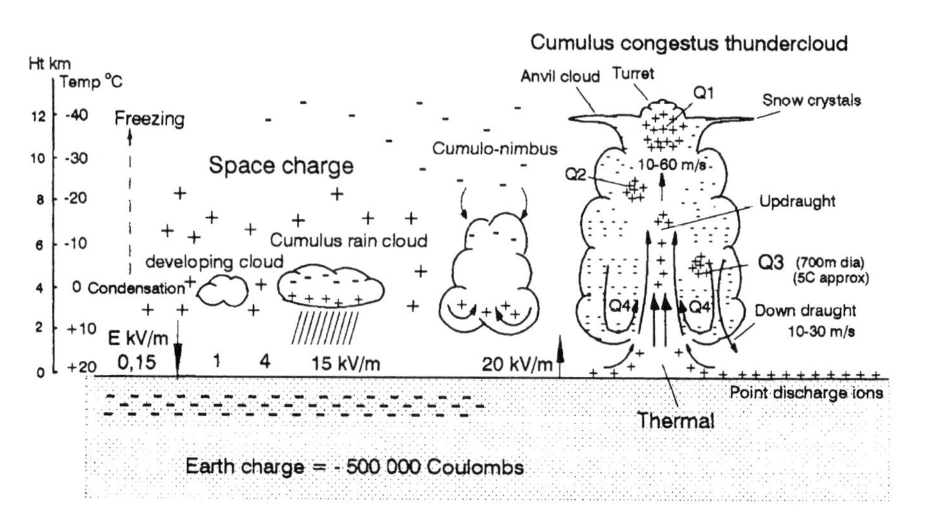

Figure 12.1
Development of thunderclouds

The condensation of water in the updraught releases latent heat which continues to drive the air upwards. A convergent flow of warm moist air flows into the updraught from a considerable area under the cloud. Updraught velocities of 10–60 m/s and downdraughts of 10–30 m/s have been inferred. Usually these downdraughts do not extend far into the subcloud air, but occasionally burst through to give dangerous 'wind shear' effects.

The diameter of the updraught cell is from 300 m to 2000 m, and the diameter of the thermal cloud system is from 3 km to 50 km, whereas frontal systems can extend for 200 km. The duration of such systems varies from 2 h to 48 h. Mountainous areas also assist in directing horizontal air flows into the vertical direction.

In the presence of suitable nuclei the liquid water droplets can freeze to form hail, at altitudes where the temperature is below 0 °C, but these nuclei appear to be limited in number. The majority of the water droplets rise to altitudes where the temperature falls to −40 °C where they freeze onto crystalline aerosol particles to form snow. Vigorous cumuli with strong updraughts often continue to grow in volume and height until they reach a thermally stable layer that serves as a ceiling on their further vertical development usually at about 12–20 km in the tropopause. Then the snow crystals spread out to form the anvil (c.incus) cloud. The updraught can penetrate through this ceiling to form turrets of cloud up to 1 km higher. At this stage the cumulo-nimbus cloud has become a cumulus congestus cloud.

Electrification of Cumulus Congestus Cloud

The bottom of the cumulo-nimbus cloud is in a region of predominantly positively charged ions, and the top is growing in a region where the ions are negatively charged. These ions become trapped amongst the water particles and can become attached to them. This severely limits their mobility so that they move with the convection currents in the clouds. The negative ions are swept downwards and the positive ions are swept upwards. The negative charge accumulating on the base of the cloud eventually causes a reversal of the electric field at the surface of the earth. The field strength increases to values of 10–20 kV/m in times of the order of 2 to 5 minutes.

At field strengths of the order of 2 kV/m, the air surrounding elevated points at the surface of the earth becomes ionized (owing to the field concentration at small radii) and an electric current flows from them. These currents, called point discharge currents, can reach values of 10 nA/m^2 in a field of 10 kV/m over small plantations of isolated trees and bushes about 2 m high. The point discharges carry positive charge into the updraught which is swept up to the cloud top. This results in a positively charged cloud top, Q1 in Figure 12.1, which attracts more negative ions (which continue to be swept downwards), over a large negatively charged region, Q4. Pockets of positive charge, Q2 and Q3, of the order of 5 C can also become trapped in small regions of the order of 700 m diameter within the negatively charged regions and have been detected by rocket and balloon sondes. The field strengths near the cloud base are about an order of magnitude greater than those at the surface of the ground.

Electrical Discharges

Clouds are even poorer conductors than air owing to the low mobility of the ions, and although some charge may diffuse through the cloud, discharges between cloud and ground or within the cloud can only take place along pre-ionized channels. Apart from lightning strokes to exceptionally high structures (of the order of 30–300 m high towers), the initiation of these self-propagating channels always takes place inside the cloud. The initiation is caused by corona from small radius water drops, or from electrodynamically or mechanically distorted drops. Ice particles may also cause initiation of corona above −15 °C. The field strengths necessary for initiation are about 250 to 400 kV/m. Such field strengths have been found in thunderclouds.

Lightning dicharges may be from cloud charge centres to space charges outside the cloud (known as air discharges), or between Q1 and Q4, or from Q2/Q3 to Q4 (called cloud flashes), or from Q1, Q2, Q3 or Q4 to ground (called ground flashes). About 90% of ground flashes are from Q4 to ground.

Lightning flashes are composed of low-luminosity, low-current leaders followed by high-luminosity, high-current strokes. A flash may consist of multiple strokes each preceded by a leader. From the initiation channel in the cloud the discharge advances toward the ground in steps of about 50 m each with about 100 μs between each step. This is called a stepped leader and it takes about 20 ms to reach the ground. As the tip of the leader arrives near the ground an upward discharge leaves the earth and initiates the main return stroke which travels up the prepared ionized channel at a speed of between 10 and 200 m/μs. This is a high current discharge with high luminosity. It neutralizes the charge brought down by the leader. This high current can come only from the earth owing to the poor conductivity of the cloud preventing such a high rate of redistribution of charge. The return stroke neutralizes the charge centre from which the leader developed. A discharge to this centre from another charge centre in the cloud may now occur, initiating, within about 40 ms, a dart leader to the ground following the same ionized path left by the main return stroke. This enables a second return

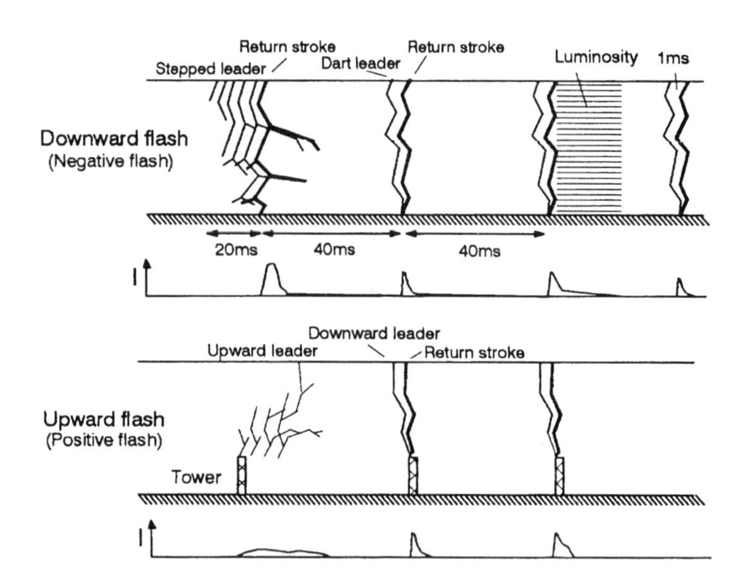

Figure 12.2
Temporal development of lightning flashes (Reprinted from *Lightning*, Vol. 1, Golde 1977, by permission of Academic Press Ltd., London)

stroke to leave the ground, and so on to produce a multiple-stroke flash. The current pulse which flows in the return strokes has a short rise time and a long tail. Often the current remains at a few amps between multiple strokes. Sometimes luminosity occurs after the return stroke corresponding to a higher continuing current. This downward or negative flash is illustrated in Figure 12.2 as it would be recorded by a rotating camera. The potential between the charge centre in the cloud and earth is estimated to be of the order of 50 to 100 MV. In temperate climes, 75% of flashes, and in sub-tropical or tropical climes 90% of flashes, are of this type.

High towers can be struck by these downward (negative) flashes, but in more than 25% of flashes to towers the flash is initiated by point discharges creating an upward (positive) leader. This may be followed by a downward leader and upward return stroke as shown also in Figure 12.2. However 50% to 60% of these positive leaders do not result in ground strokes, but can carry discharge currents of several hundred amps for several hundred milliseconds (accompanied by a hissing noise).

Characteristics of Ground Strokes

The core of the return stroke from ground is estimated to be about 30 mm diameter, surrounded by a corona envelope of about a few metres. The ionized air in the core is compressed to 2 MN/m^2 (20 atm) by magnetic forces and is heated to about 25 000 °C. The internal impedance of the 'lightning current source' is estimated to be a few thousand ohms, so that the magnitude of the stroke current is little affected by the resistance of the earth path (some hundred ohms).

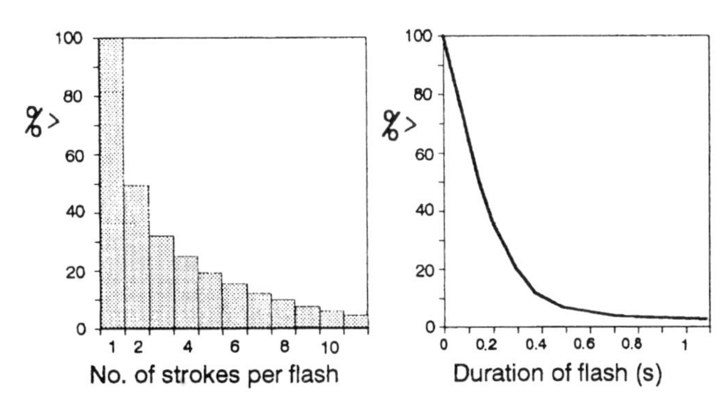

Figure 12.3
Strokes and durations of negative flashes (Compiled from *Lightning*, Vol. 1, Golde 1977, by permission of Academic Press Ltd., London)

Electrical Characteristics of Ground Strokes

Cumulative distributions in Figures 12.3 to 12.6 show the various characteristics of lightning flashes and strokes. Typically 50% of flashes are single stroke with a duration of about 60 ms. The current pulse shape is a double exponential form with typical rise time, for negative strokes, of 5 μs and time to half amplitude of 65 μc. For positive strokes the durations are about three times longer.

Peak currents in first negative strokes are greater than in subsequent strokes and have a median value of about 30 kA and a 1 percentile of about 120 kA. Positive strokes have a median value of about 35 kA but the 1 percentile can be as high as 250 kA. The peak current to buried cables is higher than to electrical transmission lines owing to the attractive range increasing with the value of the current. The variation between the distributions is due to the different measuring techniques and structures considered. The charge in first negative strokes has a median of about 7 C but in

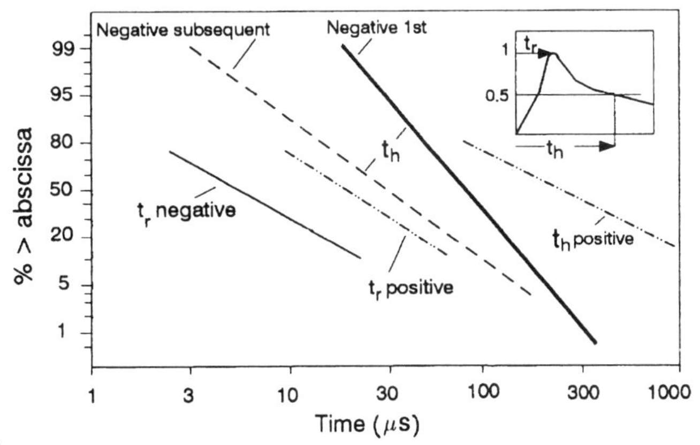

Figure 12.4
Rise and fall times of lightning pulses (Compiled from *Lightning*, Vol. 1, Golde 1977, by permission of Academic Press Ltd., London)

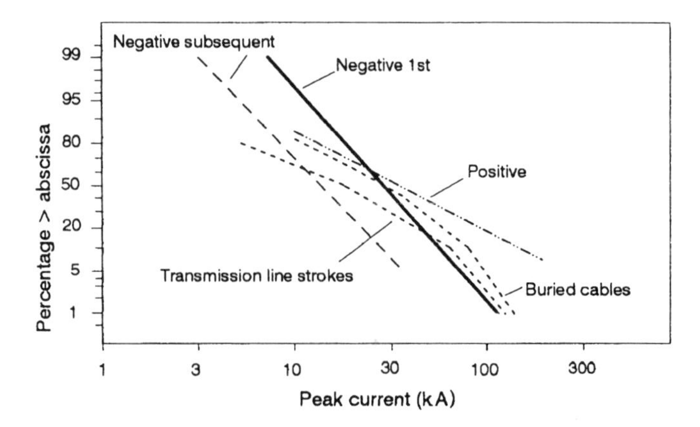

Figure 12.5
Distribution of stroke currents (Compiled from *Lightning*, Vol. 1, Golde 1977, by permission of Academic Press Ltd., London)

positive strokes the median charge is much higher at 80 C showing the greater severity of positive discharges.

For the purposes of engineering calculations for protection requirements, the current pulse shape is taken as the double exponential

$$I(t) = I^*(\varepsilon^{-at} - \varepsilon^{-bt}) \tag{12.1}$$

with the constants a and b chosen to give a true time to peak of $t_t = 5\,\mu\text{s}$ and a total time to half amplitude of $t_h = 65\,\mu\text{s}$. This is known as the 5/65 pulse shape. Also for protection calculation purposes the current in the first stroke at the 1 percentile point is generally used.

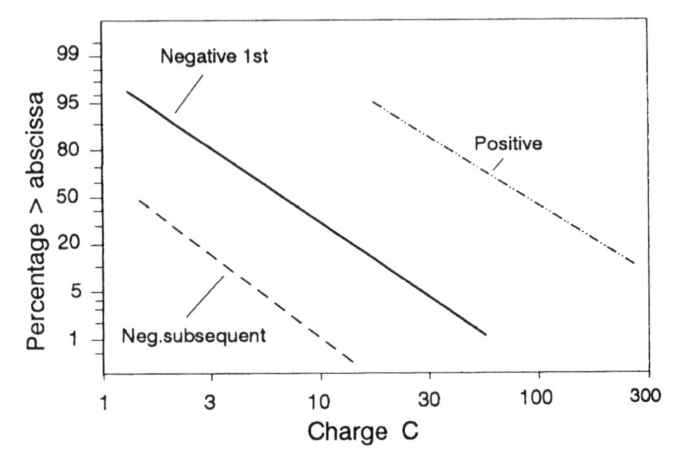

Figure 12.6
Distribution of charge in strokes (Compiled from *Lightning*, Vol. 1, Golde 1977, by permission of Academic Press Ltd., London)

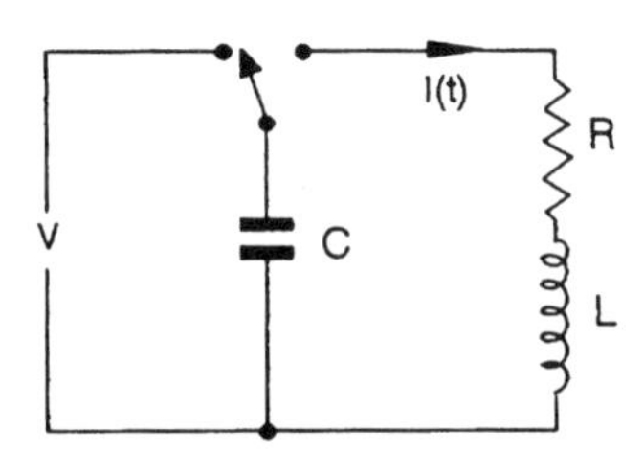

Figure 12.7
Circuit of impulse generator (Reprinted from [31] by permission of ITU)

Equivalent Diagram of Lightning Current Circuit

The double exponential pulse can be generated by the circuit shown in Figure 12.7. The following equations give the time constants

$$\frac{1}{t} = \frac{R}{2L} \pm \sqrt{\frac{R^2}{4L^2} - \frac{1}{LC}} \tag{12.2}$$

where the positive sign is used for t_1 and the negative for t_2, whence

$$C = (t_1 + t_2)/R$$
$$L = Rt_1t_2/(t_1 + t_2) \tag{12.3}$$

The lightning discharge can be represented in a very simplistic form. For the 5/65 impulse, $t_1 = 1.14\,\text{ms}$ and $t_2 = 84.9\,\text{ms}$. For a peak current of 50 kA, the following values are obtained for a lightning channel resistance of $5000\,\Omega$

$I^* = 53.7\,\text{kA}$ $\quad Q = I^*(t_2 - t_1) = 4.5\,\text{C}$

$C = 17.2\,\text{nF}$ $\quad L = 5.63\,\text{mH}$

$V = Q/C = 261\,\text{MV}$

Cloud Discharges

The investigation of in-cloud disharges is even more difficult than for ground strokes. It is generally considered that cloud discharges are between the upper positive charge and regions of denser negative charge lower down in the cloud. The discharge is often inclined to the vertical. The mechanism of the discharge is envisioned as a developing streamer of continuing luminosity, forming branches which also progress as a continuing current streamer. After about 0.25 s the luminosity disappears for 0.01–0.1 s and is followed by a series of short-duration, bright events at intervals of 3–30 ms. These are recoil streamers, following the previously ionized streamer channels and are the analogue of the return stroke in ground flashes. The recoil streamer thus discharges the region of concentrated negative charge.

The heights of the charge centres average 6 km (upper) and 4 km (lower), the charge transferred is 33 C, the streamer current is about 100 A max. and the recoil streamer 1400 A. Mean streamer duration is 250 ms and the total duration is about 500 ms.

Precipitation

One of the earliest observations on thunderstorms, was of the occurence of heavy rain and hail precipitation within about 5 min of a nearby lightning discharge. This can be of the order of 7 mm/h or heavier, after which the rate tails off to about 3 mm/h. This might be associated with acoustic shock waves from the disharge channel, or more probably with the presence of ionized channels remaining in the cloud after the discharge, facilitating the coalescence of small water droplets into heavier rain drops.

Hail is due to the growth in size of ice particles. Generally they are of about 10 mm diameter, but for ice particles which have traversed up and down several times in the strong air currents, multiple growth layers can be identified in hail stones of up to 60 mm diameter.

The rain and hail precipitation also transfers charge to the ground. The sign of the charge is generally the same as the lower regions of the cloud through which it passes. The charge concentrations in precipitation are of the order of 10–1000 nC/m^2 and current densities are of the order of 10–60 nA/m^2.

Lightning Storm Incidence

In the absence of lightning discharges to the earth, it has been calculated that the earth would lose its −500 000 C of charge at the rate of 27% each 10 min owing to the positive ion space charge. Thunderstorms at a rate of 1000 in progress at any one time over the earth's surface replenish the earth's negative charge continuously. Owing to many storms requiring thermals from the earth's surface, the storms are largely concentrated in the sub- and tropical bands, and occuring near midday result in a wave of thunderstorm development moving westward each day.

Since the characteristic 'click-tearing rattle-deep rumble' sound of thunder has no other natural source, the hearing of thunder (generally from up to 10 km distance for humans) defines a 'thunderstorm day'. The incidence of thunderstorms at a given location is defined by the number of thunderstorm days per year. Meteorological offices throughout the world maintain statistics of thunderstorm incidence. A map showing the variation of thunderstorm incidence is called an isokeraunic map. Broadly speaking the isokeraunic levels of the world are as shown in Table 12.1 with the higher values being associated with warm to hot and mountainous regions.

Table 12.1
Incidence of thunderstorm days and ground strokes

Latitudes	T (days/yr)	N_G (strokes/km^2/yr)
90 to 60	0 to 1	0 to 4
60 to 30	1 to 50	1 to 10
30 N to 30 S	30 to 200	2 to 30

Incidence of Ground Strokes

The statistics of ground strikes are expressed as ground strokes per km^2 per year; these vary according to the isokeraunic level of the location. Table 12.1 shows the range of values for some regions of the world. The number of ground strikes is directly proportional to the number of thunderstorm days, and is about 15 per km^2 per 100 thunderstorm days i.e.

$$N_G = 0.15T \text{ per km}^2 \text{ per year} \tag{12.4}$$

where N_G and T are the number of ground strokes and the isokeraunic level respectively.

Striking Range

As a negative leader approaches the ground, the electric field increases until breakdown occurs. A striking range can be inferred from laboratory measurements of rod/plane breakdown, which give an average critical field strength $E = 500\,\text{kV/m}$. For a leader tip potential of V, the striking range can be calculated as $S = V/E$. For $V = 50\,\text{MV}$ therefore $S = 100\,\text{m}$. In fact for short negative impulse breakdown the linear relation does not hold beyond 5–10 m, so that the apparent E falls below the 500 kV/m figure and S can be considerably longer than 100 m.

ATTRACTIVE RANGE OF CABLES

Attractive Range of High Structures

Golde has determined theoretically the attractive ranges of high structures and these seem to agree with field observations. The proportion of strikes likely to hit the structure falls rapidly with distance, so that the attractive range can be estimated to be two to three times the height of the structure. The protective range of a high structure is that of a cone of half-angle 45°, i.e. a radius at ground level equal to the height of the structure. For an earthed aerial wire, the attractive range will probably be less from the smooth wire than from a sharp edged structure. An attractive range of 1.5 to 2 times the height of the line has been suggested.

Attractive Range of Buried Cables

Sunde [32] has analysed the effects at the point of strike on the ground, Figure 12.8, by conductive mechanisms, to give the following. The electric field strength, E V/km, in ground of resistivity $\rho\,\Omega\cdot\text{m}$, at the point of strike is

$$E = \rho J \tag{12.5}$$

where J A/m^2, is the current density in the ground. If the field strength exceeds the breakdown value of the ground E_0, a breakdown hemisphere forms around the point of strike (or in stratified soil, a breakdown disc) of radius

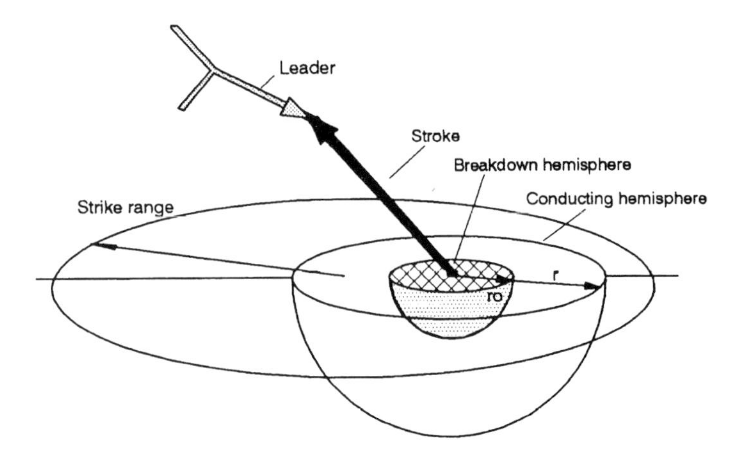

Figure 12.8
Point of strike to ground

for hemisphere

$$r_0 = \sqrt{\frac{I\rho}{2\pi E_0}} \quad \text{and} \quad R_0 = \sqrt{\frac{\rho E_0}{2\pi I}} \tag{12.6}$$

for disc

$$r_0 = \sqrt{\frac{I\rho}{4E_0}} \quad \text{and} \quad R_0 = \sqrt{\frac{\rho E_0}{4I}} \tag{12.7}$$

where R_0 is the resistance of the breakdown zone to remote earth.

For $\rho = 100\,\Omega\cdot\text{m}\ldots E_0 = 250\,\text{kV/m}$ approx. and for $\rho = 1000\,\Omega\cdot\text{m}\ldots E_0 = 500\,\text{kV/m}$ approx. For a current peak of 100 kA, for the two resistivities respectively, one has

for hemisphere

$$r_0 = 2.5\,\text{m and } R_0 = 6.3\,\Omega \quad \text{and} \quad r_0 = 5.6\,\text{m and } R_0 = 28\,\Omega$$

for disc

$$r_0 = 3.1\,\text{m and } R_0 = 7.9\,\Omega \quad \text{and} \quad r_0 = 7.1\,\text{m and } R_0 = 35\,\Omega$$

A sharp drop in potential occurs beyond the breakdown hemisphere or disc, forming a 'potential funnel' with the voltage to remote earth V_0, and the voltage between the breakdown hemisphere and a hemisphere of radius r, V_{0r}, being given by

$$V_0 = IR_0 \quad \text{and} \quad V_{0r} = \frac{I\rho}{2\pi} \times \left(\frac{1}{r_0} - \frac{1}{r}\right) \tag{12.8}$$

Thus in the examples above, V_0 reaches peak values between 630 kV and 3.5 MV. If a conductor such as a metal cable sheath is in contact with the soil and remote earth, and cuts the potential funnel at a radius of r, the potential at that point is reduced to less than 20% of its value in the absence of the conductor. The voltage between the breakdown hemisphere and the cable sheath is

$$V_{0c} = \frac{I\rho}{2\pi} \cdot \frac{1}{r_0} \tag{12.9}$$

However the presence of the conductor can increase the length r_0, and a conducting zone may also form around the conductor. Taking these effects into account, the arcing distance r, to the conductor is given by Sunde's analysis as

$$\text{for } \rho = 100\,\Omega\cdot\text{m} \quad r = 0.056\sqrt{I\rho}\,\text{m}$$

$$\text{for } \rho = 1000\,\Omega\cdot\text{m} \quad r = 0.04\sqrt{I\rho}\,\text{m} \tag{12.10}$$

However, an alternative analysis based on electrostatics given by Coleman [33] indicates that the arcing distance may be independent of earth resistivities less than 10 000 $\Omega \cdot$ m, leading to an arcing distance of

$$r = 1.5\sqrt{I}\,\text{m} \tag{12.11}$$

for a crest current I in kA. The arcing distance is interpreted as the attractive range of buried bare metal cable sheaths.

If the sheath is protected by a plastic covering, which may have a breakdown voltage of the order of 100 kV, the potential of the soil in the 'funnel' is not changed, but the plastic sheath will break down when the cable is at a distance from the breakdown hemisphere where the potential relative to remote earth is greater than 100 kV. From Equations (12.6) and (12.8) the critical radius for a sheath breakdown voltage of V_b is

$$r_b = \frac{I\rho}{2\pi V_b}\,\text{m} \tag{12.12}$$

where V_b is in kV, I is in kA and the earth resistivity is in $\Omega \cdot$ m. This may be considered the 'attractive range' of a buried insulated conductor and is shown in Figure 12.9 for $V_b = 100$ kV.

LIGHTNING EFFECTS

Lightning Effects on Aerial Lines

Even in the absence of strikes to ground, the changing electric field at the ground as an electrified cloud approaches can electrostatically induce a voltage in an aerial line at a height above the ground of h m, of

$$V = h\frac{C}{G} \cdot \frac{dE}{dt} \tag{12.13}$$

For a 3 km line, 6 m above ground, the capacitance is 7 nF/km and the leakance 0.3 nS/km. If a cumulus cloud is approaching at 18 k.p.h. (5 m/s) the rate of change of

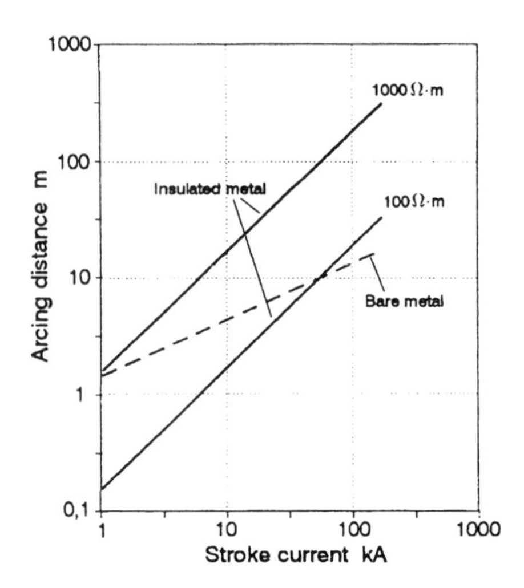

Figure 12.9
Attractive range of buried cable (Reprinted from [31] by permission of ITU)

the electric field can be 15 kV/m in the time taken to travel say 1 km, i.e. 200 s, so that $dE/dt = 0.075$ kV/m/s. Thus $V = 10.5$ kV if the line is isolated from earth. However the current flowing if one end of the line is earthed will be small, of the order of 2 mA, owing to the low energy and long time involved (1 J in 200 s = 5 mA).

Nearby cloud or air discharges, particularly where there is appreciable parallelism, can induce high voltages on the line, of the order of 10 kV/km for earth resistivities of 1000 Ω · m.

Lightning strokes to ground more than 30 km away from the line can induce peak voltages of the order of tens of volts and may cause some noise in unbalanced circuits. Ground strokes at 3–5 km from the line will normally induce less than a kilovolt in lines of appreciable length but again the currents to earth are relatively small since the surge impedance of a single line is about 500 Ω.

Ground strokes beyond the attractive range out to about 3 km can induce large currents in the line. From a point on the line opposite the strike point, a surge is propagated in both directions and repeated reflections from both ends modify both the size and duration of the current surges as shown in Figure 12.10 for a strike current peak of 150 kA (5/65) in very low resistivity ground. The surge magnitude is roughly proportional to the strike peak current. When there is more than a single wire, the current in each wire reduces approximately as the square root of the number of wires.

Where the earth resistivity is higher, the potential of the ground at radial distances from the strike point is given by Equation (12.12) and is shown in Figure 12.11. For lines whose ends are situated in the potential funnel, at different distances from the strike point, much larger voltage and current surges are induced as shown in Figure 12.12 (for 1000 Ω · m resistivity). If the voltage surge peak rises above 30–40 kV, flashover occurs between the line and the support structures owing to the small clearances, and the surge is dissipated to earth within the length of a few supports. The same happens for ground strikes within the attractive range of the line and its supports (about 3 × height of support) when the strike will most probably hit the supports and the

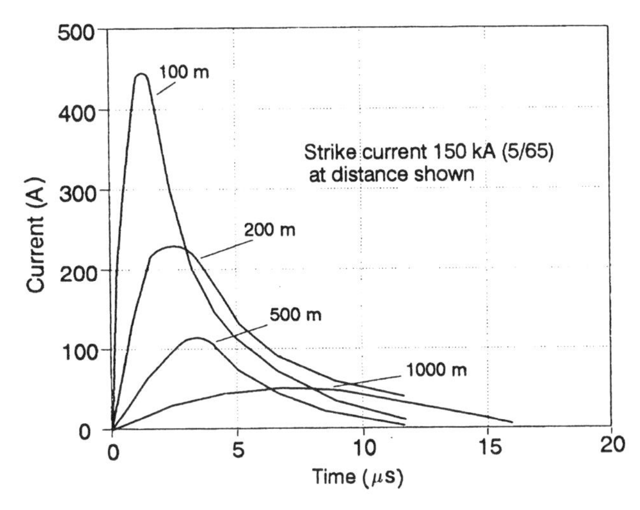

Figure 12.10
Induced current in single wire (Reprinted from [31] by permission of ITU)

stroke current will divide between the two directions of the line. There is some evidence that short lines experience higher currents than long lines owing to end reflections reducing the surge impedance.

The same considerations apply to aerial cables supported on a suspension wire, either by lashing or in a figure-of-8 sheath. Induced currents from nearby strikes will flow in both directions in the suspension wire and the cable conductors in proportion to their surge impedances. Direct strikes to supporting structures will break down to the suspension wire at the clamps and the stroke current will divide into both directions.

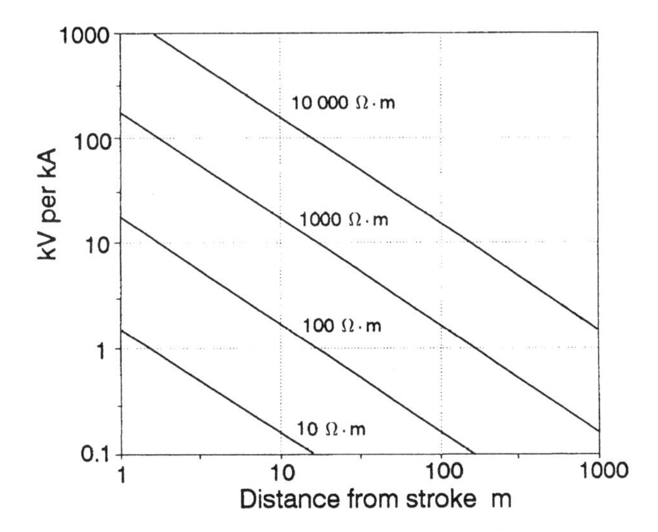

Figure 12.11
Potential in ground near strike point (Reprinted from [31] by permission of ITU)

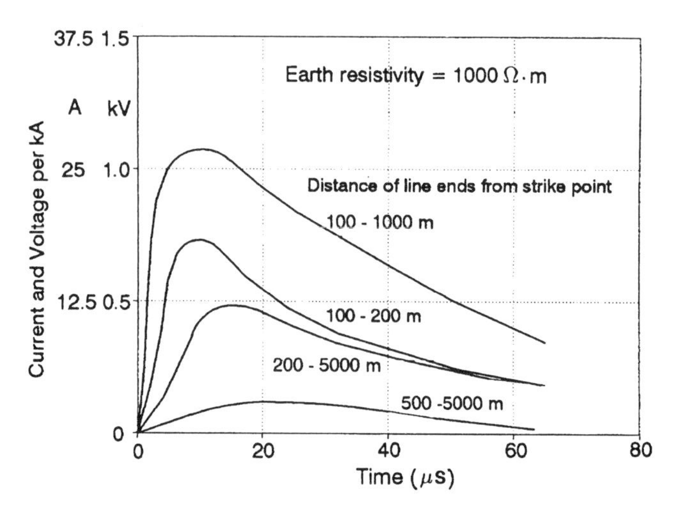

Figure 12.12
Voltage and current surges due to earth gradient (Reprinted from [31] by permission of ITU)

Lightning Effects on Buried Cable

The attractive ranges for buried cable are given in Figure 12.9 and it will be seen that the arcing distance to a bare metal sheath is considerably less than for an insulated sheath in high resistivity ground even for the average stroke current of 30 kA. When breakdown to the metallic sheath occurs the stroke current divides between the resistance of the breakdown hemisphere and the sheath resistance to ground. Since the latter is usually much lower than the former, most of the stroke current flows into the sheath, dividing into the two directions. These surge currents travel considerable distances before being attenuated.

The current surges travelling in the metal sheath appear as impulse voltages between the inside of the metal sheath and the cable conductors which are statistically at remote earth potential. If the voltage surge is large enough it will cause breakdown to the conductors.

The magnitude of the voltage surge is proportional to the sheath transfer impedance, the stroke current and the earth resistivity. Sunde gives this surge voltage as

$$V_p = 7.4 \cdot I_p \cdot Z_T \cdot \sqrt{\rho}\,\mathrm{V} \tag{12.14}$$

where I_p, Z_T and ρ are in kA, Ω/km and Ω · m respectively. This is plotted in Figure 12.13. If the sheath breaks down to the conductors, part of the surge current will flow in the conductor/sheath circuit whose attenuation is much less than the sheath/earth circuit. Hence the voltage between the conductors and sheath increases with distance from the strike point and further breakdown damage to the insulation can occur several kilometres away.

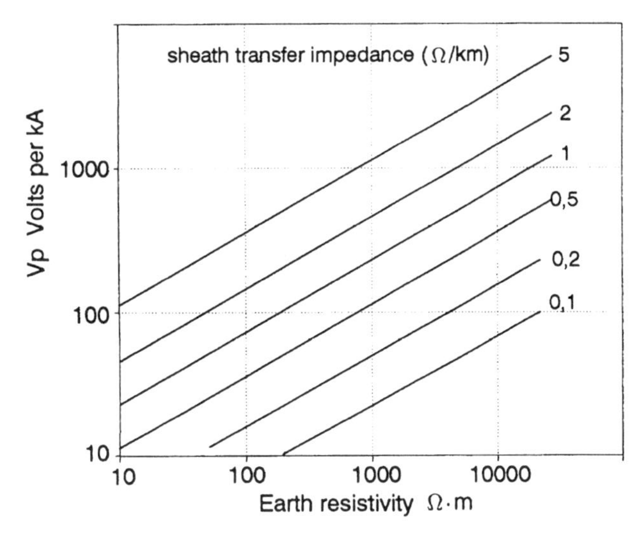

Figure 12.13
Peak induced voltage between sheath and conductors at strike point (Reprinted from [31] by permission of ITU)

The transfer impedance of the sheath can be calculated from Equation (10.24) for cylindrical metal sheaths. A sufficient approximation is to take the effective frequency of the surge as

$$f = \frac{1}{4 \cdot t_h} \tag{12.15}$$

where t_h is the time in seconds to half amplitude of the surge. The effective frequency is thus about 0.5 kHz to 3.5 kHz depending on the shape of the pulse. Hence for a non-magnetic metal sheath the transfer impedance is equal to its d.c. resistance.

The breakdown strength of paper insulation is of the order of 2 kV and for plastic insulation of the order of 10 kV (see Chapter 2). Thus the cable itself can act as protection for terminal equipment. However the resulting pinholes in the insulation are undesirable from the point of view of water penetration.

Other Damage Mechanisms of Lightning Strokes [31]

It is rare for extensive melting of metal sheaths to occur even at the point of impact. The thermal energy at the point of impact is

$$W = v \int I(t) \cdot dt = v \cdot Q \tag{12.16}$$

where v is the voltage drop where the arc contacts the metal and is about 50 V. Hence for a stroke with a charge of 30 C the energy is 1500 J or about 360 cal. To melt m g of a material with a specific heat of s cal/g and a latent heat of fusion of F cal/g requires

$$W = m(s \cdot \Delta T + F) \text{ cal} \tag{12.17}$$

For iron, $s = 0.113$, $F = 49$ and $\Delta T = 1530\text{–}20 = 1510\,°\mathrm{C}$ thus $m = 1.64\,\mathrm{g}$ or about 0.208 ml. This would be the volume of a sphere of diameter 7.5 mm For lead, $s = 0.0305$, $F = 6$ and $\Delta T = 327\text{–}20 = 307\,°\mathrm{C}$, thus $m = 23.4\,\mathrm{g}$ or 2.07 ml or a bead diameter of 15.8 mm. Similarly for aluminium the bead diameter would be 10.3 mm. For both lead and aluminium therefore, a charge of even 1 C would puncture a sheath of these materials.

A stroke of charge 30 C would raise the temperature of the PVC around a suspension wire (including the web of a figure-of-8 sheath) to the softening point over a distance of about 100 mm thus possibly allowing the cable to tear away from its suspension for a short distance. The total heat energy liberated in the arc and the conductor is proportional to $\int i^2 \cdot dt$ which on average has a value of $10^5\,\mathrm{A^2 \cdot s}$ and a maximum of about $5 \times 10^6\,\mathrm{A^2 \cdot s}$.

A 7/1.60 mm steel suspension strand has an area of $14\,\mathrm{mm^2}$, a mass of 108 kg/km, a resistance of $10.2\,\Omega$/km and a specific heat of 0.113 cal/g. The heat energy liberated in L km is

$$W = \frac{\int i^2 \cdot dt \times R \times L}{4 \cdot 2}\ \mathrm{cal} = m \times L \times s \times \Delta T$$

$$\Delta T = \frac{W}{mLs} = \frac{\int i^2 \cdot dt \times R}{4.2 \times m \times s} \tag{12.18}$$

Thus for an average stroke (say, $I = 32\,\mathrm{kA}$), the temperature rise in the suspension strand is 20 °C. For five times the energy however ($I = 70\,\mathrm{kA}$) the temperature would rise above the softening point of its figure-of-8, PVC covering, allowing separation of the cable over a considerable distance. For the maximum value of the surge energy, the temperature rise would exceed the melting point of the steel. To reach the melting point requires a surge of $7.6 \times 10^6\,\mathrm{A^2\,s}$ corresponding to a current of about 279 kA. This does not include the latent heat of fusion; including $F = 49$ gives a current of about 316 kA for fusion. As the stroke current divides into approximately equal amounts into both directions, this is a rather remote probability.

An empirical formula [31] which gives the fusing current of a conductor, is

$$I = \frac{S \times A}{\sqrt{t_h}} \tag{12.19}$$

where I is in amps, A is the cross-sectional area in $\mathrm{mm^2}$, the time to half-value is in seconds and S is an empirical constant for the material, typical values being

Copper:	310	Aluminium:	170	Iron:	78
Bronze:	240			Lead:	24

Thus for the 7/1.60 steel suspension strand, the fusing current would be about 135 kA (5/65) which is not in agreement with the earlier value calculated (the two methods of calculation are in adequate agreement if the mean value of $\int^2 \cdot dt = 10^5$ corresponds to the mean stroke current of 16 kA applicable to aerial transmission lines).

Even metal sheathed buried cables have been observed to be considerably crushed by nearby strokes. This is compatible with the compression caused by the explosive vaporization of moisture around the cables when the very short duration lightning current discharges to earth. Similarly non-conducting materials such as wooden poles, beams, trees and walls crack with explosive violence when struck, owing to the vaporization of moisture, resins etc. in the materials.

PROTECTIVE MEASURES

For aerial lines and cables an effective method of protection is to run an earthed wire of between 2.5 mm and 3 mm diameter at about 0.5 m above the lines or cable. Also it is desirable to run an earthing conductor down the support structures, particularly if these are wooden poles. This will at worst divide the stroke current by a factor of four, and increase the rate of dissipation of surges to earth. For induced surges it will reduce the magnitude by a factor of about $\sqrt{2}$.

High structures such as those in urban areas will also provide a great deal of protection to aerial lines, aerial cables and buried cables. The same is true for extensive plantations of trees, but isolated trees near the cable route can be a considerable danger owing to stroke currents being led to the cable via the root system.

For buried cables in isolated ground, the provision of two shield wires of 3 mm diameter copper, spaced 0.3 m apart and 0.3 m above the cable can be very effective in limiting damage to the cable. This is due to the screening effect of the earth wires (see Chapter 9 Equation (9.6)) and to the reduction in earth potential gradient near the cable. If there is no arcing between the shield wires and the cable screen the cable is completely protected. For this reason the cable screen should not be connected to remote earth. If arcing does occur, which is probable for stroke currents of the order of 15 kA, the puncturing of the cable sheath connects its screen to local earth potential and some current will flow in the screen. At worst, the stroke current will divide equally between the screen wires and the cable screen so that the voltage surge, screen to core, will be one third of that without the screen wires. In practice, Boyce [30] reports that fault incidence can be reduced by two orders of magnitude.

For a small polythene insulated, foil-screened and polythene sheathed junction cable buried with screen wires in earth resistivity of 1000 to 5000 $\Omega \cdot$ m in an isolated area with 7 ground flashes per km^2 per year, Boyce reports that only four proven lightning faults occured during an exposure of 750 km $\cdot$ years. This can be compared with an expected incidence of about 750 km $\cdot$ year $\times$ 0.2 km $\times$ 7 = 1050 ground strokes within the attractive range, of which 30% (315) might be expected to cause faults. This is a reduction factor of over 100.

The ability of a cable to withstand lightning damage can be improved in two ways. Firstly by reducing the transfer impedance of the cable screen, e.g. by reducing its d.c. resistance and by increasing its self-inductance by using lappings of steel tapes. The transfer impedance becomes

$$Z_T = |k| \times R_{dc} \tag{12.20}$$

where $|k|$ is given by Equation (9.14). Secondly the voltage withstand between the cable screen and the core can be increased by suitable extra insulation between conductors and screen. If the impulse breakdown voltage is V_b, a quality factor can be defined for the cable as

$$Q = \frac{V_b(\text{kV})}{Z_T(\Omega/\text{km})}\ \text{kA} \cdot \text{km} \tag{12.21}$$

Thus an APL sheathed plastic insulated cable would have a Q of about 20 and an aluminium sheathed plastic insulated cable would have Q of about 150. The required

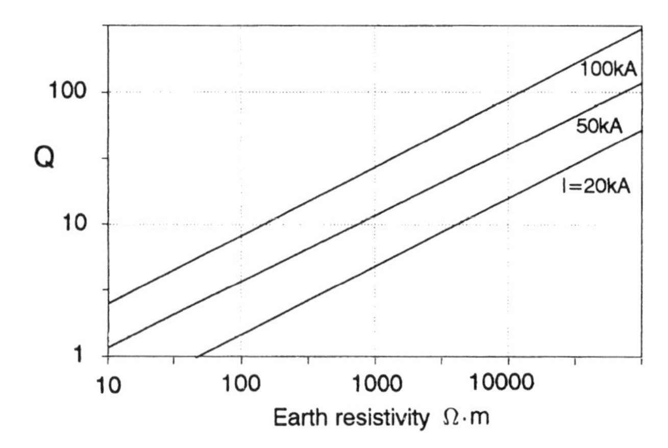

Figure 12.14
Required cable quality factor (Reprinted from [31] by permission of ITU)

quality factor for various strike currents and earth resistivities can be calculated from Equations (12.14) and (12.21) as

$$Q = \frac{V_p}{Z_T} = 7.4 \times 10^{-3} \cdot I_p \sqrt{\rho}\,\text{kA} \cdot \text{km} \tag{12.22}$$

and is shown in Figure 12.14. For the highest protection, the transfer impedance can be lowered by the use of steel tape lappings (with k of the order of 0.1) or by installing the cables in iron pipes. The screens must be well bonded across the joints in all cases. Such expensive methods are only rarely economically justifiable.

The puncturing of plastic oversheaths in the potential funnel leaves the metal sheaths open to corrosion. A method of combating this is to use semi-conducting polythene oversheaths. These have a fairly high resistance but will reduce the voltage between the ground and the metallic sheath sufficiently to prevent voltage breakdown and hence puncturing.

Protecting Equipment and Subscribers

For cable routes that do not carry d.c. currents, i.e. other than subscriber distribution networks, it is possible to isolate the conductors from local exchange earths by the use of transformers having an adequate voltage rating. Good results have also been experienced by isolating cable screens from earth, and in the case of buried-repeater systems by isolating the repeaters from earth also (completely floating systems to levels of 50 kV).

For distribution networks where direct current must be carried, protection relies on lightning arresters. These are of many forms including spark-gaps, gas-tube protectors, varactor devices and Zener diodes. The requirements of fast operation, high current carrying capacity at low voltage, low capacitance and reliable repeated operation are not always available in one type of device and multiple protection by different devices may be necessary. (Primary protectors are generally designed to limit longitudinal surges to less than 1 kV and transverse surges to less than 150 V.) This is particularly so in the case of semiconductor equipment which generally has Zener diode protection built in. Protective devices are connected between each conductor and an adequate earthing system (see Chapter 9).

PROBABILITIES OF DAMAGE AND SURGES

Estimating Probable Incidence of Lightning Damage on a Cable Route

For a route length of L km and an attractive range of r km the attractive area is $2L \cdot r\,\text{km}^2$ and the number of ground strokes in this area from Equation (12.4) is

$$N_g = 0.15T \cdot 2L \cdot r \text{ per year}$$

Let the probability of a stroke current exceeding I_p kA be p, which can be determined from Figure 12.5. The voltage surge to be expected between screen and conductors, from Equation (12.14), is

$$V_p = 7.4 \times 10^{-3} I_p Z_T \sqrt{\rho}$$

This will exceed the breakdown voltage of the insulation, Equation (2.2)

$$V_b = \frac{E_b}{2} d \cdot \cosh^{-1}\left(\frac{D}{d}\right)$$

when

$$I_p \geq \frac{V_b}{0.0074 Z_T \sqrt{\rho}} \tag{12.23}$$

The probability of this per ground stroke is p, or per year is

$$P = p \cdot N_g = 0.3pT \cdot L \cdot r \tag{12.24}$$

For a ground resistivity of $1000\,\Omega \cdot \text{m}$ an insulation impulse breakdown voltage of 20 kV, and an APL sheathed cable with $Z_T = 1\,\Omega/\text{km}$, $I_p = 86$ kA and $p = 3\%$. Thus for a route length of 10 km with $r = 0.08$ km (Figure 12.9),

$$P = 0.3 \times 0.03 \times T \times 1.6 = 0.014T$$

For an isokeraunic level of $T = 46$ thunder days per year, the lightning damage incidence is thus 0.66 times per year, or once in 1.5 years per 10 km of buried cable.

Optical Fibre Systems

Although optical fibre cables are not subject to induced or conductive surges in the fibres themselves, the cables may still be damaged by nearby lightning strokes, depending on their construction. Cables relying on metallic strength members are very susceptible to lightning damage like any other metallic conductor, and can be severely damaged to the extent of also breaking the fibres. Even non-metallic cables can be damaged by close discharges owing to thermal effects in the ground. Aerial non-metallic fibre cables can also be damaged by strikes to supports or metal cable clamps.

Surveys of Distribution Cable Surges

So far consideration has been limited to point-to-point cable routes. Distribution cables however are typified by many spurs and drops to subscribers. The result of this is that the core of the distribution cable does not maintain the same potential on all pairs in the event of a lightning surge. Thus it is not feasible to improve Q in Equation (12.21) by increasing V_b using an extra sheath of polythene beneath the screen. This was confirmed by a survey carried out by Bell Telephone Laboratories in Munson, Fa. [34] on a 14.5 km buried cable route with 13 simulated drops (both aerial and buried), which indicated that the distributions of surges measured from pair-to-pair and pair-to-screen showed no significant differences.

The cable was a 50-pair/0.63 mm solid polythene insulated filled construction with an Alpeth sheath. The measured earth resistivity on the route was 3350 Ω · m. The isokeraunic level of the region in 1977 was reported to be 93. Since the surge voltages are proportional to the square root of the ground resistivity the lightning hazard of the area may be expressed as $T\sqrt{\rho} = 93 \times \sqrt{3350} = 5383$, stated to be about the highest in USA. The investigators recorded 22 surge days in 152 days and interpreted these as thunderstorm days. The number of ground strokes would thus have been 3.3 per km^2 over this period (Equation (12.4)).

The results indicated a small number of surges up to 15 kV but the majority of surges were below 5 kV. There was one direct strike to the route which was recorded as a surge greater than 17 kV. The surges were measured at the mid-point of the route and the author estimates that the average surge attenuation could be 7 dB. Thus at the

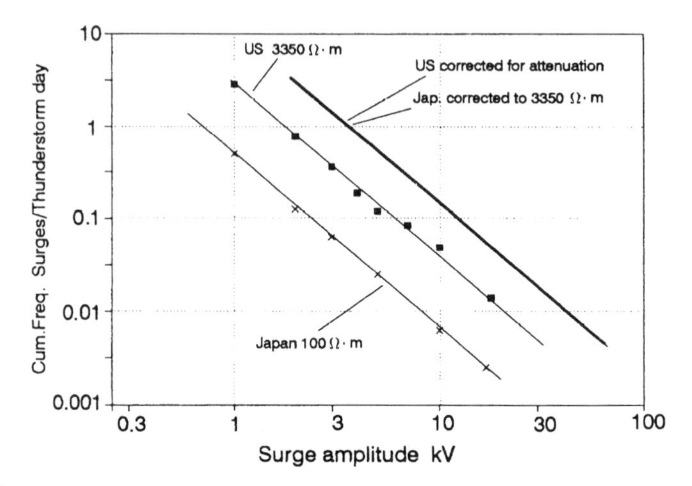

Figure 12.15
Comparison of measured surges on distribution cables

closest point to the strike the actual induced voltages could be approximately twice those recorded at the mid-point. The author has demonstrated, at least to his own satisfaction, that these results are in accord with the theory.

A five-year survey was reported on similar aerial cables of average length 1 km, at 7 locations in Japan [35]. The earth resistivity was 100 Ω · m average. 72 thunderstorm days were recorded over this period. The overall results expressed as a cumulative frequency of surges per thunderstorm day gave a good fit to a straight line on a log–log display, with a slope of approximately −2 as shown in Figure 12.15. The Munson results, when similarly expressed, are also found to fit a straight line of the same slope.

When the Japanese results are corrected by the square root of resistivity to 3350 Ω · m, and the Munson results are corrected for the average surge attenuation of 7 dB, both sets of results fall on the same line as shown. This is somewhat surprising since the aerial cable would be expected to show higher induced voltages than the buried cable. This could be due to the screening effect of the suspension strand which is estimated to be 0.7 (from Equation (9.10)). This would increase the measured surge voltages by 40% in the absence of the suspension strand.

The measurements in these surveys can be expressed by the empirical formula

$$\text{Cumulative frequency} = 0.00062\left(\frac{\sqrt{\rho}}{V_p}\right)^{1.9} \text{ surges per thunderstorm day per km} \tag{12.25}$$

where V_p in kV is the peak surge voltage opposite the point of strike (i.e. without attenuation) and is the required impulse voltage withstand of the conductor insulation.

13

Electrical Measurements

MEASUREMENT PRINCIPLES

Measurements on cables have four main objectives, namely that they

Confirm suitability for purpose

Confirm the design

Assure the user of consistency of manufacture

Provide information for system design

The suitability for purpose is generally assessed in relation to a specification drawn up in consultation between the cable designer/manufacturer and the system designer or user. The values of the parameters covered by the specification and the tolerances allowable should be realistic to the requirements and not just the best achievable, in order that an economic design and manufacture can be used.

The accuracy of the measurements should be sufficient to assure the user that his required tolerances are not being exceeded, which generally means that the measurement accuracy should be about a tenth of the tolerance, so that for a 10% tolerance an accuracy of 1% is desirable.

Since most parameters are specified in terms of cable length, this means that the cable length must also be known to this accuracy, which cannot always be assumed from normal manufacturing methods. In addition, when confirming the design or manufacture, measurements of conductor diameter and insulation diameter must also be accurate to better than the electrical measurement accuracy. This requires micrometer accuracy.

Design methods are based on straight parallel conductors, hence the pair or quad, stranding and lay-up twist lengths must be allowed for in designing for a given cable length, i.e. the product of the various take-up factors (Equation (3.1)) must be evaluated and the cable length, multiplied by this overall take-up factor, is the conductor length to be used in design. In other words, all unit length design parameters will appear larger when measured on the cable and divided by the cable length.

Other mechanical parameters of the cable are also of concern to the user, such as diameter, tensile strength, peel strength of foil laminates, and the bend, crush, impact and abrasion performance. Most of these are assessed by type tests at the design proving stage or are performed on samples taken periodically from manufacture. However cable diameter is measured on all cables and should be the mean of a number of measurements at different orientations to allow for non-circularity. For cables larger than about 15 mm, it is possible to use a diameter tape to measure mean cable diameter. This tape has graduations which are π times the tape length so that the average diameter can be read straight from the tape when wrapped around the cable circumference.

Since high-frequency measurements covering all frequencies in the band of use specified are time-consuming and require a higher degree of skill from the operators, once the design has been confirmed by detailed measurements on early production it is often possible to reduce the measurements to low-frequency tests. For example, if the conductor diameter is accurately controlled, measurements at low frequency of capacitance, the d.c. resistance and insulation resistance of the conductors will enable production control to be exercised on conductivity, insulation diameter, dielectric properties, the primary parameters, impedance, attenuation and phase at all moderate frequencies. At higher frequencies where the cable is several wavelengths in length, irregularities, especially if periodic, can affect primarily the impedance, so that it becomes necessary to perform point-by-point or swept-frequency measurements on all production.

Crosstalk for voice frequency working is generally specified in terms of allowable capacitance unbalances, since the electric and magnetic couplings between pairs, either separately or within quads, are statistically related to this unbalance. Similarly for common-mode or pair-to-earth couplings, the conductor-to-earth capacitance unbalance is specified. But for higher frequencies, again where the cable length is several wavelengths long, it is necessary to make direct, high frequency measurements of crosstalk, generally at the highest frequency of use. On large cables, since all pair–pair combinations must be measured, both capacitance unbalance and crosstalk measurements involve large numbers of measurements and automated measuring instruments are desirable.

All electrical measurements on cables involve either voltage and current measurements using suitable instruments, or voltage ratio measurements using Wheatstone-type bridges or attenuator comparison measurements. When the parameter to be measured is complex, such as for example an impedance or propagation constant, it is necessary that these measurements should provide both magnitude and phase information.

MEASUREMENT OF IMPEDANCE

The methods used to measure impedance are illustrated in Figure 13.1. In (a) is shown the basic definition of impedance. This method is used for pure resistive impedances when V and I are measured in magnitude only. When the resistance to be measured is very low, as in the measurement of 1 m samples of wire or metal sheath, the four terminal connection shown is necessary. Also the current must be limited to avoid heating the resistance under test. If the phase of V relative to I can be measured then

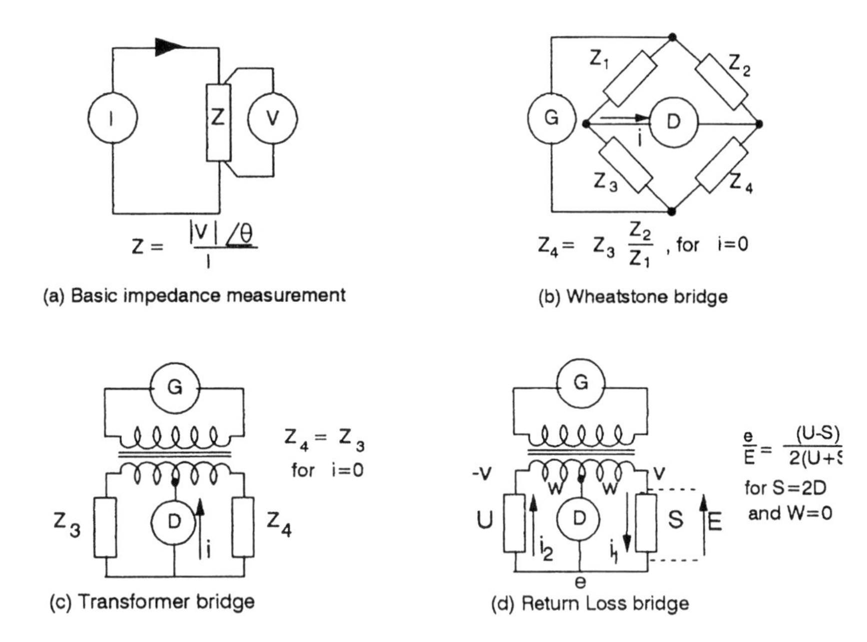

Figure 13.1
Impedance measurement

general complex impedances can also be measured in this way. The accuracy of measurement depends on the accuracy of the voltage, current and phase measurements. The voltmeter used in this sort of measurement must be of much higher impedance than the impedance being measured. Thus in measuring very high resistances, such as insulation resistance, an electrometer-type of instrument is required, and to increase sensitivity a higher voltage is used, usually 500 V.

Traditionally, the measurement of the phase of the voltage relative to the current was difficult and the bridge method was developed to compare the unknown impedance with a standard impedance. This is shown in Figure 13.1(b). The impedances Z_1 and Z_2 are known as the ratio arms, Z_4 represents the unknown impedance and Z_3 is the standard. For fixed ratio arms the standard impedance is varied until balance is obtained, i.e. for $i = 0$. Then the equation shown in (b) gives the value of the unknown. If the ratio arm impedances are equal in magnitude and phase, then at balance the unknown equals the standard in magnitude and phase.

By constructing the ratio arm impedances in exactly the same way it is possible to make both the resistive and reactive components equal to high accuracy over a wide frequency range. However owing to inevitable stray impedances being present, it is necessary to include zeroing components in the bridge which can be used to initially balance the bridge under standard conditions, e.g. with both Z_3 and Z_4 set to zero or open circuited, or to equal value. In general this needs to be done at each frequency of measurement.

The standard impedance needs to be variable over a range to suit the unknown values to be measured. The standards are usually made from decades of fixed pure resistances paralleled by decades of fixed pure capacitors. Both the resitive and capacitive decades need to have a continuously variable element to determine the exact value. By making this a small portion of the whole, the required accuracy of the variable can be reasonable.

The many variants of bridge circuits are due to the choice of the arrangement of standards to achieve the highest accuracy for small, medium or large phase angles of the unkown impedance and also to achieve an arrangement where the balance condition is independent of frequency.

It is difficult to construct wide-range standards up to high frequencies and use is made of varying the ratio arms in decade steps to increase the range of the bridge. The best bridges made in this way (using a lot of double-screening of components to stabilize stray elements) can achieve accuracies of 1% up to 1.6 MHz for both balanced and unbalanced unknown impedances. For balanced impedances the two bridge terminals connected to the unknown should have high and equal impedances to earth which generally requires that both the generator or detector must be transformer-coupled to the bridge.

By using bifilar winding of the secondary of the generator transformer, the two very equal halves of this winding can be used in the transformer ratio arm bridge shown in Figure 13.1(c). The voltages at the terminals of this winding are exactly equal because they depend only on the number of turns, which must be integral. Since the stray components are also very equal, such bridges can be constructed to operate accurately up to 100 MHz or more. There seems to no reason why such bridges should not also be constructed for balanced impedance measurement, using two bifilar secondary windings. To use an unbalanced bridge to measure a balanced impedance a suitable balanced-to-unbalanced transformer, or 'balun' is required. By using decade taps, or 10 : 1 autotransformers, the ratio of the voltages can be changed (changing the strays in the same ratio) to extend the range of the standard impedances.

Return Loss Bridge

In either bridge circuit, if the unknown impedance is not too different from the standard impedance (say within ±25%), the unbalanced detector current can be calibrated in terms of the unknown impedance. In the case of the transformer bridge this leads directly to the configuration of Figure 13.1(d) which is called a return loss bridge.

If the standard impedance, S, and the detector impedance, D, are purely resistive, the secondary winding impedances are each W, and the unkown is denoted by U, then the mesh equations are

$$V = i_1(S + W) + (i_1 - i_2)D \quad \text{and} \quad -V = -i_2(U + W) + (i_1 - i_2)D$$

from which the voltage across the detector can be shown to be

$$e = (i_1 - i_2)D = \frac{i_1 D}{(U + W + 2D)}(U - S)$$

If this is compared with the voltage across the standard resistance, $E = i_1 S$, then

$$\frac{e}{E} = \frac{D(U - S)}{S(U + W + 2D)}$$

If now $D = S/2$ and W is negligibly small, i.e. if the generator G is a constant voltage source of very low impedance,

$$\frac{e}{E} = \frac{(U - S)}{2(U + S)} = \frac{\mathrm{r}}{2} \tag{13.1}$$

where r is the return loss of U with respect to S. If this ratio is measured in magnitude and phase and S is purely resistive, then U can be determined in magnitude and phase.

$$\mathrm{U} = S \cdot \frac{(1 + \mathrm{r})}{(1 - \mathrm{r})} \tag{13.2}$$

$$= S \cdot \left(\frac{1 + r\cos\theta + jr\sin\theta}{1 - r\cos\theta - jr\sin\theta} \right)$$

$$|U| = S \cdot \sqrt{\frac{1 + 2r\cos\theta + r^2}{1 - 2r\cos\theta + r^2}} \tag{13.3}$$

$$\angle U = \frac{1}{2}\left[\tan^{-1}\left(\frac{r\sin\theta}{1 + r\cos\theta}\right) - \tan^{-1}\left(\frac{-r\sin\theta}{1 - r\cos\theta}\right)\right] \tag{13.4}$$

The common practice of measuring only the magnitude of r and using Equation (13.2) in a scalar fashion results in an error in $|U|$ which increases as r increases. For $r < 0.1$ and $\theta \leq 25°$ the error is less than about 2%. Fortunately for terminated cables this is true above about 500 kHz if S is within 25% of U.

The transformer can be regarded as a directional coupler between S and U with a loss of 6 dB (Equation 13.1). Such directional couplers generally achieve a directivity of better than −40 dB (2% deviation of impedance).

Impedance Standards

The standard should preferably be of the same kind as the unknown. For a largely capacitive unknown a standard capacitor should be used. There is little difficulty in making a pure capacitor with a very low loss angle. For a largely inductive unknown however it is very difficult to make a pure inductance standard owing to the resistance associated with a copper winding. This can be overcome in some cases by using a mutual inductance as a standard or by using a Maxwell bridge variant of the Wheatstone bridge. In this variation, the standard in the form of parallel-connected R and C standards, Z_3, is interchanged with the resistive ratio arm Z_1, in Figure 13.1(b). Then at balance

$$R_u = \frac{Z_2 Z_1}{R_s} \quad \text{and} \quad L_u = Z_2 Z_1 \cdot C_s \tag{13.5}$$

Stable resistance standards were traditionally made by winding thin resistance wire in various sectionalized bifilar forms to minimize the associated inductance and capacitance strays. With the advent of metal or metal oxide films deposited on stable ceramic formers, it became possible to make stable resistive standards which were almost purely resistive up to very high frequencies.

Consider the cylindrical film resistor in Figure 13.2 deposited on a ceramic former with plated copper ends for connection. The resistance will be

$$R = \frac{l \cdot \rho}{\pi d \cdot t}\ \Omega \tag{13.6}$$

where the film length, diameter and thickness are l, d and t in m and the resistivity ρ is in $\Omega \cdot$ m. The inductance associated with the film is given by

$$L = 0.2\left(\ln \frac{4l}{d} + \frac{d}{2l} + \ln \zeta - 1\right)\ \mu\text{H/m} \tag{13.7}$$

where $\ln \zeta = 0$ for the very thin film. The capacitance between the edges of the copper film connectors is estimated from

$$C = \frac{\varepsilon \varepsilon_0 A}{l} = \varepsilon \times 8.9 \times \frac{A}{l}\ \text{pF} \tag{13.8}$$

where A is the area of the annular edge of the copper films in m^2 and ε is the permittivity of the ceramic. The resistance, inductance and capacitance form a parallel resonant circuit with an admittance given by

$$Y = \frac{1}{Z} = \frac{R}{R^2 + \omega^2 L^2} + j\omega\left(C - \frac{L}{R^2 + \omega^2 L^2}\right) \tag{13.9}$$

which resonates at a frequency which makes the imaginary part of this impedance zero, i.e. at

$$f_0 = \frac{1}{2\pi}\sqrt{\frac{1}{LC} - \frac{R^2}{L^2}} \tag{13.10}$$

However, if C is augmented, the imaginary part of the impedance can be controlled to be negative at all frequencies, reducing the rise in value of the real part.

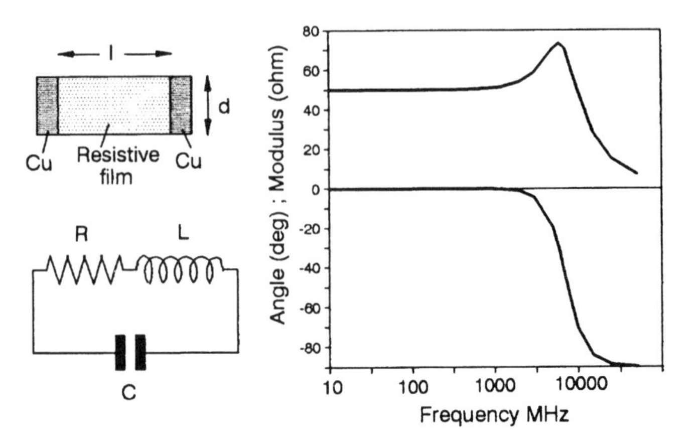

Figure 13.2
Resistive cylindrical film as a standard resistance

To estimate the possibilities, let the length of the cylindrical film be 3 mm with 1mm copper connections and the diameter be 2 mm. The thickness of the film for a 50 Ω resistance would be 9.5 μm at a resistivity of 1 μΩ · m. The inductance and capacitance are calculated to be 1.1 nH and 0.0014 pF. The resonant frequency would be 126 GHz, but if the stray capacitance of connecting the resistor to the circuit is controlled to be 0.45 pF, then the impedance will be as shown in Figure 13.2. The impedance rises by 1% up to 700 MHz and 2% at 1000 MHz with the angle remaining less than 0.2°. Although this requires some well-controlled micro-engineering, it shows what can be achieved, particularly with printed circuit techniques using linear resistive film deposition.

Swept-Frequency Impedance Measurements

The advantage of measuring impedances via the return loss to a fixed standard is firstly that the precision required is concentrated in two or three components and secondly the saving in time involved, particularly when looking for rapid variations in the impedance, due for example, to periodic reflections in the length of a cable. The disadvantages are that if the modulus and angle of the impedance are required then the return loss needs also needs to be measured as a vector at each frequency and Equations (13.3) and (13.4) must be solved. With a built-in microprocessor this can be done in real time. Another disadvantage is that the range of the unknown impedance should not deviate more than about 25% from the standard, if scalar measurement of the return loss is required. Consequently cables being measured need to be properly terminated (see Chapter 7). For swept frequency measurements it is necessary to perform a calibrating run prior to the measurement sweep and to apply corrections as required.

Characteristic, Iterative and Image impedances

Characteristic impedance is a mathematical concept defined in Chapter 7, and only applies to line sections that are perfectly uniform. For lines which are only moderately uniform the characteristic impedance can still be calculated from input impedance measurements with the line successively open- and short-circuited at a given frequency, using Equations (7.14) and (7.15). The input impedance of a semi-infinite line is its iterative impedance. If the line is reasonably uniform this approximates to the characteristic impedance. If the line is perfectly uniform the iterative impedance is equal to the characteristic impedance. If the line is not semi-infinite, terminating the line accurately in its iterative impedance (magnitude and phase) enables the input impedance to equal the iterative impedance. This allows the measurement of iterative impedance of finite length lines, by using two sets of identical variable standards, one for the bridge and one for the termination. When the two sets of standards are equal and the bridge also balances, then both are equal to the iterative impedance, and for reasonably uniform lines, to the characteristic impedance. If both ends of the cable are close together, the standards can be 'ganged' and the arrangement is then known as an iterative impedance bridge. However, as the length approaches an integral multiple of the half-wavelength of the frequency it acts as a 'half-wavelength transformer', when the input impedance equals the terminating impedance, and the bridge balance becomes indeterminate. Thus from Equation (7.12)

$$\frac{Z_{in}}{Z_0} = \frac{Z_T/Z_0 + \tanh\gamma l}{(1 + Z_T/Z_0(\tanh\gamma l))} \tag{13.11}$$

If the attenuation is small, $\tanh\gamma l = \tanh(\alpha l + j\beta l) = j\cdot\tan\beta l$. For $\tan\beta l = 0$, $l = n\lambda/2$ and $Z_T = mZ_0$ then $Z_{in} = mZ_0 = Z_T$ and the iterative bridge balances at all settings since the bridge standard equals the termination. (m is a general multiplying factor.) For $\tan\beta l = 1$, $l = (2n+1)\lambda/4$ and $Z_T = Z_0$ then $Z_{in}/Z_T = (Z_0/Z_T)^2$ which is the 'quarter-wavelength transformer' condition. In this case when

$$Z_T = Z_0 \quad \text{then} \quad Z_{in} = Z_0$$

and the bridge has a unique balance.

For an end-to-end asymmetrical network or cable, the characteristic impedance degenerates into two image impedances such that the input impedance at one pair of terminals equals the image impedance of that end of the network when the second pair of terminals is terminated with the image impedance at that pair of terminals. An asymmetric network is correctly terminated when it is terminated in its image impedances. For a symmetric network the image impedances become the characteristic impedance.

Resonant Impedance Measurements

When the use of semi-rigid coaxial cables for long distance analogue TV-signal transmission was started in about 1948, there was a requirement for the return loss at joints to be less than −54 dB. This is equivalent to 0.3 Ω difference in impedance between two 75 Ω coaxials. Thus it was necessary to be able to measure the impedances to better than 0.2% up to 6 MHz. At that time there were no available bridges or standards to meet this requirement. However making use of open- and short-circuit resonances in the input impedance of a reasonably short length of the coaxial line enabled the characteristic impedance, attenuation and phase to be calculated to the required accuracy.

From Equations (7.14) and (7.15)

$$\begin{aligned} Z_{sc} &= Z_0 \tanh\gamma l = (Z_R + jZ_I)\cdot\tanh(\alpha + j\beta)l \\ &= (Z_R + jZ_I)\left(\frac{\sinh 2\alpha l + j\sin 2\beta l}{\cosh 2\alpha l + \cos 2\beta l}\right) \end{aligned} \tag{13.12}$$

$$\begin{aligned} Z_{oc} &= Z_0 \coth\gamma l = (Z_R + jZ_I)\cdot\coth(\alpha + j\beta)l \\ &= (Z_R + jZ_I)\left(\frac{\sinh 2\alpha l + j\sin 2\beta l}{\cosh 2\alpha l + \cos 2\beta l}\right) \end{aligned} \tag{13.13}$$

Separating the real and imaginary parts of these equations enables the resistive and reactive components to be calculated with results as shown in Figure 13.3 for a 200 m length of 1.2/4.4 mm coaxial. The frequencies of resonance are where the reactance

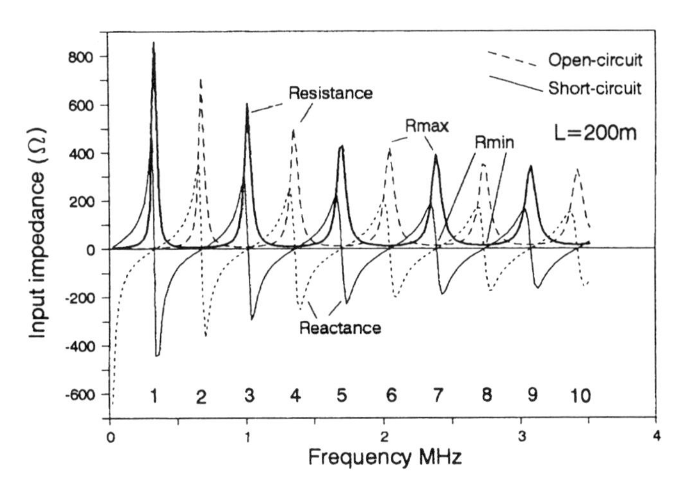

Figure 13.3
Impedance resonances on a 4.4 mm coaxial pair

curves cross the zero of the y-axis, and the first ten resonances have been covered by the calculation. The resonances are nearly but not quite harmonic owing to the phase constant being not quite linear with frequency. Also Z_I contributes to the reactance at the lower frequencies.

At any resonance there is a resistive component which is a maximum R_{max}, or a minimum R_{min}, depending on the order of resonance, n, and whether the cable is open- or short-circuited. From Equations (13.12) and (13.13), with $\beta l = n\pi/2$ when $n = 1, 3, 5, \ldots$

$$Z_{sc} = (Z_R + jZ_I) \cdot \tanh \alpha l = R_{max} \quad \text{and} \quad Z_{oc} = (Z_R + jZ_I)/\tanh \alpha l = R_{min}$$

when $n = 2, 4, 6, \ldots$

$$Z_{sc} = (Z_R + jZ_I) \cdot \tanh \alpha l = R_{min} \quad \text{and} \quad Z_{oc} = (Z_R + jZ_I) \cdot \tanh \alpha l = R_{max}$$

Thus, if $Z_I < Z_R$ then

$$\alpha l = \tanh^{-1}\left(\frac{R_{min}}{Z_R}\right) \text{ nepers} \tag{13.14}$$

for whichever condition (oc or sc) gives R_{min} for the real part of the input impedance. The error in the assumption is a factor $(1 \pm (Z_I/Z_R)^2 \tanh^2 \alpha l)$ on Z_R which is usually negligible.

Since R_{min} is of the order of 5 to 25 Ω it was possible to use small carbon rod resistors as standards, since they could be assumed to be flat with frequency up to about 25 MHz, and their actual values could be determined by d.c. resistance measurements. The bridge was usually a simple transformer bridge.

Now consider Equation (7.6)

$$\frac{\gamma}{Z_0} = G + j\omega C = \omega C(\tan \delta + j \cdot 1)$$

Since the loss angle of the dielectric is generally less than 0.0003, G can be neglected to within 0.03% and

$$\gamma = (\alpha + j\beta) = (Z_R + jZ_I) \cdot j\omega C = j\omega C \cdot Z_R - \omega C \cdot Z_I \tag{13.15}$$

whence

$$\beta = \omega C \cdot Z_R = 2\pi f_0 C \cdot Z_R = \frac{n \cdot \pi}{2l} \quad \text{at a resonance } f_0$$

and so

$$Z_R = \frac{n}{4lC \cdot f_0} \tag{13.16}$$

For good dielectrics (particularly with highly air-spaced constructions) the high-frequency value of lC will be identical with the low-frequency capacitance which can be measured to high accuracy (better than 0.05%). The resonant frequency can be measured to better than 0.05% with a heterodyne or digital frequency meter and the order of resonance, n, is an integer. Thus Z_R can be determined to an accuracy of better than 0.1% without having to know the value of R_{min} or the cable length to the same accuracy.

The other cable parameters can then be determined from Equations (13.15) and (7.5) with the assumptions of negligible G and $Z_I^2 \ll Z_R^2$

$$Z_I = -\frac{\alpha}{\omega C} \qquad Z_\infty = Z_R + Z_I \tag{13.17}$$

$$R = 2\alpha \cdot Z_R \qquad L = \frac{nZ_R}{4l \cdot f_0} \qquad V_p = \frac{1}{\sqrt{LC}} = \frac{1}{CZ_R}$$

$$\varepsilon = \left(\frac{300 \cdot n}{4 \cdot l \cdot f_0}\right)^2 \frac{1}{1 + \dfrac{4\alpha l}{n\pi}} \quad \text{for } f_0 \text{ in MHz and } l \text{ in m} \tag{13.18}$$

Also, just for interest, from (13.15) and (13.17)

$$\frac{Z_I}{Z_R} = \frac{-\alpha}{\beta} \tag{13.19}$$

In all these equations the attenuation α is in nepers/km and the phase constant β is in rad/km.

Direct Measurement of *R*, *L*, *G* and *C* on Short Cable Lengths

The primary parameters can be measured directly on a short length of cable provided the length is such that the first resonant frequency is at least 10 times the highest measurement frequency. This will ensure that the parameters are not changed by more than 1% owing to the onset of resonance.

The longest length that can be used for 1% accuracy with a top frequency of measurement of f MHz is thus given by

$$l = \frac{7.5}{f\sqrt{K}} \text{ m} \tag{13.20}$$

where K is the effective permittivity of the insulation. This principle also applies to the simple low-frequency measurement of pair capacitance, where even at 1 kHz a length of 5300 m would produce a 1% error in the measurement.

The use of short lengths has the disadvantage that the length may not be typical of the whole cable length and that the sample length will probably represent scrap after measurement. Also the shorter the length, the smaller are the impedances to be measured, which may lead to loss of bridge accuracy. On the other hand, the use of short lengths for oc/sc measurement, when using the method of analysis of Chapter 7, which recognizes the possible onset of resonance, removes $n\pi/2$ ambiguities in the determination of the phase constant. Another advantage of short length measurements, is that physical measurements of diameters on the ends of the sample length are much more likely to be representative of the whole, when they are used for design verification.

ATTENUATION AND CROSSTALK MEASUREMENT

As with characteristic impedance, the attenuation constant is a mathematical concept and should be calculated from the oc and sc input impedances of the line. The comparison of the input voltage with the output voltage, as shown in Figure 13.4(a), gives the insertion loss between the particular termination conditions used. If the line is accurately terminated in the characteristic impedance at the frequency of measurement then the insertion loss will approximate closely to the attenuation.

For installation measurements, the insertion loss between nominal system resistive terminations is measured using a transmission test set. This comprises a stable generator which feeds the line with a stabilised voltage and a detector in the form of an

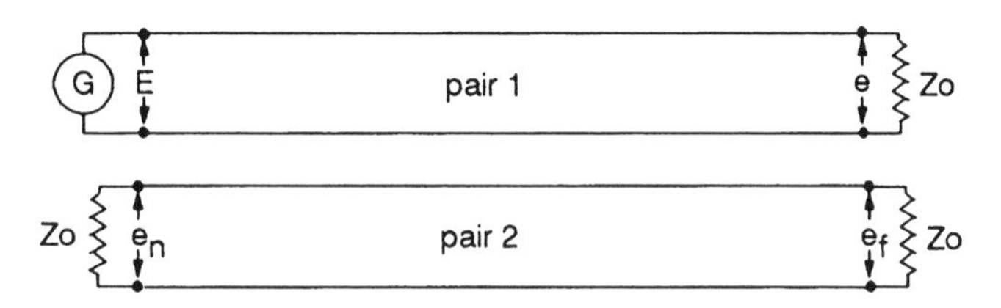

(a) Direct voltage measurements

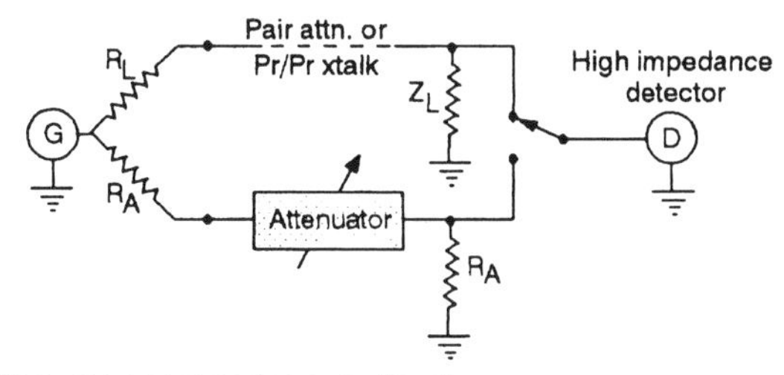

(b) Attenuator comparison method

Figure 13.4
Attenuation and crosstalk measurements

amplifier and rectifier voltmeter, which measures the received voltage. The termination is usually a resistor equal to the nominal line impedance. The insertion loss is expressed as the logarithmic ratio of the sent and received voltages, in N or dB. It is necessary to standardize the receiver and generator before measurement. Both are frequently calibrated in dB simplifying the determination of the loss to subtracting the readings.

The effect of the termination not accurately matching the line impedance is to produce an error

$$\text{termination error} = 20 - \log \frac{2Z_T}{Z_0 + Z_T}\ \text{dB} \tag{13.21}$$

Thus a 10% error in termination would produce an 0.4 dB difference between the measured insertion loss and the attenuation. This would not be very significant on long cable lengths where an attenuation of 30–40 dB is expected, but for short cable lengths with low attenuations it could be serious.

The measurement of crosstalk attenuation is similar to line attenuation measurement except that the loss is usually much larger and the signal may approach the noise level. Thus it becomes essential to reduce the bandwidth of the receiver to limit the noise. When both ends of the cable are available, a comparison method of measurement may be used as shown in Figure 13.4(b) with a calibrated variable attenuator.

Usually a resistive splitter is used from the generator, feeding the line from its nominal impedance value, R_L, and the attenuator from its impedance, R_A. By using a high-impedance detector both the line and the attenuator can also have their correct terminations. The splitter produces a 6 dB loss from the generator to both the line and the attenuator when correctly adjusted. If the line and the attenuator are both reasonably well terminated to avoid reflections, the splitter is not strictly required, and a simple parallel connection gives the same voltage on both the line and attenuator inputs.

If it is required to measure the phase constant as well as the attenuation of the line, then the phase of the voltage e must be measured relative to the input voltage E. The phase will be measured as a lag between 0 and $-360°$ (0 and -2π) Thus each time the signal frequency passes through a cable resonance point, the measured lag must be augmented by $-360°$ (or -2π rad).

For very high crosstalk attenuations, e.g. greater than 120 dB, it is necessary to use a higher input voltage and sometimes extreme measures to isolate the generator and detector from stray and earth-loop couplings, even to the extent of running the generator from batteries.

When the phase of the coupling between pairs is required as well as the magnitude, then the attenuator is replaced with a complex coupling network (calibrated in terms of capacitive C, and conductive G/ω, couplings) and the comparator switch is replaced with a differential transformer to compare the line and network outputs directly. This is known as a complex coupling bridge.

When measuring crosstalk attenuations it is necessary that both ends of both pairs are terminated to avoid reflections giving rise to extra crosstalk paths. Also in some circumstances it is necessary also to terminate third circuits. For common-mode rejection measurements, the generator is used to excite the common mode via a centre-tapped tranformer and the received balanced-mode voltage across the pair is measured and compared with the generator voltage (allowing for the transformer loss).

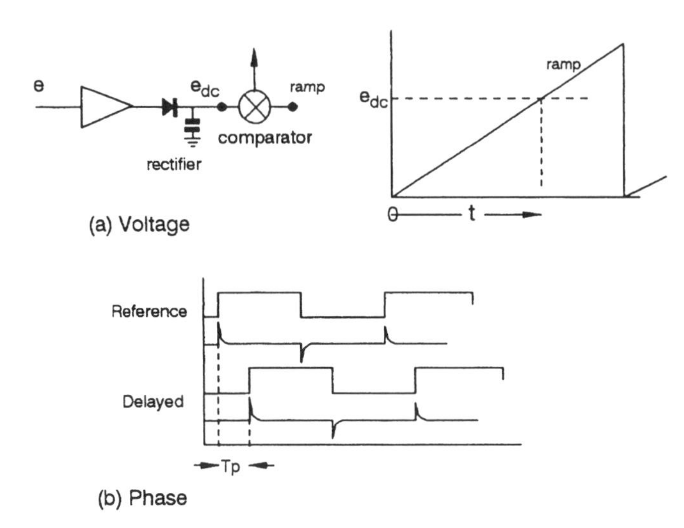

Figure 13.5
Attenuation and crosstalk measurement principles

DIGITAL VOLTAGE AND PHASE MEASUREMENT

It is desirable to have a visualization of the methods used to measure voltage and phase digitally. Possible methods are sketched in Figure 13.5. In (a) the voltage measurement is carried out by comparing the amplified and rectified signal, e_{dc}, with a linear time-ramp voltage. The time required for the ramp to equal e_{dc} is measured by a counter and displayed. By suitable calibration the voltage is related to the count. High resolution and accuracy is possible with self-calibration facilities. In (b), the delay between the signal and the reference waveforms (suitably squared and differentiated) is also measured by a counter. The count, relative to a full period, is calibrated in degrees or radians lag.

NETWORK ANALYSERS

Much use is now made of general-purpose network analysers for cable measurements. These use a swept-frequency frequency-synthesizer generator covering a very wide band up to 500–1000 MHz. A resistive splitter is used to feed the cable (or differential transformer return loss bridge) and a reference port, usually at 50 Ω or 75 Ω impedance level.

For impedance measurement, the cable is fed from the return loss bridge via a suitable impedance matching and balancing transformer, and is terminated at its far end. The output of the RL bridge is measured in magnitude and phase and compared with the reference port voltage. An in-built microprocessor calculates the input impedance of the terminated pair and displays it versus frequency on the built-in digital oscilloscope.

Preliminary calibration (or normalization) runs with the test port open- and short-circuited and terminated are carried out to allow for frequency variations. These are stored and automatically correct the measurements. For best accuracy a suitable simulator (Figure 7.14) should be used to terminate the pair.

For insertion loss and phase constant measurement a second input port is used to measure the received voltage in magnitude and phase, compare it with the reference port voltage, calculate the loss in dB and correct the phase for $2n\pi$ transitions. The attenuation and phase can then be diplayed on alternate oscilloscope channels simutaneously. Crosstalk losses can be measured in a similar fashion.

The oscilloscope display can be frozen and digitally stored, or displayed on a printer. Markers can be deployed along the traces to give digital read-outs at desired frequencies, etc. Network analysers are very accurate, versatile and convenient measuring devices, but the costs are high (US$100 000 or so).

PULSE TESTING

Another technique for testing cables, much used with precision coaxial pairs, is pulse testing. A short pulse generated by the pulse generator PG (Figure 13.6) is applied to the differential transformer centre tap of a return loss bridge. The cable is connected to one port of the bridge and an impedance simulator to the opposite port. The output of the bridge is the return loss spread out in time owing to the the delay time of the pulse traversing the pair. When displayed on the CRT against a suitable linear timebase the display represents the pair impedance, relative to the simulator impedance, at progressive distances down the pair.

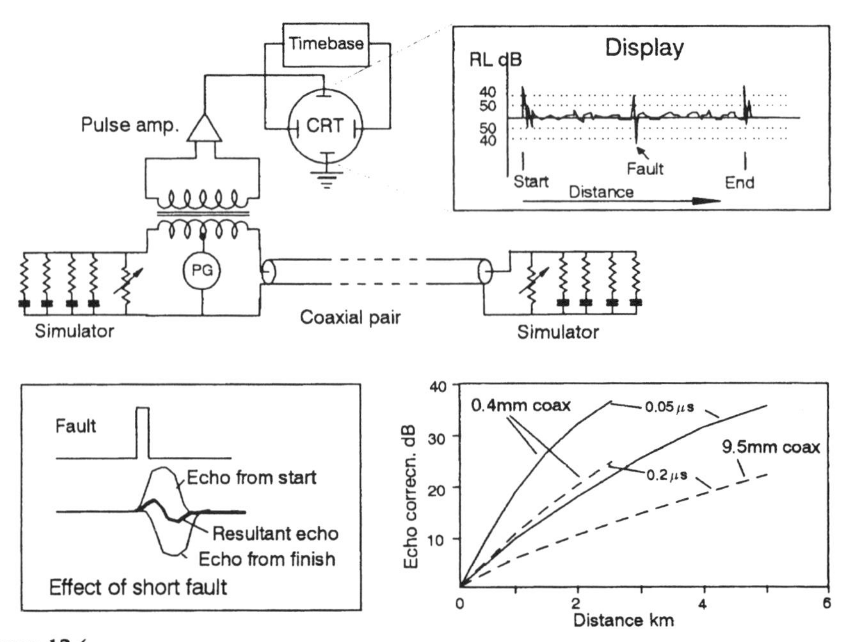

Figure 13.6
Pulse testing of cables

The near-end simulator is adjusted to the best match at the input end of the pair. This match is generally limited by the transformer to about −40 dB (or 2% impedance mismatch). At the far end however a match of −60 dB is generally achievable with a precision simulator, representing a 0.2% impedance match. A four-branch simulator can be designed to match the coaxial pair impedance to better than 0.05% over a frequency range of 30 kHz to 8 MHz.

A discrete fault on the pair will be displayed at a point on the trace corresponding to its distance down the length. The size of the fault, at its location on the pair, is given by the pulse amplitude of the displayed fault corrected by twice the pulse amplitude attenuation from the start of the pair to the location of the fault. Thus a displayed fault of −50 dB at 1000 m would be corrected to −40 dB (for a 0.05 μs pulse on a 9.5 mm coaxial). The typical correction curves for 4.4 mm and 9.5 mm coaxials for 0.05 μs and 0.2 μs pulses are shown in the insert of Figure 13.6. The pulse reflection from an impedance fault shorter than half the pulse width is the resultant of the echoes from the beginning and end of the fault as shown in Figure 13.6. The size of the resultant reflection reduces rapidly (at 20 dB per decade of length ratio) as the fault becomes shorter than the pulse. This is somewhat dependent on pulse shape. For precision pulse testers the pulse is controlled to have a sine-squared shape. Pulse testing of this nature, but of less precision in terms of impedance, is used on a wide variety of cables for fault location.

SCREENING FACTOR AND PERMEABILITY

In order to measure the intrinsic screening factor, defined in Chapter 9, on short cables say of the order of 1 m length, it is necessary to simulate the earth return inductance of 2 μH/m. As it is virtually impossible to achieve this with a lumped inductance without adding considerable resistance, the artifice of using mutual inductance is used. As shown in Figure 13.7 the potential lead used for measuring E is taken very close to (or inside)

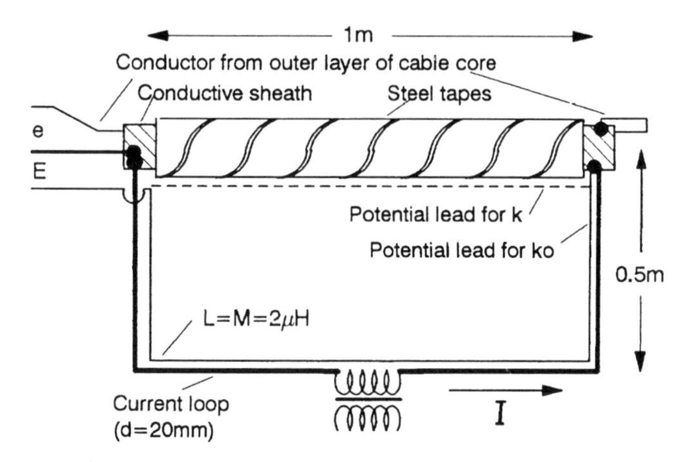

Figure 13.7
Measurement of screening factor on short cable length

the current loop formed from 20 mm diameter copper tubing. This loop is arranged to be 0.5m from the cable sample and has an inductance of

$$L = 0.0002p\left(\ln\frac{4p}{d} - 3.04\right)\text{mH} \tag{13.22}$$

where p = perimeter of the loop (m) and d is the conductor diameter (m). For the dimensions shown in the figure this results in $2\,\mu$H added inductance.

Since the potential lead is very close to this path, the mutual inductance, M, is equal to the inductance, L, and the voltage measured on this lead is

$$E = I\{(R_{dc} + R_S) + j\omega(\mu + L_S)\} \tag{13.23}$$

If however the potential lead is run very close to the cable sample, as shown by the dotted line, the simulated earth return inductance is not added to the cable inductance and the voltage becomes

$$E' = I\{(R_{dc} + R_S) + j\omega L_S\} \tag{13.24}$$

The voltage e, on a conductor in the outer layer of the cable is measured in magnitude and phase relative to E to determine the intrinsic screening factor, k_0 (or relative to E' to determine the basic screening factor k).

The measurement of the basic screening factor, k, is used to calculate the real and imaginary parts of the magnetic pearmeability of the steel tapes. Since

$$k = \frac{R_{dc}}{(R_{dc} + R_S) + j\omega L_S}$$

$$R_S + j\omega L_S = R_{dc}\left(\frac{1}{k} - 1\right) = j\omega\left(\frac{0.4lmnt}{\text{D}}\right)(\mu_R - j\mu_I)\,\mu\Omega/\text{m} \tag{13.25}$$

from Equation (9.12). Since $k = |k|\angle\theta$, equating real and imaginary parts gives

$$\mu_R = -Q\cdot\frac{1}{|k|}\cdot\sin\theta \qquad \text{and} \qquad \mu_I = Q\left(\frac{1}{|k|}\cdot\cos\theta - 1\right) \tag{13.26}$$

where

$$Q = \frac{R_{dc}\cdot D\cdot 10^6}{\omega\cdot 0.4\cdot lmnt} \tag{13.27}$$

Since θ is negative, both parts of the permeability are positive. D is the mean diameter of the steel tape layer in m, l is the sample length in m, n is the number of tapes of thickness t m and m is the gap factor defined in connection with Equation (9.12).

Because of the low values of resistance involved, the connections should preferably be soldered or welded. Also at the higher field strengths of 400 V/km, the current is such as to dissipate approximately 50 W in the loop, so that unwanted heating may occur. This may be an additional reason for the measured intrinsic screening factors deviating from the calculated values in Figure 9.7 at the higher field strengths.

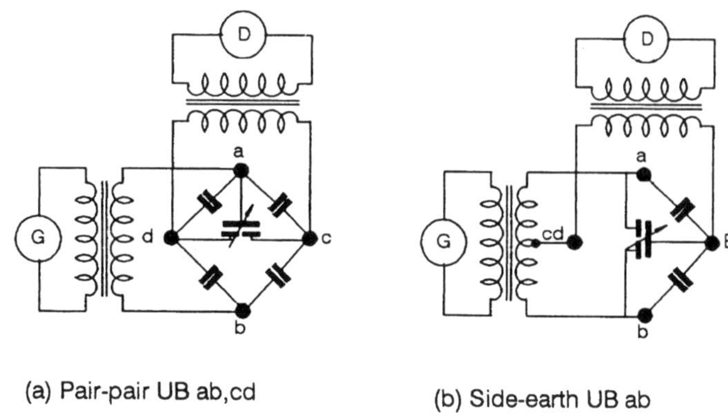

(a) Pair-pair UB ab,cd (b) Side-earth UB ab

Figure 13.8
Capacitance unbalance measurements

CAPACITANCE UNBALANCE MEASUREMENT

In principle, capacitance unbalances are measured by setting up a generator and detector, usually at low frequency, to detect the voltage unbalance produced. A calibrated differential capacitor is then connected and adjusted to compensate for the voltage unbalance. This is illustrated in Figure 13.8 for two important unbalances, viz. pair-to-pair and wire-to-earth unbalances. A differential resistor may also be required to sharpen the balance adjustment by compensating for any conductance unbalance. Suitable zeroing components are also usually included.

Checking the Balance of Capacitance Meters

Small hand-held digital capacitance meters are often used to measure pair capacitances. Apart from the basic accuracy of these devices, it is important to know whether they are sufficiently balanced to earth not to affect the measurement of balanced pairs. This can be checked by carrying out a three-capacitance measurement as shown in Figure 13.9, where C_a, C_b and C_{ab} are the partial capacitances from the conductors to earth and between the conductors respectively. The instrument is used to measure the three capacitances (typical measurements are shown in brackets).

$$\begin{array}{lll} \text{from a to b} + \text{E} & C_1 = C_a + C_{ab} & (7.006\,\text{nF}) \\ \text{from b to a} + \text{E} & C_2 = C_b + C_{ab} & (6.985\,\text{nF}) \\ \text{from a} + \text{B to E} & C_3 = C_a + C_b & (7.535\,\text{nF}) \end{array} \qquad (13.28)$$

These measurements are not affected by the balance of the instrument. Then

$$\begin{array}{rl} (C_1 C_2) + C_3 = 2C_a & (= 2 \times 3.778\,\text{nF}) \\ C_3 - (C_1 - C_2) = 2C_b & (= 2 \times 3.757\,\text{nF}) \\ C_1 - C_a = C_2 - C_b = C_{ab} & (= 3.228 \text{ and } 3228\,\text{nF}) \end{array} \qquad (13.29)$$

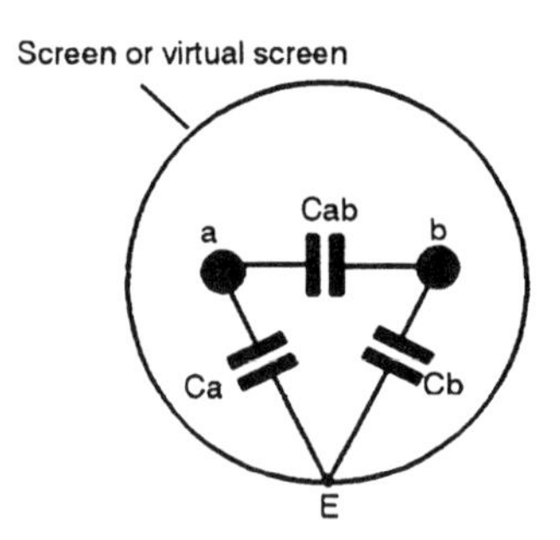

Figure 13.9
Three-capacitance measurement

The balanced pair capacitance is then

$$C_{bal} = C_{ab} + \frac{C_a C_b}{C_a + C_b} \quad (= 5.112\,\text{nF; cf. } 5.107\,\text{nF measured}) \tag{13.30}$$

The agreement to within 0.1% between the calculated and measured balanced pair capacitance shows that the instrument is well balanced.

Voltage Breakdown Measurement

These are straightforward measurements but the safety aspects should be observed. The current from the test voltage source should be limited to less than 4 mA to avoid heart fibrillation in case of accidental contact. Remember to discharge the cable adequately after test since the charge stored in the cable capacitance can be lethal.

Other measurements

The measurement methods of some parameters have been included in the text as they occurred. These are

- Earth resistance and resistivity — Chapter 9
- Transfer impedance, screening efficiency/factor of braids — Chapter 10
- Permittivity by derivation from impedance measurements — Chapter 7

14

Optical Fibre Transmission

HISTORY

The quest for greater bandwidths on copper transmission lines was accompanied by greater attenuation at higher frequencies (proportional to the square root of frequency). This resulted in amplitude/frequency distortion, the correction of which limited the distance of transmission, until for 60 MHz bandwidth the distance between regenerators reduced to a few kilometres even with large diameter coaxial constructions.

By modulating the wideband signal on to a very-high-frequency carrier the spread of the resultant signal could be made a small percentage of the carrier frequency. This greatly reduced the amplitude/frequency distortion if the change of attenuation with frequency was small. If then a suitable guiding medium for the high frequency carrier could be found, with a low loss, a much improved transmission line could be constructed.

At first, in the late 1960s, efforts were concentrated on 50 mm diameter circular waveguides which could be constructed to have a loss of about 0.6 dB/km at about 100 GHz. For a 1 GHz bandwidth this would give a fractional frequency spread of 1% with a negligible amplitude/frequency distortion and a regenerator spacing of about 15–20 km. However these low attenuations only applied to waveguide sections of good commercial straightness. For bends there was an attenuation penalty of

$$\Delta\alpha = 2.2\sqrt{\frac{\theta}{R}} \qquad \text{dB for a helix waveguide} \qquad (14.1)$$

and

$$\Delta\alpha = 6.5\left(\frac{\theta}{R}\right) \qquad \text{dB for a dielectric waveguide} \qquad (14.2)$$

where θ is the angle turned in rad at a radius of curvature of R m. So that a 90° turn at a radius of 10 m incurred a penalty of 0.87 dB for the helix guide and 1.02 dB for the dielectric guide. Also the tilt between two waveguide sections had to be limited to less than 1°. All this proved to be too much of a limitation for realistic installations.

Meanwhile attention had turned to the possible use of light as a carrier supported by a suitable guiding medium. Light with a wavelength of 0.5 μm has a frequency of about 600 THz (600×10^{12} Hz). This would thus give almost unlimited bandwidth possibilities.

The use of light beams for line-of-sight communication in the form of the heliograph had been well known for centuries, even the guiding of light beams along liquid jets has been described by Tyndall in 1870. Glass fibres were used for guiding light in gastroscopes in the 1940s and sintered fibre faceplates for CRTs were in use in the 1950s. In 1962, Eaglesfield of STL proposed the use of internally silvered optical pipelines of about 25 mm diameter for communication purposes and considered the attenuation, mode conversion and bandwidth properties. This led to consideration of the use of very small diameter glass fibres for telecommunication in the mid-1960s. However even the best optical glasses at that time had attenuations of the order of 2000 dB/km (equivalent to 99% transmittance/cm).

The invention of optical fibres for telecommunication is generally attributed to Kao and Hockman [36] of STL(UK) in 1966. A joint study by STL and the British Post Office defined the minimum aim of 20 dB/km and 100 Mb/s transmission rate. A material study sponsored by BPO and STL at Sheffield university had elucidated the sources of the high attenuation in fused silica glasses by 1967. The contamination of the glass lattice by ions of the transition metals (Ti, V, Cr, Mn, Fe, Co, Ni, Cu and Zn) together with boron oxides and hydroxyl ions, even at concentrations as low as parts-per-million was found to cause overlapping resonant peaks of high attenuation, producing attenuation coefficients somewhat like that shown in Figure 14.1. Nevertheless, by careful purification, alkali glasses were made which could be melted in, and drawn into fibre from, platinum crucibles. Such fibres were eventually made with losses as low as 10 dB/km at 800 nm. But really to eliminate all these contaminants, pure silica glass has to be made by oxidation of the vapour of silicon tetrachloride and depositing it on a suitable surface, usually a natural quartz rod or tube. This produces a rod preform from which fibres can be drawn. The liquid silicon tetrachloride can be prepared by chemical and distillation techniques to a very high purity of 99.9999% (six-nines purity).

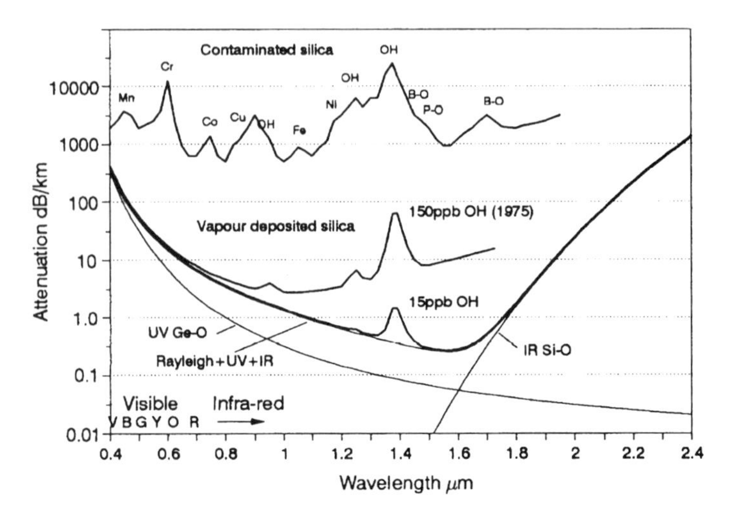

Figure 14.1
Absorption and scatter losses in silica glasses

The first fibres drawn from silica prepared in this way (called Chemical Vapour Deposition – CVD) in 1970 by Corning Glass had an attenuation of less than 20 dB/km at about 800 nm wavelength. However fibre strength was a problem still to be solved. Bell Telephone Laboratories were licensed by Corning, and both BTL and Corning were producing fibres with attenuations of less than 5 dB/km (at 850 nm) in 1 km lengths by 1975. They used a method of outside deposition on to rotating (and removable) ceramic formers. Meanwhile STL had developed the inside deposition method where the silica glass was formed on the inside surface of a rotating natural quartz tube. This was known as the Modified Chemical Vapour Deposition or MCVD method. They were also producing 5 dB/km fibre in 1 km lengths by 1975.

By 1980 the Japanese had developed an alternative outside vapour deposition method where the silica was deposited on to the end of a rotating quartz rod to build up a preform. This was called the Vapour Axial Deposition (VAD) method and had the advantage of making much larger preforms but it was more difficult to control.

Pure fused silica glass itself (with a small amount of germanium doping), has a low attenuation window, reaching a minimum loss of about 0.25 dB/km at 1550 nm, bounded on the high wavelength side by infra-red absorption, on the low wavelength side by UV resonances due to Ge–O vibration, and in between limited by Rayleigh scattering as shown in Figure 14.1. However until hydroxyl contamination (which gave rise to harmonics of several resonances situated at about 2.7 μm wavelength) could be controlled, two 'windows' around 850 nm and 1100 nm at about 3.5 dB/km each were used, which suited available LED and laser sources.

Control of hydroxyl contamination down to 10 parts per billion was demonstrated by about 1979, and the rise in attenuation above 1 μm wavelength was identified with the presence of a broad P–OH resonance at 3.05 μm (phosphorus had been used as a dopant and also to lower the melting point of the deposited glass). When this was removed a very close approach to the theoretical attenuation of pure silica was possible.

The world's first commercial telecommunication optical cable route (9 km from Hitchin to Stevenage in the UK) was installed by STC in 1977 and operated at 140 Mb/s using 850 nm solid state lasers and avalanche photodiode receivers. The cable was an 8-fibre construction, having 3 dB/km attenuation at 850 nm and with dispersion controlled by graded refractive indices.

Also indicated on the x-axis of Figure 14.1 is the wavelength range of the visible light spectrum (Red, Orange, Yellow, Green, Blue and Violet). From this it will be seen that the low attenuation wavelengths are in the near infra-red. The standard wavelengths used are 850 nm, 1300 nm and 1550 nm for which solid state laser sources are now readily available.

PHYSICS OF OPTICAL FIBRE TRANSMISSION

When a light ray travels across a boundary between a denser medium and a less dense medium its direction changes to make a larger angle with the normal to the surface, i.e. it is refracted, as shown in Figure 14.2(a). Snell's law relates the angle of incidence θ_1 to the angle of refraction θ_2 as follows

$$n_1 \sin\theta_1 = n_2 \sin\theta_2 \tag{14.1}$$

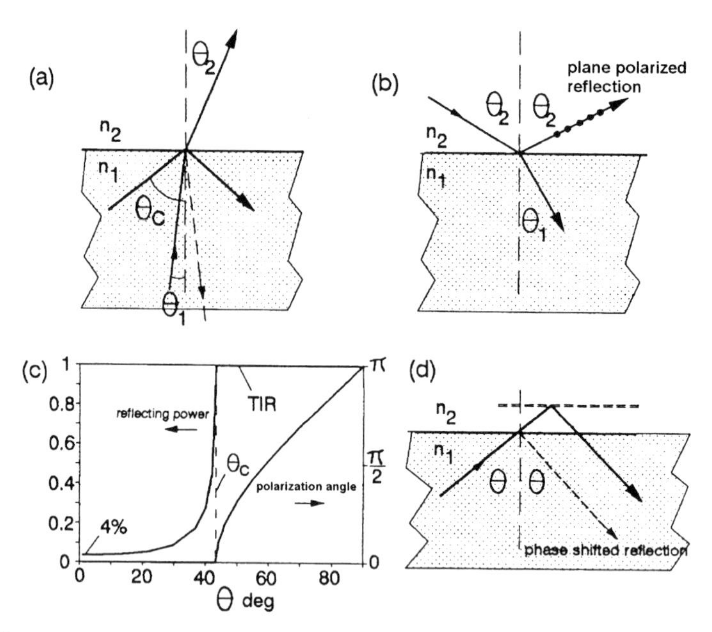

Figure 14.2
Reflection and refraction at dielectric boundaries

where n_1 and n_2 are the refractive indices of the two media. The velocities of light in the two media are different and are given by c/n_1 and c/n_2 respectively, where c is the velocity of light in free space ($n = 1$ and $c = 3 \times 10^8$ m/s). Also a proportion of the incident power is reflected with a reflected angle equal to θ_1. When the angle of incidence increases beyond a critical value θ_c, the incident light is totally reflected. This corresponds to $\theta_2 = 90°$ and from Snell's law

$$\sin \theta_c = \frac{n_2}{n_1} \tag{14.2}$$

The situation for a light ray approaching the boundary from the less dense medium is the same, but at a particular angle of incidence θ_B called the Brewster angle, such that $\theta_1 + \theta_2 = 90°$, the reflection from the surface (with a reflected power of about 15% of the incident power), is plane polarized perpendicularly to the plane of incidence, even though the incident ray is unpolarized. This is shown in Figure 14.2(b). Brewster's law is

$$\tan \theta_B = \frac{n_1}{n_2} \tag{14.3}$$

Fresnel first studied the phenomenon of reflection at boundaries and his results for internal reflections are shown in Figure 14.2(c) for rays polarized perpendicularly to the plane of incidence.

The reflecting power of the boundary increases from about 4% at normal incidence to 100% at the critical angle. Beyond the critical angle the reflecting power stays at 100% but the plane of polarization changes from 0 to π radians (180°). At normal incidence the reflecting power of the boundary between a medium of refractive index n and air, for both perpendicular and parallel polarizations, is given by

$$r = \frac{(n-1)^2}{(n+1)^2} \tag{14.4}$$

For a refractive index of 1.5 this reflecting power is 4%. Since the refractive index varies with wavelength, so will this reflecting power. This is generally the reflecting power referred to as 'Fresnel reflection'.

A further phenomenon at the internal reflection point on the boundary is that of a phase shift of the reflected wave. This is equivalent to the reflection point actually being outside the boundary as shown in Figure 14.2(d). This has been confirmed experimentally and the penetration into the less dense medium is about two or three wavelengths. This means that some power flows on the surface just outside the boundary.

When a light ray approaches the end face of a circular fibre, it is refracted as it enters the fibre and travels on in the fibre until it meets the boundary wall. If the angle of incidence is greater than the critical angle it is then totally internally reflected and travels further until it is again reflected at the fibre wall and so on. In general the ray is guided into an approximately helical path down the fibre as shown in Figure 14.3(a) for the skew ray B. If the ray happens to lie in a meridional plane as it enters the fibre, i.e. the ray and a diameter of the fibre lie in the same plane, it is refracted into the fibre and stays in the same plane as it is totally reflected from side to side, e.g. the ray A. Looking into the end face of the fibre, the meridional ray bounces back and forth along the one diameter whilst the skew ray traces out a polygon. Generally there are very many more

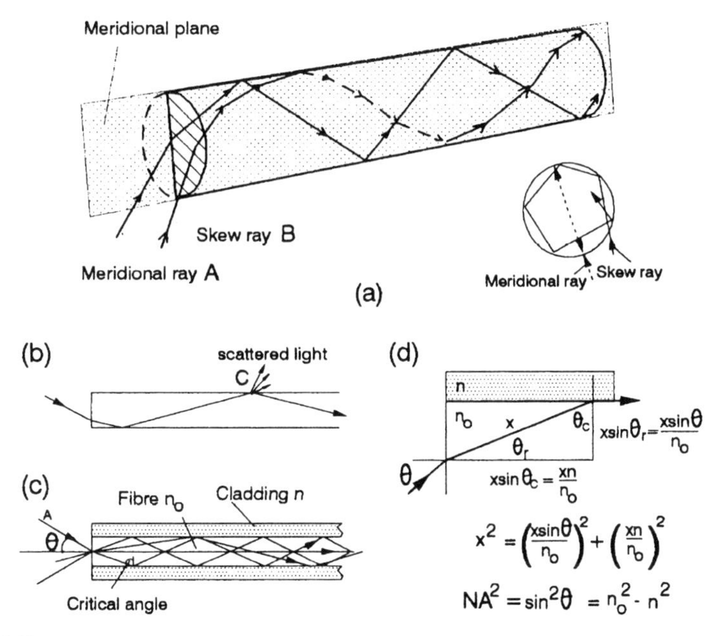

Figure 14.3
Ray paths in step index fibres

skew rays than meridional rays, but the latter are easier to draw and conventionally this is how fibre ray transmission is shown.

If the surface of the fibre is damaged (for example by a scratch) or is contaminated by oil, grease or water on the surface at the reflection point, power will be lost by scattering through the surface as at C in Figure 14.3(b). To avoid this power loss the clean, undamaged fibre is coated with a cladding of lower refractive index than the fibre, as shown in Figure 14.3(c). This cladding may be of plastic or of a lower refractive index glass. Alternatively the fibre may be drawn from a preform made with the core having a dopant which raises the refractive index above that of the outer layers. Because the boundary penetration in total internal reflection results in some power flowing on the outside of the boundary, any losses due to the cladding material affect the total losses of the guided rays. Consequently the cladding material should be of equally low loss material as the core.

When the angle of incidence at the fibre wall is exactly the critical angle, the ray incident on the fibre face is at an angle of θ to the fibre axis. This angle defines a cone at the fibre face. If a ray is at a greater angle than this it will meet the fibre wall, after the entrance refraction, at less than the critical angle, will be largely refracted out of the fibre and will not be guided down the fibre. The sine of the half-angle of this cone of acceptance defines the Numerical Aperture (NA) of the fibre. The derivation of NA is shown in Figure 14.3(d)

$$NA = \sin\theta = \sqrt{n_0^2 - n^2} \approx n\sqrt{2\Delta} \tag{14.5}$$

where $\Delta = (n_0 - n)/n = dn/n$ is small and is called the fractional index difference (often also expressed as a percentage). dn is the actual index difference.

Fibres in which the core index is uniform with the radius are known as 'step index' fibres. Notice that the path followed by the axial ray is shorter than the path followed by any other ray and that the extreme rays entering at an angle θ take a much longer path. Since the refractive index across the core is constant, the velocity for all rays is the same hence the time taken by the extreme rays to go from one end to the other of the fibre is longer than that taken by the axial ray. This leads to the phenomenon of dispersion which can seriously widen the shape of a transmitted pulse, thus limiting the bandwidth of the fibre typically to about 10 MHz · km.

This dispersion can be reduced by varying the refractive index of the core with the radius so that the rays near the axis travel though a higher refractive index, and travel slower than the rays which spend most of their time remote from the axis (typically the skew rays). Such fibres are known as 'graded index' fibres and by proper choice of the index profile shape, can achieve bandwidths of the order of 500 MHz · km. This will be discussed in more detail later.

Due to the electromagnetic wave nature of light, a further limitation is imposed on the guided rays. The waves associated with the rays must set up a standing wave pattern across the width of the fibre core if they are to be guided. This means that the vector component of the ray propagation coefficient normal to the axis must cause a half-wavelength, or multiples of a half-wavelength, of the field to exist across the core width.

The propagation constant of the ray is $k = 2\pi n/\lambda$ rad/m, and the component normal to the axis is $k_r = k\sin\theta = 2\pi n/\lambda \sin\theta$, where n is the core index and λ is the free space wavelength. For standing waves to exist across the width ($2a$) of the core in a *planar* guide there must be a field node at each core boundary. Thus

$$\cos(mak_r) = \cos\left(m\,\frac{2\pi an}{\lambda}\sin\theta\right) = \cos\left(m\,\frac{2\pi a\cdot \mathrm{NA}}{\lambda}\right) = 0 \tag{14.6}$$

This is shown on the left of the fibre in Figure 14.4. The cosine function is periodic, with the nodes of the harmonics occuring together at the core boundary.

For *circular* guides the cosine function is replaced with the Bessel function of the first kind and order m, so that for standing waves

$$J_m(ak_r) = J_m\left(\frac{2\pi an}{\lambda}\sin\theta\right) = J_m\left(\frac{2\pi a\cdot \mathrm{NA}}{\lambda}\right) = 0 \tag{14.7}$$

This is shown on the right side of Figure 14.4. The Bessel function is quasi-periodic, but the nodes adjust to the core boundary by a slight change of θ for each m. In both cases the rays that cause standing waves are restricted to values of θ and m which satisfy Equations (14.6) or (14.7). These are called the transmission modes of the fibre.

In circular waveguide theory for three dimensions of the ray vector, the argument of the Bessel function is usually written as V, so that

$$J_{l,m}(V) = 0 \tag{14.8}$$

where

$$V = \frac{2\pi a\sqrt{n_0^2 - n^2}}{\lambda} = \frac{2\pi a\cdot \mathrm{NA}}{\lambda} \tag{14.9}$$

This takes account of radial m, and azimuthal l, modes. V is called the normalized frequency and n_0 and n are the core and cladding indices respectively. If dn is very small these modes are linearly polarized and are designated the LP_{lm} modes.

The field pattern across the fibre core, illustrating the azimuthal and radial modes, is shown for a few modes in Figure 14.5. Since these modes can exist in two orthogonal polarizations, there are four modes for each value of l (except for $l = 0$ when there are only two).

The number of modes existing in a step index fibre can be estimated as follows. From diffraction theory it can be shown that each ray exiting the fibre occupies a solid angle of $\pi\delta^2$ where $\delta = \lambda/\pi a$. The total exit cone of the fibre has a solid angle of $\pi\theta^2$

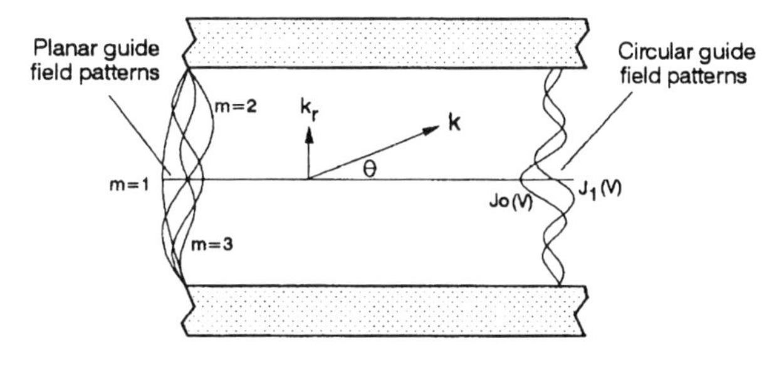

Figure 14.4
Transmission modes in rectangular and circular fibres

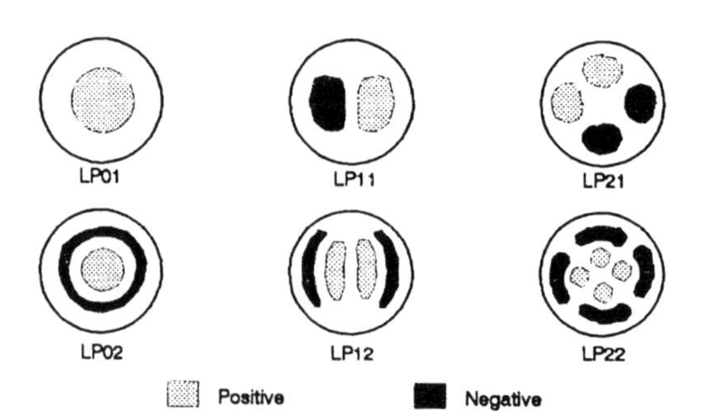

Figure 14.5
Field patterns of modes

where $\theta \approx \sin\theta = \surd(n_0^2 - n^2)$. Since each ray represents two orthogonally polarized modes, the total number of modes which can fit into the exit cone is

$$N \approx 2\left(\frac{\theta}{\delta}\right)^2 = 2\left(\frac{\pi a}{\lambda}\sqrt{n_0^2 - n^2}\right)^2 = \frac{V^2}{2} \tag{14.10}$$

This only holds for a large number of modes. Thus for a fibre with a silica core of radius $a = 50\,\mu\text{m}$ and an index difference $dn = 0.015$ the $NA = 0.209$, $V = 77.29$ at 850 nm wavelength and there will be approximately 3000 modes.

As V decreases fewer modes can be supported. In Figure 14.6 the field patterns along a radius for two orders of the Bessel function, namely J_{0m} and J_{1m}, are shown. At $V = 10$ the requirement that $J_{l,m}(V) = 0$ indicates that about $6 \times 4 - 2 = 22$ modes can coexist.

The requirement that there should be a field node at the core/cladding boundary is necessary for the onset of the mode but the mode can continue to exist as V is reduced, by the field spreading into the cladding (where it dies away exponentially). This can only occur to a limited extent since it drains power into the cladding; too much and the ray ceases to be guided. In other words the mode reaches cut-off, which occurs at

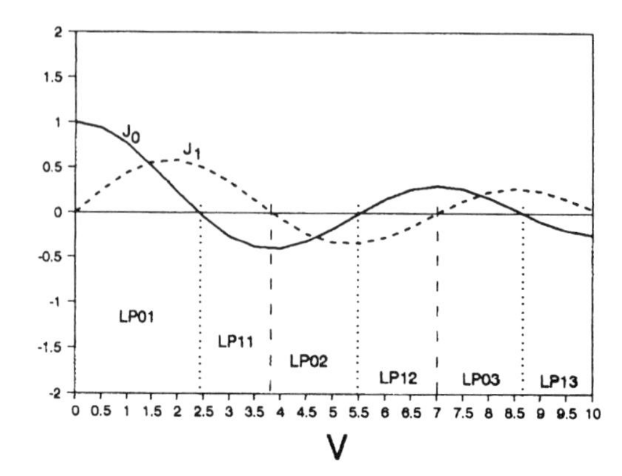

Figure 14.6
Radial fields in circular fibres

the next lowest root of Equation (14.8). Thus in Figure 14.6, the LP_{13} mode can exist until V reduces from 10.1 to 8.65 at which point it reaches cut-off, ceases to be guided and its power is rapidly dissipated into the cladding. Similarly the LP_{01} and LP_{11} modes can coexist below $V = 3.85$ down to $V = 2.405$ where the LP_{11} mode cuts off, leaving only the two polarizations of the LP_{01} mode being guided. This then is the condition for a single-mode fibre, that $V \leq 2.405$. The LP_{01} mode does not have a sharp cut-off as V decreases further, but more and more power spreads into the cladding and the mode becomes more and more weakly guided as the wavelength increases away from the LP_{11} mode cut-off. For a silica fibre with $n_0 = 1.453$ and $n = 1.45$, a radius of 5 µm will cause the LP_{11} mode to cut off at a wavelength of 1.22 µm, so that light at a wavelength of 1.3 µm will be transmitted as a single mode. Since both polarizations of this mode travel at the same velocity there will be no modal dispersion, and the bandwidth of the fibre will be very large (tens of GHz · km), limited only by chromatic dispersion.

The effects of dispersion are shown in Figure 14.7(a). Short pulses transmitted down the fibre equally in all allowable modes are broadened by the dispersion until they start to overlap and it becomes difficult to discriminate between them. If the pulses are Gaussian in shape, it is possible to determine the dispersion from the half-height widths of the sent and received pulses

$$\delta t = \sqrt{T^2 - t^2}\ \text{ns/km} \tag{14.11}$$

and for Gaussian pulses the bandwidth, B can be estimated from

$$B = \frac{440}{\delta t}\ \text{MHz} \cdot \text{km} \tag{14.12}$$

This becomes more inaccurate as the dispersion becomes significant to the width, T, since the modal dispersion is not Gaussian in shape. In general the sent and received

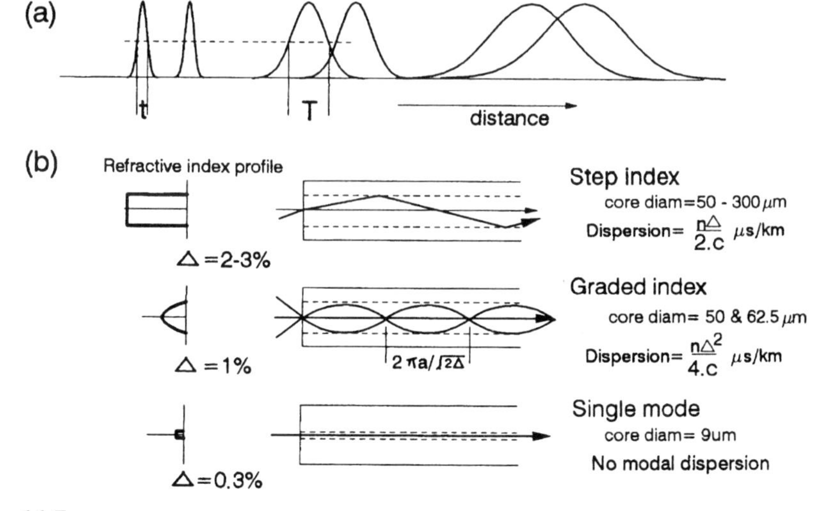

Figure 14.7
Modal dispersion and index profiles of fibres

pulse shapes should be fast Fourier transformed into the frequency domain and subtracted to obtain the frequency response. The optical bandwidth is then taken as the -3 dB point of the optical power.

In Figure 14.7(b) the two ways of minimising the modal dispersion are illustrated. Firstly by grading the refractive index of the core with radius, light travelling further from the axis travels faster than axial rays. By choosing the index profile correctly all modes arrive at nearly the same time. This profile turns out to be a truncated parabola. The meridional rays of all modes cross the axis at the same nodal points, separated by

$$\delta l = \frac{2\pi a}{\sqrt{2\Delta}} \tag{14.13}$$

Skew rays follow quasi-helical paths between caustic radii of R_{min} and R_{max} with the minimum radii corresponding to the nodes of the meridional rays. The theoretical half-height dispersions for the step index and the ideal parabolic graded index are indicated in Figure 14.7(b), (where n is the core index and c is the free space velocity of light 0.3 km/µs) and are functions of the fractional index difference between the axis and the cladding. In practice the ideal profiles are difficult to achieve, which improves the step index bandwidth and degrades the graded index bandwidth. Secondly the core radius and the index difference of the step index profile can be reduced until V is less than 2.405 when all but the fundamental modes cut off and a single-mode fibre is produced. There is no modal dispersion in such fibres but the bandwidth is somewhat limited by the chromatic dispersion acting on the fractional wavelength spread of the light source. Typical core diameters are also shown in Figure 14.7. The cladding diameters for telecommunication fibres have been standardized at 125 µm.

The core refractive index and the refractive index of deposited cladding layers are controlled by doping the deposited silica with suitable dopants. The most usual dopants are germanium from germanium tetrachloride, phosphorus from phosphorus oxychloride and fluorine generally from 'Freons'. The effects of these dopants at various concentrations is shown in Figure 14.8. Germanium and phosphorus increase the refractive index and boron and fluorine reduce the index. Aluminium doping has also been used to increase the index. Often phosphorus and fluorine are used together

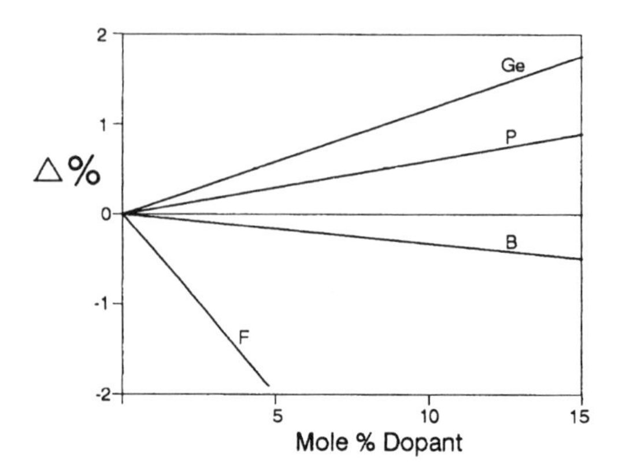

Figure 14.8
Effect of dopants on refractive index

to reduce the melting point of the deposited glass without changing the resultant index, as for example in depositing cladding layers of low optical loss which match the index of the quartz support tube.

For graded refractive index the amount of germanium doping is progressively increased from outer to inner layers. Control of the doping is difficult and also dopant can be lost or can migrate during sintering and vitrification of the deposited silica, or in subsequent operations such as collapse and fibre drawing. This makes the achievement of theoretical profiles quite difficult.

MULTIMODE FIBRES

Profile Analysis for Dispersion and Other Parameters

The effect of profile shape has been well analysed for analytically defined profiles. Although these profiles are difficult to manufacture they do indicate the profiles to be aimed for and the results that theoretically can be achieved. The profiles analysed are generally the set of 'power profiles' defined by

$$n_r = \left\{ \begin{array}{ll} n_0\sqrt{1 - 2\Delta(r/a)^\alpha} & \text{for } r < a \\ n_0\sqrt{1 - 2\Delta} & \text{for } r > a \end{array} \right\} \tag{14.14}$$

where $\Delta = (n_0 - n)/n$. These profiles are graphed in Figure 14.9 for a few values of the profile exponent α and are assumed to be circularly symmetrical.

An exponent of 0.5 gives a cusp profile, 1 gives a conical profile, 2 gives the parabolic profile and 1000 is near enough to the step index profile. Thus the analysis for dispersion covers both step index and graded index fibres. Olshansky and Keck [37] showed that the r.m.s. pulse broadening is given by

$$\sigma = \frac{LN_g\Delta}{2c}\left(\frac{\alpha}{\alpha+1}\right)\sqrt{\frac{\alpha+2}{3\alpha+2}} \times \sqrt{c_1^2 + \frac{4c_1c_2\Delta(\alpha+1)}{(2\alpha+1)} + \frac{4c_2^2\Delta^2(2\alpha+2)^2}{(5\alpha+2)(3\alpha+2)}} \tag{14.15}$$

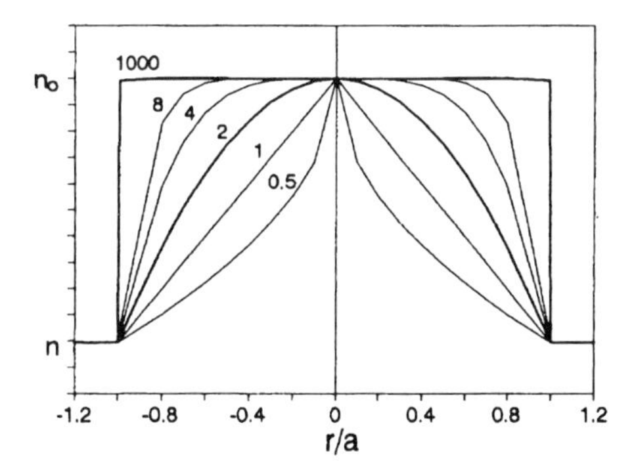

Figure 14.9
Power profiles

where

$$c_1 = \frac{\alpha - 2 - 2y}{\alpha + 2} \qquad c_2 = \frac{3\alpha - 2 - 2y}{2(\alpha + 2)} \qquad y = \frac{2n_0}{N_g}\frac{\lambda}{\Delta}\left(\frac{d\Delta}{d\lambda}\right) \tag{14.16}$$

and the group index is given by

$$N_g = n_0 - \lambda\frac{dn_0}{d\lambda} \tag{14.17}$$

L is the fibre length in km and $c = 300\,\text{km/ms}$ is the velocity of light in free space.

Equation (14.15) has an optimum value, giving the minimum pulse broadening, for

$$\alpha_{opt} = 2 + y - \Delta\left[\frac{(4+\alpha)(3+y)}{(5+2y)}\right] \tag{14.18}$$

For $y = 0$, that is in the absence of material dispersion $d\Delta/d\lambda$,

$$\alpha_{opt} = 2 - \tfrac{12}{5}\Delta \tag{14.19}$$

So that for $\Delta = 0.01$, $\alpha_{opt} = 1.976$ which is approximately the parabolic profile.

To evaluate Equation (14.15) for a range of profiles in the presence of material dispersion, it is necessary to know the index on the axis n_0, and $dn_0/d\lambda$ at the wavelength of interest. These can be obtained by using the Sellmeier coefficients measured by Fleming [38], scaling for dopant mole percentage, and using numerical differentiation. Figure 14.10 shows the dependence of σ on α. As expected there is a very sharp dip around $\alpha = 2$ for both 850 nm and 1300 nm wavelength giving a 200-fold improvement over the step index value.

In order to convert these pulse broadenings into bandwidths, they are interpreted as the standard deviations, σ, of a Gaussian pulse. The equation for a Gaussian pulse and its Fourier transform $F(\omega)$ are

$$f(t) = \frac{1}{\sigma\sqrt{2\pi}}\exp\left(\frac{-x^2}{2\sigma^2}\right); \qquad F(\omega) = \frac{1}{\sqrt{2\pi}}\exp\left(\frac{-\omega^2\sigma^2}{2}\right) \tag{14.20}$$

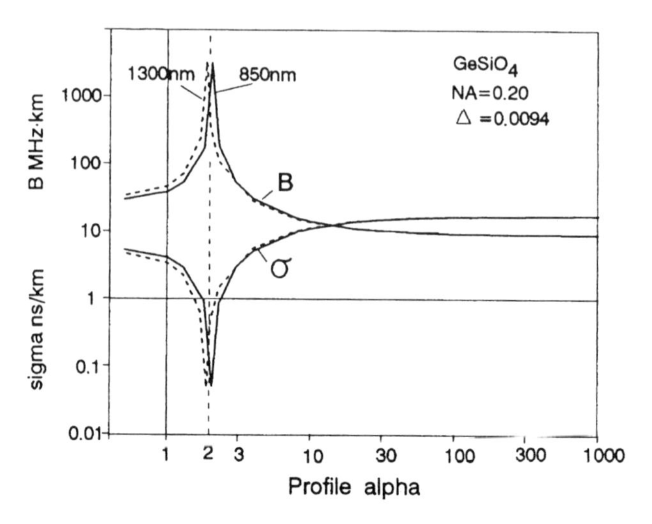

Figure 14.10
Effect of profile on dispersion and bandwidth

from which it may be calculated that the half-power (−3 dB) bandwidth is

$$B = \frac{0.1873}{\sigma}\ \text{GHz} \cdot \text{km for } \sigma \text{ in ns/km} \tag{14.21}$$

The dispersion, D, is defined as the full width of the Gaussian pulse at half-height,

$$D = 2 \times 1.1774\sigma\ \text{ns/km} \tag{14.22}$$

and

$$B \cdot D = 441\ \text{ns} \cdot \text{MHz} \tag{14.23}$$

The bandwidths evaluated in this way are also shown in Figure 14.10. It will be seen that the bandwidth increases from about 8 MHz · km for the step index to values in excess of 1 GHz · km at the optimum profile. Notice also that a slight rounding of the step index profile to $\alpha = 6$ increases the bandwidth to 20 MHz · km.

Achievement of the exact profile exponent for the maximum bandwidth at a given wavelength is very difficult, notice how sharp the peaks are. In order to achieve better than 1 GHz · km bandwidth, at 850 nm $\alpha = 2.05 \pm 0.04$ and at 1300 nm $\alpha = 1.87 \pm 0.06$, a control of better than 3%.

A better way of ensuring saleable production would be to aim for a double window fibre having adequately high bandwidths at both wavelengths. Figure 14.11 shows the trade-off between 850 nm and 1300 nm bandwidths for various achievements in exponent value. To obtain better than 600 MHz · km at either wavelength the exponent to be aimed at is 1.97 (for this NA and material). Then for a control of 5% almost 100% of production fibres will be above 400 MHz · km for either wavelength and a good proportion will be above 800 MHz · km at one or other wavelength.

The refractive indices used in the calculations for Figures 14.10 and 14.11 for a fibre with a germanosilicate core having a NA of 0.20 were derived from Fleming's Sellmeier

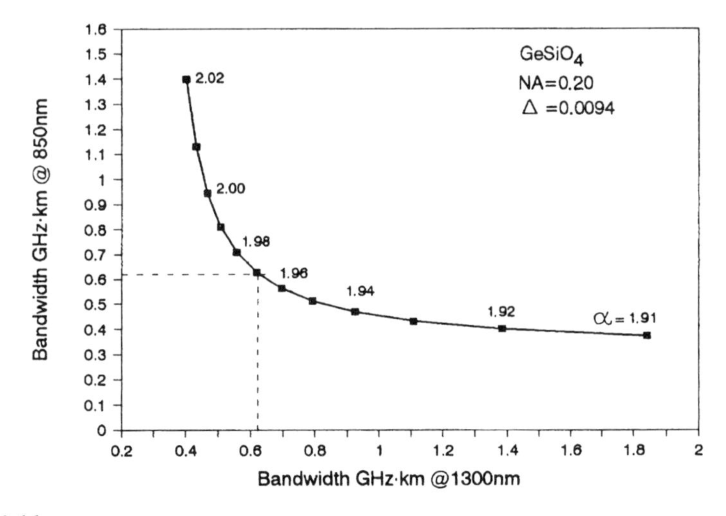

Figure 14.11
Double window bandwidths

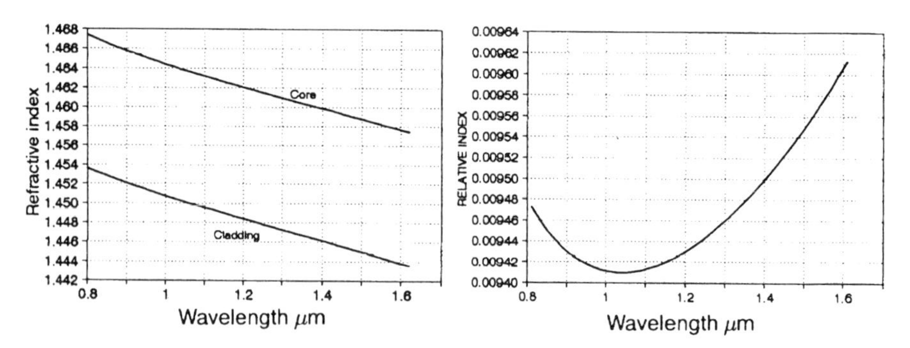

Figure 14.12
Variation of refractive index and relative index difference with wavelength

coefficients [38], scaled for a germanium concentration of 8.85 mole percent. The core and cladding indices and the relative index difference Δ are shown in Figure 14.12.

The first and second derivatives of the core and cladding indices, $dn/d\lambda$ and $d^2n/d\lambda^2$ are shown in Figure 14.13. The first of these is the material dispersion and the second is the chromatic dispersion. The derivative $d\Delta/d\lambda$ and the group index N_g, used in Equation (14.16), are shown in Figure 14.14. The group velocity of transmission down the fibre is $V_g = c/N_g$ and is about 203 m/µs.

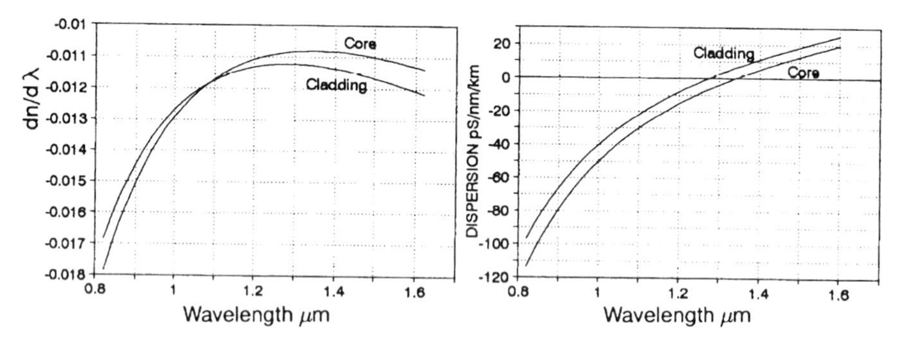

Figure 14.13
Material and chromatic dispersions versus wavelength

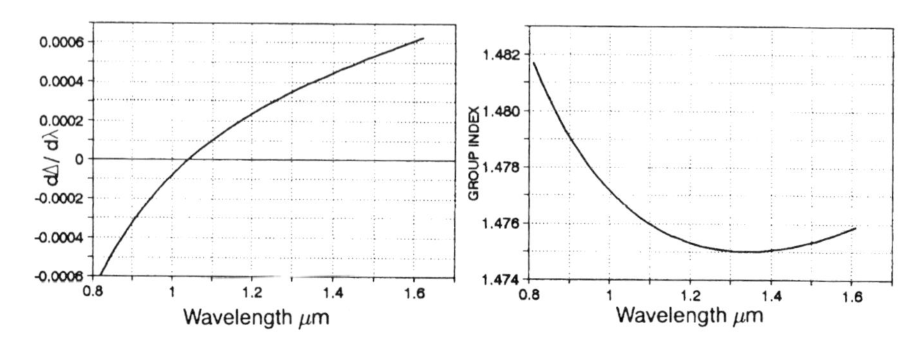

Figure 14.14
Derivative of delta and group index versus wavelength

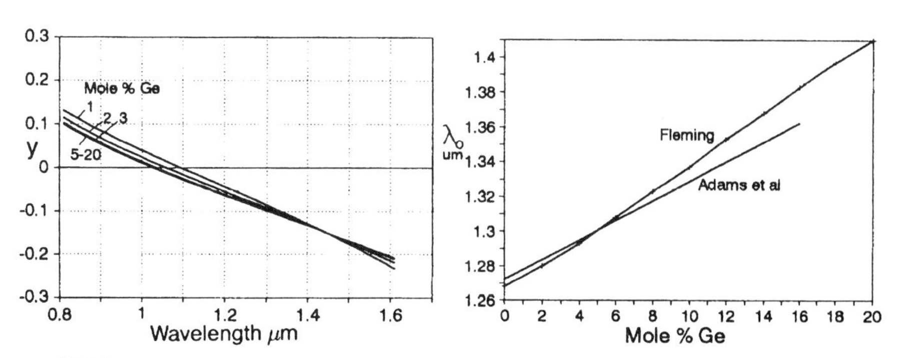

Figure 14.15
Test of Fleming's refractive index data

There was some discussion by Adams et al. [39, 40] of Fleming's measurements and other sources of refractive index information which were not in agreement. The author has checked the requirement that y in Equation (14.16) should be independent of the dopant concentration for Equations (14.15) to (14.19) to hold, and as shown in Figure 14.15 this is very nearly true for Fleming's data. The wavelength of zero dispersion given by $d^2n/d\lambda^2 = 0$ from Fleming's data does differ from the best fit line deduced by Adams et al., but only by 5 nm over the concentrations of interest as also shown in Figure 14.15.

The chromatic dispersion does degrade the bandwidth of the optimum profile fibre at 850 nm, but only significantly above 1 GHz · km if the source spectral width is small (e.g. about 1 nm for a laser source). At 1300 nm which is near the zero dispersion wavelength, there is no significant effect on the bandwidth.

In Figure 14.16 the ray with propagation constant $k_r = 2\pi n_r/\lambda$ rad/m has cylindrical coordinates of β in the axial direction, U_r in the radial direction and l/r in the tangential direction. Thus

$$k_r^2 = U_r^2 + \beta^2 + l^2/r^2 \tag{14.24}$$

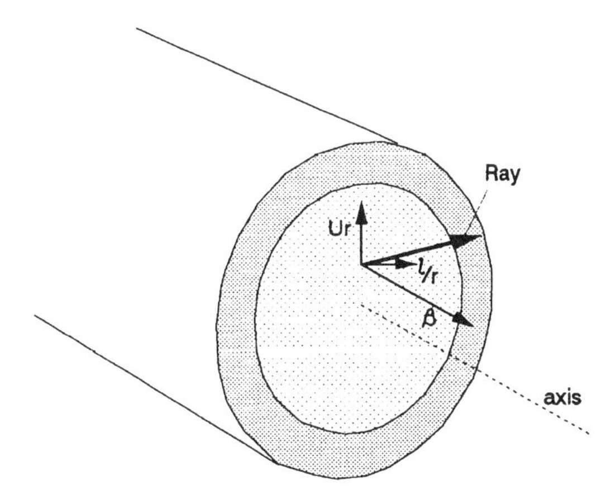

Figure 14.16
Ray vector components

and

$$U_r = \sqrt{k_r^2 - \beta^2 - l^2/r^2} \tag{14.25}$$

Following Gloge and Marcatili [41], the solution to this latter equation has two roots R_1 and R_2 within which U_r is real and the field is periodic. Outside these radii the field dies away rapidly and is thus evanescent. Within these radii a periodic field pattern is established if the phase change from R_1 to R_2 adds up to an integral number of half-periods, thus

$$m\pi = \int_{R_1}^{R_2} U_r \cdot dr = \int_{R_1}^{R_2} (k_r^2 - \beta^2 - l^2/r^2)^{1/2} \cdot dr \tag{14.26}$$

m is the radial mode number and l the azimuthal mode number.

In terms of rays the modes follow helical paths down the guide with the radius of the helix varying periodically between R_1 and R_2. When $l = 0$ the rays have no tangential component, the modes are meridional and $R_1 = 0$, consequently the rays cross the axis of the fibre and travel in a fixed plane. The fundamental mode occurs when $m = 1$, the LP_{01} mode. Equation (14.26) can be solved numerically choosing β for each mode l, m and finding the caustic radii by interpolation. The minimum value for β is $\beta_c = 2\pi n/\lambda$ when m reaches a maximum, $m_{\max}$.

For the power profiles the analysis gives the total number of guided modes (allowing for polarization and azimuthal degeneracy) as

$$M = \frac{4}{\pi} \int_0^{m_{\max}} \int_{R_1}^{R_2} \sqrt{k_r^2 - \beta_c^2 - l^2/r^2}\, dr \cdot dl$$

whence it is shown that

$$M = \left(\frac{\alpha}{\alpha + 2}\right)\left(\frac{2\pi a n_0}{\lambda}\right)^2 \Delta \approx \left(\frac{\alpha}{\alpha + 2}\right) V^2 \tag{14.27}$$

This assumes a large number of modes so that m can be considered a continuous variable. Thus a step index profile guides twice as many modes as a parabolic profile and three times as many as a conic profile. Notice that this approximation differs from that of Equation (14.10) by a factor of two.

Gloge and Marcatili also give the total modal dispersion for power law profiles as

$$dT \begin{Bmatrix} \dfrac{\alpha - 2}{\alpha + 2}\Delta & \text{except for } \alpha \approx 2 \\ \dfrac{\Delta^2}{2} & \text{for } \alpha = 2 \end{Bmatrix} \tag{14.28}$$

Numerical Profile Analysis

The analytical method of profile analysis is adequate for indicating the way transmission parameters depend on gross features of the power-law refractive index profiles, but it is inadequate for profiles with fine features such as exist in practice.

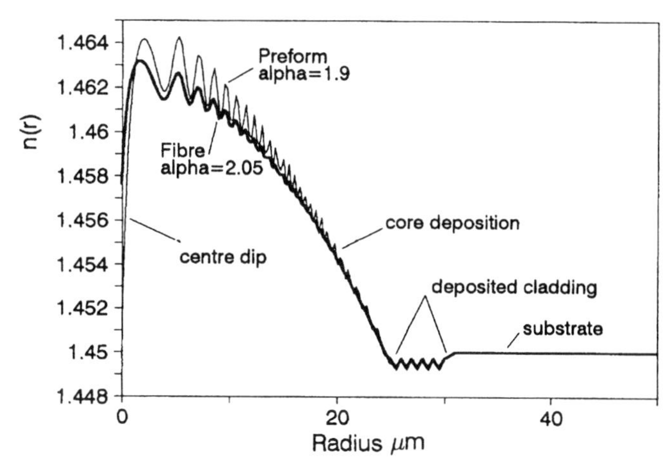

Figure 14.17
Typical preform and fibre refractive index profiles

A typical example of the profiles it is desired to analyse is shown in Figure 14.17. The profiles shown have been simulated from actual measurements on a preform and the fibre drawn from it. It will be seen that the profile changes during the fibre drawing process. The central dip diminishes, the ripples are less pronounced and the mean profile exponent increases. This is due to dopant migration within the core when the preform tip is heated to the melting point.

It is therefore desirable to find a way of dealing with generalized profiles numerically, principally to determine dispersion and bandwidth. The time delay of a mode relative to the fundamental mode is

$$\frac{\tau}{\tau_0} = \left(\frac{k_0}{\beta}\right) \frac{\int_{R_1}^{R_2} (n_r^2/n_0^2)[k_r^2 - \beta^2 - l^2/r^2]^{-0.5} \cdot dr}{\int_{R_1}^{R_2} [k_r^2 - \beta^2 - l^2/r^2]^{-0.5} \cdot dr} \tag{14.29}$$

The difficulty of integrating Equation (14.29) numerically is that both integrands have 'poles' (i.e. go to infinity) at R_1 and R_2. Irving and Karbowiac [42] suggested the use of a Gauss–Chebychev approximation to overcome this difficulty resulting in Equation (14.29) being transformed into

$$\frac{\tau}{\tau_0} = \left(\frac{k_0}{\beta}\right) \frac{\sum_{i=1}^{N} (n_r^2/n_0^2) \cdot f(r_i)}{\sum_{i=1}^{N} f(r_i)} \tag{14.30}$$

where

$$r_i^2 = \frac{R_1^2 + R_2^2}{2} + \frac{R_1^2 - R_2^2}{2} \cdot \cos\frac{(2i-1)\pi}{2N} \tag{14.31}$$

and

$$f(r) = l \cdot \sqrt{\frac{(r^2 - R_1^2)(R_2^2 - r^2)}{r^2(k_r^2 - \beta^2) - l^2}} \tag{14.32}$$

where $k = 2\pi/\lambda$, $k_r = 2\pi n_r/\lambda$ and β is the radial propagation constant that satisfies Equation (14.26) for a radial mode number m, and the caustic radii R_1 and R_2. The azimuthal mode number is l. N is the number of sampling points on the profile between the caustic radii, and should be large enough to resolve the fine details of the profile. To take account of material dispersion n_r^2 in Equation (14.30) is replaced with $n_r(n_r + k \cdot dn_r \cdot dn/dk)$ where $k \cdot dn/dk$ is 0.04553 at 850 nm and −0.045 at 1300 nm. These values are derived from Fleming's Sellmeier coefficients.

Fearsome as these equations appear they can be solved numerically on a PC spreadsheet if the macro programming facilities are used, but it takes some time. Using LOTUS 123r2 on a 386DX-40 with math.co-processor typically takes 12 seconds per mode. Since there may well be 400 modes on a GI fibre (400/4 if we ignore polarization degeneracy), the programme takes about 20 minutes per profile.

Implementation on a Spreadsheet

1. *The profile*

The profile needs to be expressed analytically because Equations (14.30) and (14.32) require n_r to be calculable after determining r_i from Equation (14.31). For an arbitrary profile this can tax mathematical ingenuity. For the profiles of Figure 14.17 the following were used

$$n_r = n_0 - dn\left(\frac{r}{a}\right)^{\alpha} - D \cdot dn \cdot \exp\left(-4\,\frac{r}{w}\right) + A \sin\left\{2\pi S\left(\frac{r}{a}\right)^{p} \cdot \exp\left(\frac{-F \cdot r}{a}\right)\right\} \quad \text{for } r < a \tag{14.33}$$

$$n_r = n_c + A_c \sin 2\pi T\left\{\frac{r - R_c}{R_c - a}\right\} \quad \text{for } a < r < R_c \tag{14.34}$$

where

α = the mean profile exponent
n_0 = the projected on-axis index
dn = index difference
a = core radius
D = fractional centre dip size
w = centre dip radial width
A, A_c = ripple amplitude (core or cladding)
S = number of ripples within the core
T = number of ripples in the deposited cladding
p = exponent to control core ripple spacing

F = exponent to adjust core ripple decay
n_c = mean deposited cladding index
R_c = outer radius of deposited cladding

In order to adequately define the profile, it is necessary to list the refractive index in 0.25 μm steps.

2. *Finding the caustic radii*

Having listed n_r against r, the square of the integrand, I^2 in Equation (14.26), is calculated in the adjacent column. This is calculated for $l = 0$ and $m = 1$ starting with the maximum possible value of β, namely kn_0. The next column calculates the square root conditionally, (i.e. if $I^2 < 0$ it enters 0 in the column). This column is then summed to give $\sum I \cdot dr$ and this value is compared with $m\pi$. If it does not match, β is decreased until equality is achieved to within less than 0.1%. The I-column is then searched to find the two radii at which I goes to zero, and the caustic radii R_1 and R_2 are determined by interpolation of the radius column. The values of β, R_1 and R_2 are stored in an array against the values of l and m.

3. *Gauss–Chebychev sub-routine*

The values of r_i are listed using Equation (14.31) and the corresponding $n(r_i)$ are now calculated from Equations (14.33) and (14.34) as appropriate and $f(r_i)$ is calculated from Equation (14.32). Equation (14.30) can now be calculated using simple column summations to determine τ/τ_0. Then the actual time delay difference from that of the fundamental mode is

$$dt = \left(\frac{n_0}{c}\right)\left(\frac{\tau}{\tau_0} - 1\right) \tag{14.35}$$

This will be in ns/km if the velocity c is 0.0003 km/ns. The values of dt are stored in a separate array versus l and m.

4. *Next mode*

m is now incremented by one, and steps 2 and 3 are repeated and so on until $\beta_{lm} < \beta_{\min} = (2\pi/\lambda)n_{\min}$. When this occurs m is reset to 0 and l is incremented by one, and the procedures are repeated. Finally when both m and l are too large the programme stops.

5. *Statistical procedures*

The first array listing the caustic radii and β is available for graphing or printout as required. The second array needs processing to determine the dispersion and bandwidth. The LOTUS statistical algorithms are used to derive and plot a frequency polygon of t versus a suitable 'Bin' range of time about zero. (It is necessary to do this in two steps in order to weight the frequency of dt_{lm}, $l \neq 0$ by 2 to account for polarization degeneracy).This gives a representation of the shape of output pulse which would be received after sending an infinitely short pulse down a kilometre of the fibre. The mean and standard deviation of dt can also be obtained. If the simulated output pulse is clean and can reasonably be termed Gaussian, then the bandwidth can be estimated from the standard deviation using Equation (14.21). However this can lead to gross errors if the pulse is not clean or Gaussian and the bandwidth should be determined by applying a suitable fast Fourier transform algorithm (e.g. in 'Statgraphics') to the output pulse frequency polygon.

Tests of Numerical Programme

The programme can be tested against the theoretical analysis of power profiles. Equations (14.27) and (14.28) can be checked by running the numerical analysis on the series of power profiles in Figure 14.18 with $a = 25\,\mu\text{m}$, $dn = 0.014$, $NA = 0.202$, $\Delta = 0.0097$ and at $\lambda = 850\,\text{nm}$, $V = 37.33$. The number of modes recorded by the programme, multiplied by 4 to account for polarization, is given in the table as M_n and is within 10% of half the value given by Equation (14.27), M_{gm}. It was noted that Equation (14.10) differed from Equation (14.27) by a factor of two. This is probably due to the analysis assuming that m is a continuous variable. The bracketed numbers in the figure show the calculated spread of dT calculated from Equation (14.28) and show excellent agreement within the limitations of the frequency polygons. Note that Equation (14.28) is multiplied by n_0/c as in Equation (14.35) to obtain dT in ns/km.

Using the programme with greater time resolution to determine the optimum profile exponent at 850 and 1300 nm gave the results of Figure 14.19 in which it will be seen that at 850 nm and 1300 nm the narrowest pulse spreads are given by $\alpha = 2.07$ and 1.89 respectively. These compare well (within 1%) with the values of 2.05 and 1.87 given by the analytical values from Equation (14.18).

Accepting that a Gaussian shape is only an approximation, the bandwidths calculated by Equation (14.21) from the standard deviation of these pulses, are about 8 and 9 GHz · km respectively. The 850 nm bandwidth would be limited to about 3 GHz · km by chromatic dispersion however.

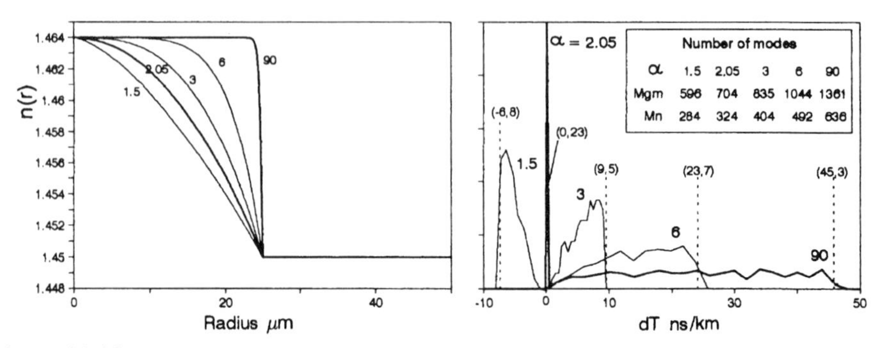

Figure 14.18
Numerical analysis of modal dispersion of power profiles

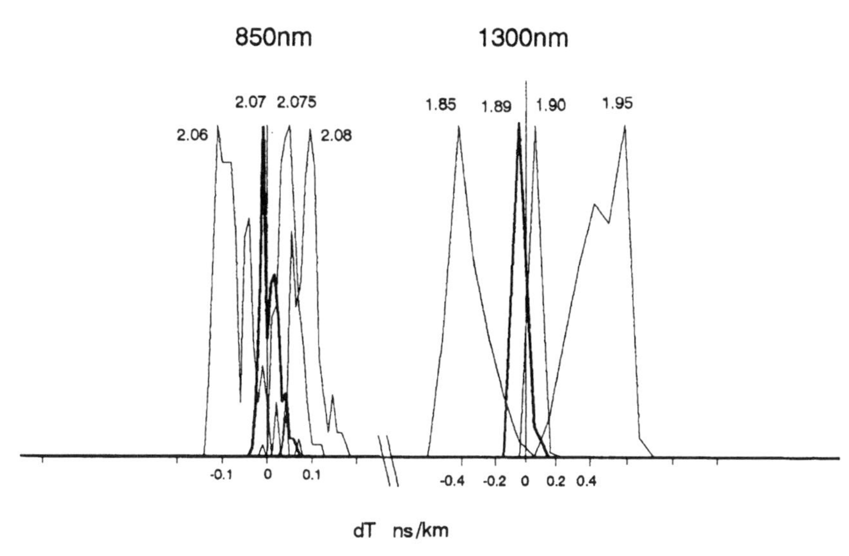

Figure 14.19
Optimum profile from numerical analysis

The pulses for 1300 nm involve fewer modes (V is inversely proportional to λ so there are only 0.43 times the number of modes) and the class interval was accordingly increased for the frequency polygons. These tests give a reasonable assurance that the numerical analysis gives believable results, with the proviso that N be increased for non-ideal profiles.

Caustic Radii

The caustic radii calculated by the numerical analysis for the typical fibre profile of Figure 14.17 are shown in Figure 14.20. If the average radius for a given l is weighted by the number of modes with that l, and the average taken it results in a weighted average radius of 12.85 μm, or normalized to the core radius, $0.514a$. This is where the transmitted light spends its average time. The refractive index difference at this radius, for an approximately parabolic profile, is given by

$$\overline{dn} = (1 - 0.514^2) \times dn_0 = 0.74 \cdot dn_0 \tag{14.36}$$

This is important when considering the loss mechanisms contributing to fibre attenuation.

Bandwidth Concatenation with Length

All the analyses above have assumed that the input light power is distributed uniformly between all allowable modes. As the modes propagate down the fibre however a phenomenon known as mode coupling begins to occur. This is due to microscattering

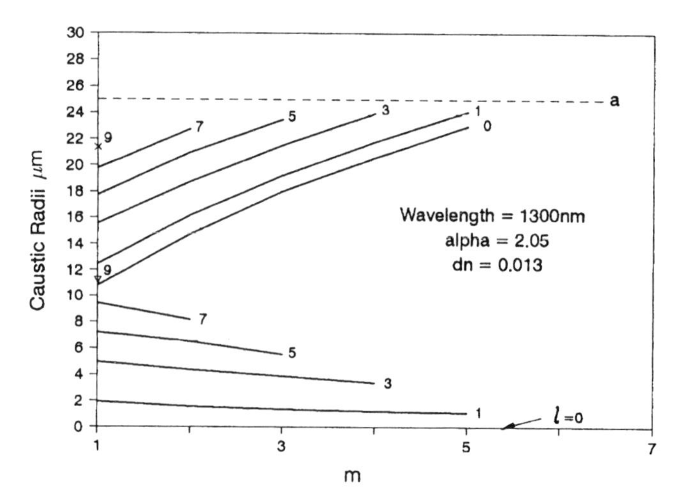

Figure 14.20
Caustic radii of modes in graded index fibre

and bends in the fibre. The effect of this is to scramble the power distribution between the modes until after some distance a statistically stable power distribution is attained. Power is still being interchanged between the faster and slower modes however. The effect of this is that the dispersion builds up for a short distance linearly with length, and then as the square root of length. To determine the length at which this transition occurs experimentally is very difficult. In order that the calculated bandwidths from the numerical analysis should agree in order of magnitude with measurements (where the linear region is truncated by deliberate mode-scrambling), the transition must occur somewhere between 200 m and 400 m as indicated for bandwidth in Figure 14.21. This means that the numerical analysis for bandwidth should be carried out for a length of

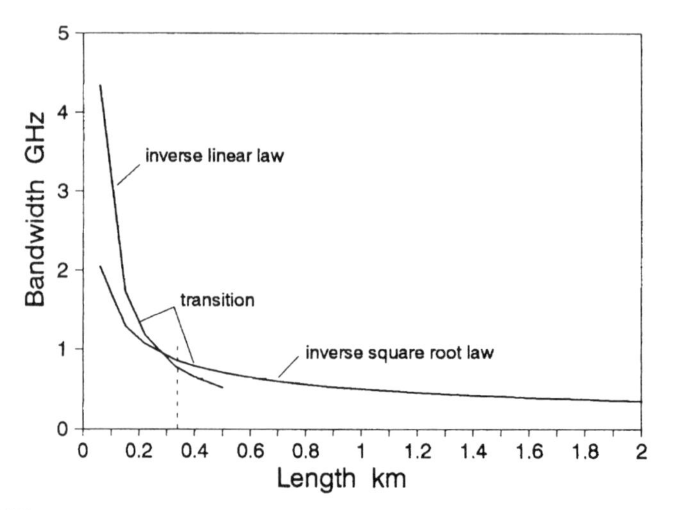

Figure 14.21
Bandwith concatenation with length

say 333 m and then the one kilometre bandwidth figure calculated as $1\sqrt{3}$ times this value. It furthermore means that the bandwidth measurements should actually be expressed as $\mathrm{MHz}\cdot\mathrm{km}^{1/2}$ and not in $\mathrm{MHz}\cdot\mathrm{km}$.

When measuring bandwidth and attenuation it is necessary to simulate this stable power distribution at the beginning of the measured length so that the measurements can be normalized to one kilometre and then extrapolated to longer lengths without having to make allowances for launch conditions. This can be done in two ways, either by a restricted launch aperture and spot size (usually 60% of NA and 80% of core diameter) or by launching the light through a 'mode scrambler' which is a much distorted short length of fibre in which the small diameter bends, or exaggerated microbends, accelerate the mode coupling effects. Either way can be considered as launching through several kilometres of fibre (but with less attenuation). Then when correcting the measurement to a one kilometre length in the case of bandwidth, the correction becomes

$$B_1 = B_L \times \sqrt{\frac{L + L_S}{1 + L_S}} = B_L \times L^{\gamma} \tag{14.37}$$

where L is the measured length and L_S is the effective length of the mode scrambler. The exponent γ is called the concatenation factor for the measurement set-up. This can be determined by making sufficient measurements on say 4 km lengths of fibre and comparing these with measurements on the same fibres cut to 2 km.

On a typical good measuring arrangement this gave a value of $\gamma = 0.19 \pm 0.02$ (the higher value applying to 1300 nm and the lower to 850 nm wavelength) for these two particular lengths. This is equivalent to a scrambler 'length' of about 4 km. Thus when converting from a 4 km or 2 km length to one kilometre the exponent becomes 0.32 or 0.55 respectively. Note that the value of the concatenation factor will vary with the actual lengths used in its determination, whereas the 'scrambler length' will not and is therefore to be preferred.

Comparison with Measurements

A measured fibre profile was simulated as closely as possible by the Equations (14.33) and (14.34) and the numerical analysis (with N in Equation (14.32) increased to 52) was applied at 850 nm and 1300 nm wavelengths. The results are shown in Figure 14.22. The class intervals used to form the frequency polygons were adjusted to the minimum value consistent with a clean pulse shape. Because of the pre-pulses, the bandwidth was derived from the full-width at half magnitude (FWHM), i.e. the dispersion, using Equation (14.23) and is shown as Bh in Figure 14.22 and compared with the measured values Bm. The measured bandwidth was obtained by the fast Fourier transform of the deconvolved output pulse of the cut-back measurement technique on 2.83 km of the fibre. The bandwidth concatenations discussed above were applied.

The agreement between the measured and numerical analysis values is reasonable considering the difficulty of simulating the profile. Note that the bandwidth values are not consistent with those to be expected from the mean profile exponent of 2.05. This is because the effects of the profile distortions mask the expected theoretical values, as will be shown below.

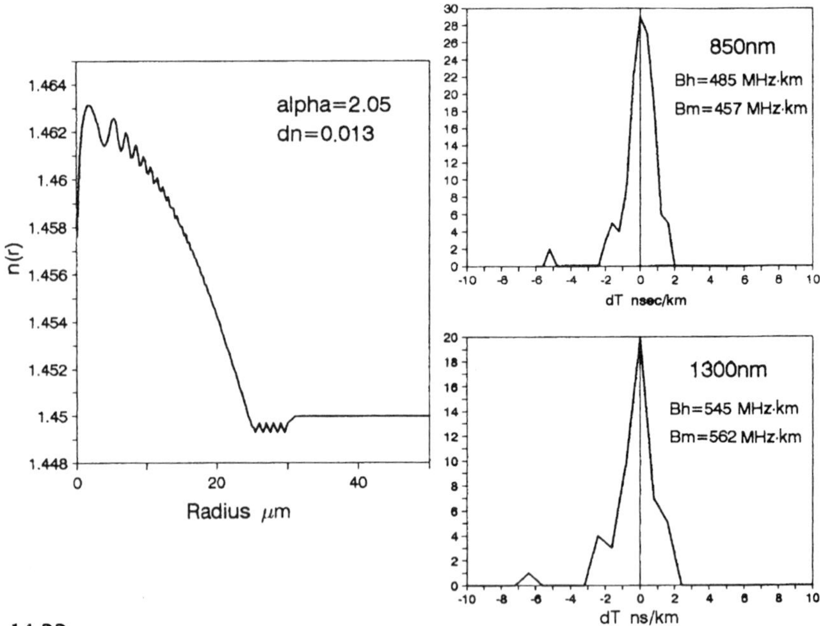

Figure 14.22
Profile and results of numerical analysis and measured bandwidths

Effect of Profile Distortions

To illustrate the effect of the core ripples on the bandwidth, the decay factor F in Equation (14.33) is increased from 5 to 10 with the result shown in the profile in Figure 14.23. All other factors are kept the same. The numerical analysis at 850 nm and 1300 nm gives the frequency polygons shown alongside, together with the bandwidths calculated from the FWHM values. Actually in this case the bandwidths calculated from the standard deviations agree well with the FWHM values. The effect of the profile alpha is now becoming apparent with the 850 nm values being significantly better than the 1300 nm bandwidth since the alpha value is much higher than the optimum at this wavelength.

The core ripple spacing and amplitude are determined by the thickness of the layers deposited during preform manufacture and are more or less fixed by the deposition rate achieved and economic considerations, but the ripple decay rate can be modified after preform collapse, by an annealing pass. Also by increasing the residence time of the preform tip during fibre drawing, more dopant migration can be encouraged thus attenuating the ripples. By increasing the amount of dopant on the inner layer of the preform it is possible to remove the centre dip of the profile after fibre drawing. However this does not have such a major effect on the core ripples, as shown in Figure 14.24.

The effect of ripples in the deposited cladding has little effect so long as the mean index matches or is slightly depressed compared with the start-tube index. If however the mean cladding index is slightly raised compared with that of the start-tube, extra radial modes can be supported for $l = 0, 1, 2 \ldots 7$. Since these modes spend a significant time at the index of the cladding they travel faster and are scattered forward in time

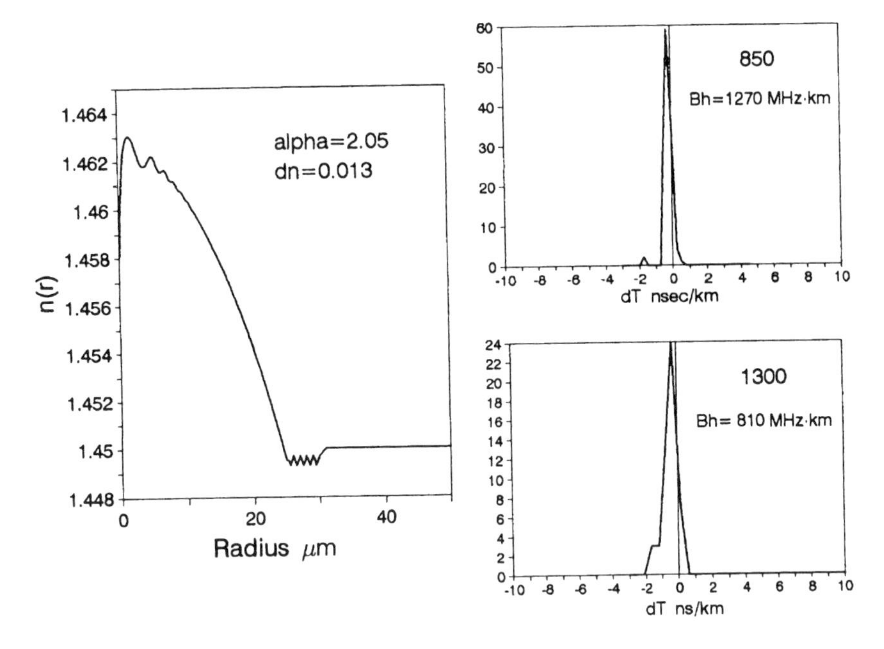

Figure 14.23
Effect on bandwidth of reducing core ripples

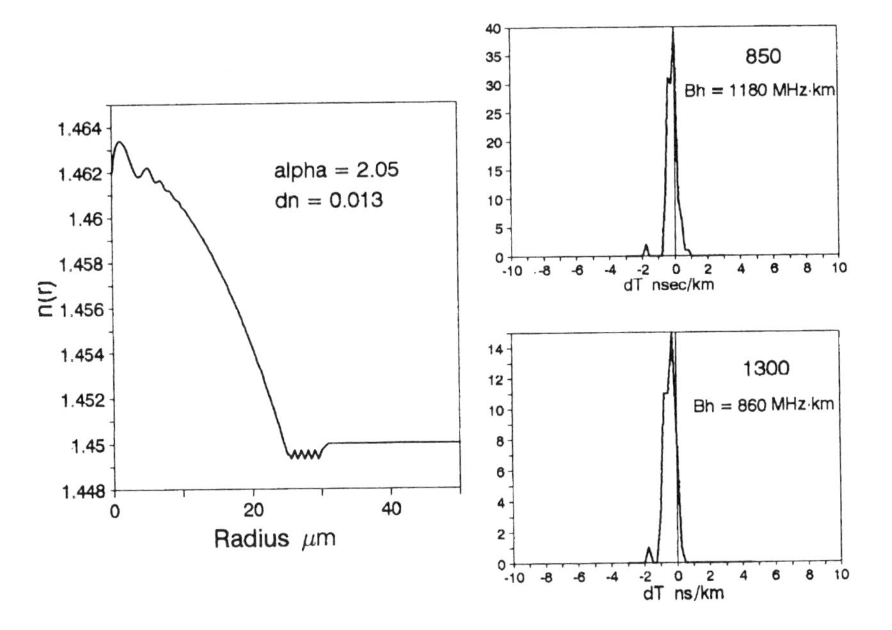

Figure 14.24
Effect of profile centre dip on bandwidth

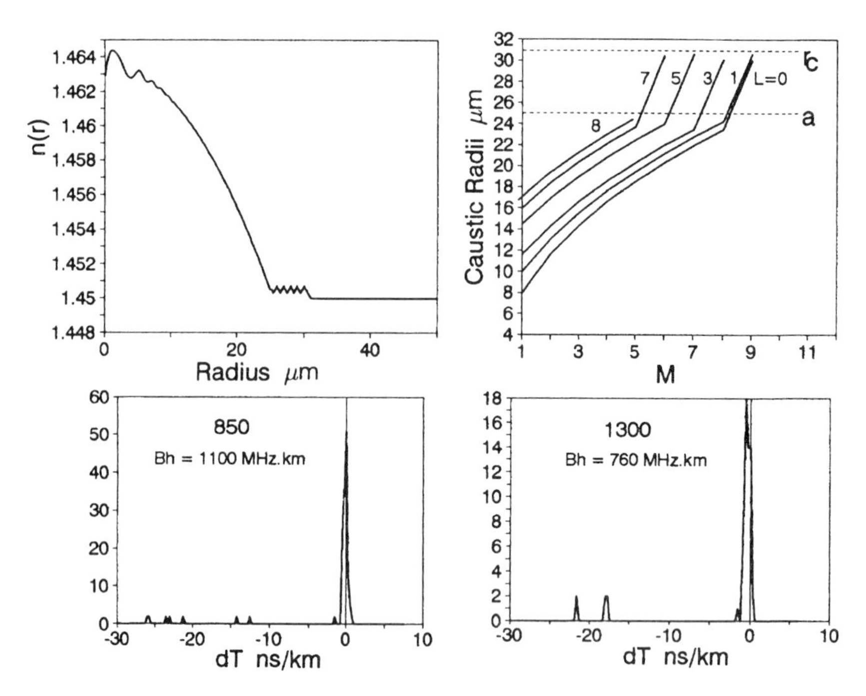

Figure 14.25
Effect of raised cladding index on caustic radii and bandwidth

compared with the majority of the modes. This is illustrated in Figure 14.25. The bandwidth, calculated from the FWHM of the main pulse, is only little affected, however, compared with Figure 14.24. The forward scattered pulses are not desirable and the profile design is generally aimed at the slightly depressed cladding form.

If all the core ripples and the centre dip are removed from the profile simulation, the bandwidths, calculated from FWHM of the output pulse, are 3.18 GHz · km and 1.33 GHz · km at 850 nm and 1300 nm respectively. The lack of agreement with the theory of Figure 14.11 is because the theory ignores the mode coupling effects which improve the bandwidth considerably at both wavelengths.

SINGLE-MODE FIBRES

As mentioned before, modal dispersion can be eliminated by proportioning the index profile to ensure that $V < 2.405$ for a rectangular step index profile at the wavelength of transmission. This is done by reducing both the core diameter and the index difference. Thus at 1300 nm wavelength, for an index difference of 0.003 and a core diameter of 10 μm, $V = 2.26$ and only the LP_{01} mode will propagate. The cut-off wavelength of the LP_{11} mode is

$$\lambda_c = \frac{2\pi a\sqrt{n_0^2 - n^2}}{2.405} \tag{14.38}$$

and for these dimensions it is 1219 nm.

Owing to the variation of refractive index with wavelength however there will still be chromatic dispersion because of the finite spectral width of the light source. For a good-quality solid-state laser this spectral width is less than 1 nm, but for an LED it is generally of the order of tens of nanometres. The dispersion of a typical single-mode fibre at 1300 nm is small and is about −5 ps/km per nm of source spectral width, but at 1550 nm it rises to about 16 ps/km/nm. Thus the bandwidth for a laser source at 1300 nm would be of the order of 100 GHz · km (Equation (14.23)) but for an LED source it would be about 10 GHz · km. Since there is no modal mixing these bandwidths will degrade linearly with length, unlike the graded index multimode fibre bandwidth.

Step Index Theory for Single-Mode Fibre

In a material of refractive index n, the phase change coefficient, or propagation constant, is given by

$$\beta = kn = \frac{2\pi}{\lambda} n \quad \text{where } \lambda \text{ is the free space wavelength} \tag{14.39}$$

For a step index fibre of core radius a and refractive index n_0 in a semi-infinite cladding of index n, the weakly guiding theory of Gloge [43] defines the following parameters

$$U = a\sqrt{kn_0^2 - \beta^2} \quad \text{and} \quad W = a\sqrt{\beta^2 - kn^2} \tag{14.40}$$

The normalized frequency and normalized propagation constant are then given by

$$V = a\sqrt{U^2 + W^2} = ka\sqrt{n_0^2 - n^2} \tag{14.41}$$

and

$$b = 1 - \left(\frac{U}{V}\right)^2 = \left(\frac{W}{V}\right)^2 \tag{14.42}$$

whence

$$\beta = k\sqrt{n^2 + b(n_0^2 - n^2)} \tag{14.43}$$

An effective refractive index for the fibre is defined as

$$n_e = \sqrt{n^2 + b(n_0^2 - n^2)} = n(1 + b \cdot \Delta) \quad \text{where } \Delta = dn/n \text{ for small } dn \tag{14.44}$$

The group time of travel and the dispersion coefficient are given by

$$T_g = \frac{1}{c} \cdot (n_e - \lambda \cdot n_e') \tag{14.45}$$

$$D = dT_g/d\lambda = \frac{-\lambda}{c} n_e'' \tag{14.46}$$

where c is the vacuum velocity of light and the primes indicate differentiation with respect to wavelength. For the step index fibre, Equation (14.44) can be differentiated to give

$$D = D_w + D_c + D_d + D_x \tag{14.47}$$

where the four terms are the waveguide dispersion, the material chromatic dispersion, a differential chromatic dispersion and a cross-product dispersion respectively (White and Nelson [44]). By using an empirical equation, due to Rudolph and Neumann [45],

$$W = pV + q = 1.1428V - 0.996 \tag{14.48}$$

which is accurate to 0.1% for $1.5 < V < 2.5$, White and Nelson showed that the waveguide dispersion is given by

$$D_w = \frac{-q^2\lambda}{nc(2\pi a)^2} \tag{14.49}$$

which depends only on the wavelength in the fibre and the core diameter. The material chromatic dispersion was derived as

$$D_c = -\frac{\lambda}{c}\, n'' + \delta D \tag{14.50}$$

where δD are terms in n, n' and n'', i.e. only the cladding material index, which amount to about -0.25 ps/km/nm for silica and are negligible.

$$D_d = D_w\left(\frac{pV}{q}\right)\left[\left(\frac{pV}{q}+1\right)^2 \frac{\lambda^2\Delta''}{2\Delta} + \frac{\lambda\Delta'}{\Delta} + \left(\frac{\lambda\Delta'}{2\Delta}\right)^2\right] \tag{14.51}$$

which varies from about -0.2 to 2.5 ps/km/nm varying with wavelength and inversely as core diameter. This is only negligible at the lower wavelengths (1 to 1.3 μm)

$$D_x = D_w\left(\frac{pV}{q}\right)^2 \left(\frac{\lambda n'}{n} \cdot \frac{\lambda\Delta'}{\Delta}\right) \tag{14.52}$$

This is about -0.002 to 0.03 and is also negligible. These dispersion terms are shown in Figure 14.26. The effect of neglecting D_d is also shown in the D curves by the lower, dashed line.

The material chromatic dispersion has a wavelength of zero dispersion at 1275 nm and the effect of adding the waveguide dispersion is to shift the zero dispersion wavelength of the fibre to longer wavelengths as shown. For a 5 μm radius core it is shifted near to the operating wavelength of 1300 nm. For a 2 μm radius it is shifted near to the other operating wavelength of 1550 nm. However such a large shift also moves the cut-off wavelength of the second mode down to about 500 nm. Consequently, light transmitted at 1550 nm would be only very weakly guided, with a lot of power travelling in the cladding, and micro- and macrobend losses would be high. Obviously the smaller core diameter would also be more difficult to joint. However special profiles can be used to make successful dispersion shifted fibres (see later).

As with graded index cores, practical step index profiles are far from perfect rectangles, usually they exhibit centre dips and ripples, as shown in Figure 14.27. A means of analysing these real profiles is needed. One method suggested by Hussey and Pask [46]

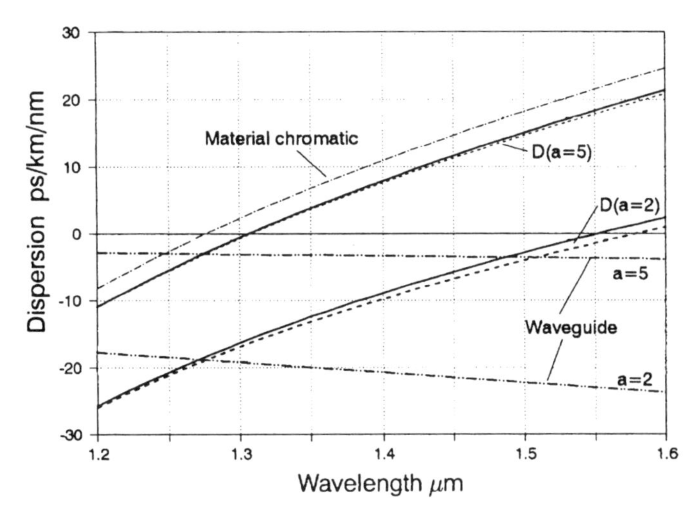

Figure 14.26
Dispersion components in single-mode fibre; $dn = 0.003$

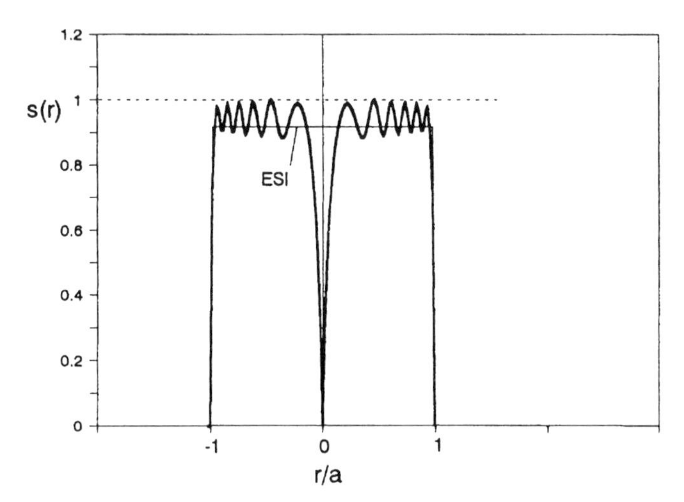

Figure 14.27
Typical single-mode fibre core profile and equivalent step index

was to construct an equivalent step index (ESI) profile from the moments of the actual profile. This approach was further extended by Martinez and Hussey [47] to give an enhanced ESI method (EESI).

ESI Method of Analysing Actual Profiles

The actual profile is first described by the shape function

$$s(r) = \frac{n(r)^2 - n^2}{n_{\max}^2 - n^2} = \frac{dn(r)}{dn_{\max}} \tag{14.53}$$

where $dn(r)$ is the index difference at radius r and dn_{max} is the maximum index difference. The moments of $s(r)$ are given by

$$M_p = \int_0^a s(r) \cdot r^{p+1} \cdot dr \tag{14.54}$$

Only the first few even moments are used, $p = 0$, 2 and 4.

The equivalent fibre parameters are given by

$$a_s = a \cdot \sqrt{\frac{2M_2}{M_0}} = a \cdot \sqrt{2M} \quad \text{where } M = M_2/M_0 \tag{14.55}$$

$$h_s = \frac{M_0}{M} \tag{14.56}$$

$$V_s = \sqrt{2M_0} \cdot V \tag{14.57}$$

where the suffix, s, indicates the equivalent step index values of radius, fractional height and V-value.

These use only the first two even moments of $s(r)$. To improve the dispersion prediction the EESI calls on the normalized difference of the fourth moments of the actual and equivalent step index profiles

$$\delta M_4 = \frac{\left(\frac{M_4}{M_0} - \frac{4M^2}{3}\right)}{\frac{4M^2}{3}} \tag{14.58}$$

and gives the dispersion parameter as

$$b(V) = h_s \cdot b_s(V_s) \cdot [1 + |\delta M_4| \cdot f(V_s)] \tag{14.59}$$

where from Equation (14.48)

$$b_s(V_s) = \left(\frac{W_s}{V_s}\right)^2 = \left(\frac{1.1428V_s - 0.996}{V_s}\right)^2 \tag{14.60}$$

and the empirical function

$$f(V_s) = 0.313V_s - 0.013V_s^2 \tag{14.61}$$

The moments are evaluated numerically, using Equation (14.54) in steps of $r/a = 0.02$, to give a_s, h_s and V_s which enables the cut-off wavelength to be calculated, using $n_0 = n_{max}$, in

$$\lambda_c = \frac{2\pi a \sqrt{n_0^2 - n^2}}{V_c} \tag{14.62}$$

where

$$V_c = \frac{2.405}{\sqrt{2M_0}} \tag{14.63}$$

Note that this is the theoretical cut-off wavelength which is generally 11.7% higher than the effective cut-off wavelength measured on a 3 m fibre length according to the recommended test method (see 'Fibre measurements' later).

$b(V)$ from Equation (14.59) is used in Equation (14.44) to derive the effective refractive index, n_e versus wavelength. To do this, a listing versus wavelength in 10 nm steps, of V, V_s, $f(V_s)$, $b_s(V_s)$, $b(V)$, n and n_e is created. n_e is then differentiated numerically twice for use in Equation (14.46) to give the total dispersion, D, versus wavelength. If required the waveguide dispersion, D_w, and the material chromatic dispersion, D_c, can then be calculated from Equations (14.49) and (14.47).

Calculation of Refractive Indices

The refractive indices are calculated versus wavelength by using a three-term Sellmeier equation

$$n^2 - 1 = \frac{A\lambda^2}{\lambda^2 - B^2} + \frac{C\lambda^2}{\lambda^2 - D^2} + \frac{E\lambda^2}{\lambda^2 - F^2} \tag{14.64}$$

Fleming [38] gives values for the constants for a range of glass compositions (Table 14.1).

The compositions are given in mole percentages. For different dopant concentrations the calculated refractive index at the given concentration, n_g, is scaled as follows

$$n_m = (n_g - n_{Si})\frac{m}{m_g} + n_{Si} \tag{14.65}$$

where m is the desired mole fraction, m_g is the given mole fraction and n_{Si} is the index of pure silica.

Mode Field Radii

Coupling of light into the fibre, fibre-to-fibre jointing, bending and microbending losses are functions of various definitions of the radial extent of the field in the fibre. For launching and jointing, the applicable radius is that at which the field amplitude is $1/e$

Table 14.1
Sellmeier coefficients for lightguide glasses (reprinted from [38] by permission of IEE)

Material	*A*	*B*	*C*	*D*	*E*	*F*
Quenched pure silica	0.696750	0.069066	0.408218	0.115662	0.890815	9.900559
13.5% Ge/SiO_2	0.711040	0.064270	0.451885	0.129408	0.704048	9.425478
9.1% P/SiO_2	0.695790	0.061568	0.452497	0.119921	0.712513	8.656641
13.3% B/SiO_2	0.690618	0.061900	0.401996	0.123662	0.898817	9.098960
1.0% F/SiO_2	0.691116	0.068227	0.399166	0.116460	0.890423	9.993707
16.9% Na/32.5% B/ 50.6% SiO_2	0.796468	0.094359	0.497614	0.093386	0.358924	5.999652

of its on-axis maximum. As it is the intensity or power of the field which is measured, this is the radius giving the $1/e^2$ value of the intensity. This is called the mode field radius (MFR) or 'spot size' of an optimally exciting Gaussian beam. Marcuse [48] has given an empirical formula, accurate to 1%, for step index fibres

$$w_0 = a(0.650 + 1.619/V^{1.5} + 2.879/V^6) \tag{14.66}$$

When using the moments method on actual profiles, not only are the ESI radius and V_s used in this formula but also a multiplier $\sqrt{2M}$ to give

$$w_{0s} = \sqrt{2M} \cdot a_s(0.650 + 1.619/V_s^{1.5} + 2.879/V_s^6) \tag{14.67}$$

For joint offset losses, a field radius called the Petermann 'strange' spot size is appropriate. This has an exact analytical value, but an approximation was derived by Hussey and Martinez [49] accurate also to 1%, namely

$$\bar{w}_s = w_{0s} = \sqrt{2M} \cdot a_s(0.016 + 1.561/V_s^7) \tag{14.68}$$

The fraction of *field* propagating in the core, η, can also be related to the Petermann spot size and the normalized propagation constant, b from Equation (14.59), by

$$\eta = \frac{2}{V_s^2 \bar{w}_s^2} + b \tag{14.69}$$

and the fraction of *power* propagating in the core is given by

$$P_{core} = 1 - (1 - \eta)^2 \tag{14.70}$$

These are illustrated in Figure 14.28. At 1300 nm in a step index fibre with core radius 5 μm and index difference of 0.003, $V = 2.71$ and about 83% of the field (or 97% of the power) is in the core, and decreases as the wavelength gets further from the second mode cut-off wavelength.

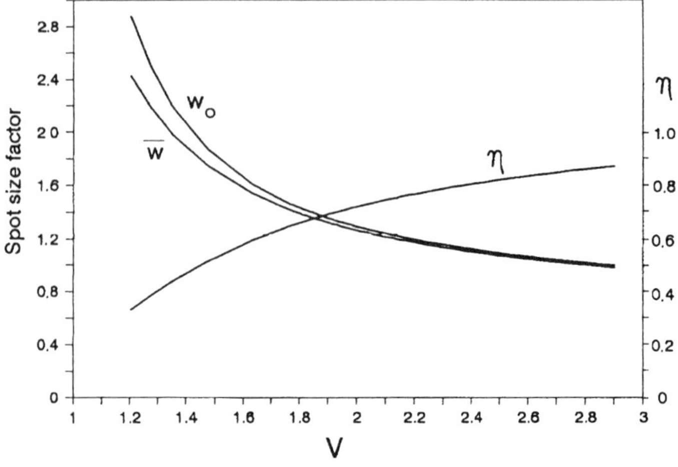

Figure 14.28
Spotsize factors and field fraction in core

Table 14.2
Mean measured and EESI-calculated values of fibre parameters

Parameter		Measured Mean	Measured Std dev.	EESI calculation numerical	EESI calculation analytical	Estimated meas. error
Core radius	μm	4.56	0.22	4.19		0.03
Index difference		0.00461	0.00042	0.00422		0.00003
Effective cut-off wavelength (3 m measurement)	nm	1202.5	40.6	1201.5		7
Mode field radius	μm	4.95	0.27	5.02		0.05
Dispersion 1285 nm	ps/km/nm	−2.30	0.32	−2.85	−2.54	0.03
Dispersion 1330 nm	ps/km/nm	1.60	0.36	1.24	1.58	0.03
Dispersion 1550 nm	ps/km/nm	16.77	0.47	17.35	17.74	0.03
Zero dispersion wavelength	nm	1311	3.81	1316	1313	0.5

Application of EESI Method

Comparison of the dispersion calculated from the analytical expressions and from the double numerical differentiation of the effective refractive index shows an agreement within better than 0.6 ps/km/nm and within 5 nm for the zero dispersion wavelength, for typical dimensions of step index single-mode fibres.

Using a database of 56 preform profiles and about 100 fibres drawn from them, which were measured for dispersion at 1285 nm, 1330 nm and 1550 nm, cut-off wavelength and MFR at 1300 nm, the average values were derived and an average profile was constructed using Equation (14.33). The latter is shown in Figure 14.27. The EESI method was applied with the results shown in Table 14.2.

The profile moments established by numerical integration were (step index values in brackets)

$$M_0 = 0.447 \quad (0.500)$$
$$M_2 = 0.218 \quad (0.250)$$
$$M_4 = 0.141 \quad (0.167)$$

Table 14.3
Effect of centre dip and core ripples on fibre parameters

Parameter	Dip and ripple	Ripples no dip	Dip no ripples	No dip no ripples
Eff. cut-off wavelength	1201.5	1203.9	1200.0	1203.8
MFR	5.02	5.00	4.85	4.79
D1285	−2.85	−2.85	−3.06	−3.03
D1330	1.24	1.24	1.06	1.08
D1550	17.35	17.34	17.22	17.23
Zero dispn. wavelength	1316	1317	1319	1317
Vc	2.54	2.55	2.47	2.47

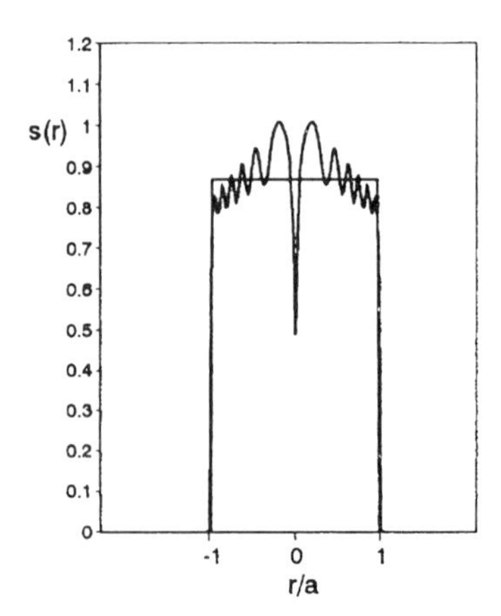

Figure 14.29
Individual profile and EESI results

It will be seen that having established the EESI parameters for the profile, the analytical calculation of the dispersions gives good agreement with the measured values and the numerical differentiation gives very reasonable agreement. The effect of the centre dip and the ripples in the profile is small, as can be seen in Table 14.3.

An individual fibre, chosen for the distinct slope to the top of its profile and small centre dip as shown in Figure 14.29 also gave reasonable agreement between the calculated and measured values, Table 14.4.

Although this ESI approach is satisfyingly instinctive and has been derived by Hussey and Pask from the scalar wave equation for the fibre, the author has not found it possible to use it for dealing with cladding profile variations or for the more complex profiles used for dispersion-flattened fibres.

Central Difference Approximation

An alternative approximate method for solving the scalar wave equation was given by Sammut and Pask [50] based on central difference approximations, as follows.

Table 14.4
Measured and EESI-calculated parameters for individual fibre

		Measured	EESI
Core radius	μm	4.87	4.70
Index difference		0.00417	0.00362
Eff. cut-off	nm	1159.5	1144
MFR	μm	4.95	5.38
D1285		−2.63	−2.44
D1330		1.24	1.65
Zero disp. wavelength	nm	1314	1312

For a circularly symmetric fibre profile with the function

$$g(r) = (n_0^2 - n_R^2)/(n_0^2 - n^2) \tag{14.71}$$

the scalar wave equation is

$$\psi'' + \psi'/R + (U^2 - V^2 g(R) - l^2/R^2)\psi = 0 \tag{14.72}$$

where the primes denote differentiation with respect to wavelength. $R = r/a$ is the radius normalized to the core radius, a. U, V are the usual eigenvalues Equation (14.40) and l is the azimuthal mode number. The boundary conditions on the field, ψ, ensure that it remains finite between $R = 0$ and $R = \infty$, and that ψ and ψ' are continuous everywhere. The derivatives are replaced by central difference approximations, using a stepping interval h and a stepping function S_R

$$\psi_{R+h} = S_R \cdot \psi_R \tag{14.73}$$

and Equation (14.72) is converted to the recurrence relation

$$(2R + h) \cdot S_R + (2R - h) \cdot S_{R-h} - 4R + 2h^2 R(U^2 - V^2 g(R) - l^2/R^2) = 0 \tag{14.74}$$

with initial conditions

$$S_0 = 1 - h^2/4 \cdot (U^2 - V^2 g(0)) \quad \text{for } l = 0 \tag{14.75}$$

$$S_h = (\tfrac{2}{3})[l^2 + 2 - h^2(U^2 - V^2 g(0))] \quad \text{for } l \geq 1 \tag{14.76}$$

A circular fibre will have a uniform refractive index beyond some normalized radius R_0 and in this region the field is proportional to $K_l(WR)$, where $W = \sqrt{V^2 - U^2}$ and K_l is a modified Bessel function of the second kind. Matching the values of ψ'/ψ on either side of $R = R_0$ gives the second boundary condition

$$S_{R_0} - \frac{1}{S_{R_0 - h}} = \frac{2h \cdot W}{R_0} \left[\frac{K_l'(WR_0)}{K_l(WR_0)}\right] \tag{14.77}$$

For the fundamental mode, $l = 0$ and $K_0'(z) = -K_1(z)$ so that the boundary condition for the fundamental mode becomes

$$S_{R_0} - \frac{1}{S_{R_0 - h}} = -\frac{2h \cdot W}{R_0} \left[\frac{K_1(WR_0)}{K_0(WR_0)}\right] \tag{14.78}$$

The Bessel functions can be evaluated from the formula

$$K_l(z) = \int_0^\infty e^{-z \cosh\theta} \cdot \cosh l\theta \cdot d\theta \qquad \text{((14.79)}$$

by numerical integration with $d\theta = 0,1$ and the limit ∞ taken as about 5.

To apply this on a spreadsheet, n_R, $g(R)$, S_R and the field $\psi_R$are listed against R at intervals of $h = 0.01$ out to a radius beyond R_0. S_0 for the fundamental mode is obtained from Equation (14.75) and subsequent values of S_R are calculated from the transposed Equation (14.74) namely

$$S_R = \left[\frac{1}{2R+h}\left\{\frac{-(2R-h)}{S_{R-h}} + 4R - 2h^2R(U^2 - V^2g(R) - l^2/R^2)\right\}\right] \tag{14.80}$$

using the calculated V for the profile, (Equation (14.41)) and a trial value of $U < V$. The field ψ_{R+h}, is then calculated using Equation (14.73) and the previous values of S_R and ψ_R.

The boundary condition comparison Equation (14.78), is set up using numerical integration of Equation (14.79) and U is adjusted manually (or using a macro command) until the difference is very close to zero. This should be carried out to about eight significant figures in U. This is a little tedious but the convergence is quite rapid. Greater accuracy and even quicker convergence is obtained with larger values of R_0. The initial choice of U is made easier by observing a graph of the field against R, which, for the fundamental mode, should never become negative and should tail away to zero at large R.

The normalized propagation constant, b, is calculated from Equation (14.42), using the final value determined for U, and is used to determine the effective refractive index from Equations (14.44) and (14.64).

To determine the dispersion by numerical differentiation the whole procedure must be carried out at three closely spaced wavelengths λ and $\lambda \pm d\lambda$ where λ is the wavelength at which the dispersion is to be calculated from

$$D = -\frac{\lambda}{c} \cdot \frac{d^2 n_e}{d\lambda^2} \tag{14.81}$$

A value of $d\lambda = 5$ or 10 nm is suitable.

To determine the cut-off value of V for the second mode where $l = 1$, U is set equal to V in the spreadsheet, S_h is determined from Equation (14.76) and the second boundary condition becomes

$$S_{R_0} - \frac{1}{S_{R_0-h}} = -\frac{2h \cdot l}{R_0} \tag{14.82}$$

A trial value of U is then adjusted to satisfy this condition as accurately as required, generally to four or five significant figures.

Applications of the Central Difference Method

For both the fundamental and second mode boundary conditions, Equations (14.77) and (14.82), the author thinks there are errors in the original paper, which have been corrected above. With these corrections the V_c values of a step index and of a truncated parabolic profile were determined and compared with published values as below

Profile	Calculated V_c	Published V_c
Step index	2.40485	2.40491
Parabolic	3.51872	3.51805

an agreement of better than 0.02%.

For the dispersion of a germanosilicate step index profile with $a = 5\,\mu m$ and $dn = 0.003$ determined as above, the author obtained the following values

Wavelength nm	Dispersion ps/km/nm Central Difference	Rudolph–Neumann	(difference)
1285	−1.69	−2.38	(+0.79)
1330	+2.32	+1.81	(+0.51)
1550	+17.32	+18.16	(−0.84)

The comparison column uses values of b calculated from the Rudolph–Neumann empirical relation Equation (14.48). An agreement of about 0.8 ps/km/nm is indicated.

For a step index, *dispersion-shifted,* germanosilicate fibre with $a = 2\,\mu m$ and $dn = 0.0209$ (13.5 mol% of germanium) the dispersions at the same wavelengths as above were respectively −14.3, −10.7 and +1.6 ps/km/nm. These values agreed to similar accuracy with a more accurate variational matrix calculation on a step index fibre of the same dimensions, published by Sharma and Banerjee [51]. The fields for this fibre are shown in Figure 14.30 together with the MFRs at $1/e$ of the axial field value. Note that the profile function plotted is $(1 - g(R))$. The effective cut-off wavelength of this fibre would be 1153 nm so that signals at both 1300 nm and 1550 nm would be single-moded. However the MFRs are small at about 2.2 and 2.4 μm which would make jointing of the fibres more difficult than with the non-shifted design.

The central difference method also enables the effect of cladding ripples to be analysed. The average profile of Figure 14.27 is accompanied by an extensive deposited cladding having the inevitable ripples as shown in Figure 14.31. The analysis shows that the fields spread far into the cladding, which carries 4% to 8% of the power at

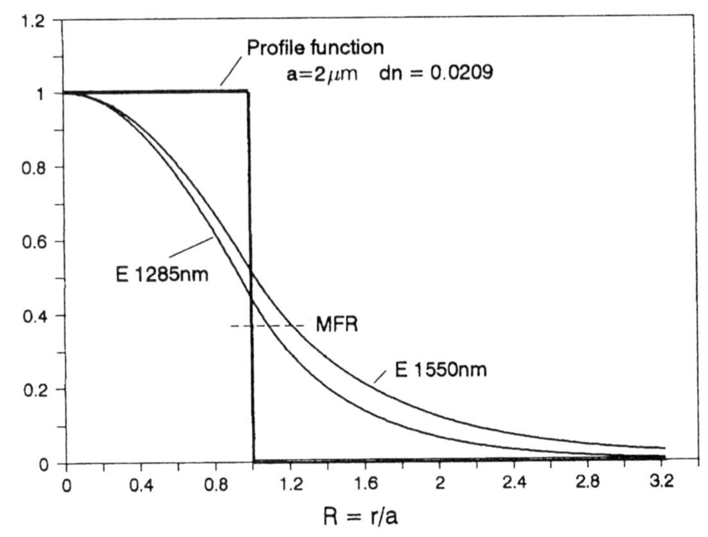

Figure 14.30
Dispersion-shifted fibre profile and fields

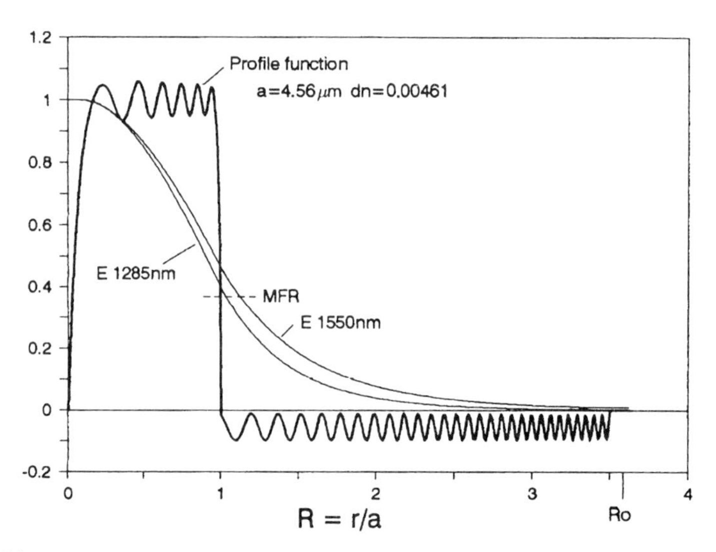

Figure 14.31
Simulated preform profile with cladding ripples

1285 nm and 1550 nm. The calculated dispersions, zero dispersion wavelength and the cut-off wavelength are given below

Wavelength	Cladding ripples With	Without	Measd
1285	−1.46	−1.80	−2.30
1330	+2.48	+2.13	+1.60
1550	+17.37	+16.95	+16.77
Zero disprsn	1302	1306	1311
Eff. Cut-off	1216	1216	1203

The effect of the cladding ripples is seen to be about +0.5 ps/km/nm. The comparison with the measured values again shows an agreement of about 0.8 ps/km/nm. This is probably acceptable since the profile used for the analysis is only approximately simulated and applies to the preform before drawing into fibre, whereas the measurements were made on fibres. Comparison with the values calculated from the EESI method (without cladding ripples) is also within 1 ps/km/nm.

Thus it is concluded that both methods of analysis are probably accurate to within better than 1 ps/km/nm for dispersion and to within 5 nm to 10 nm for zero dispersion wavelength and cut-off wavelength. Considering the difficulty of implementing a designed profile into an actual fibre profile this is considered satisfactory. More accurate methods of analysis are available to the mathematically able, but these two methods are sufficiently accurate for most purposes and can be readily applied on PC spreadsheets. The moments method is instinctively attractive but does not handle cladding variations, whereas the central difference method does, and is mathematically easier to appreciate without being too difficult to use. It also generates the field values as part of the calculation.

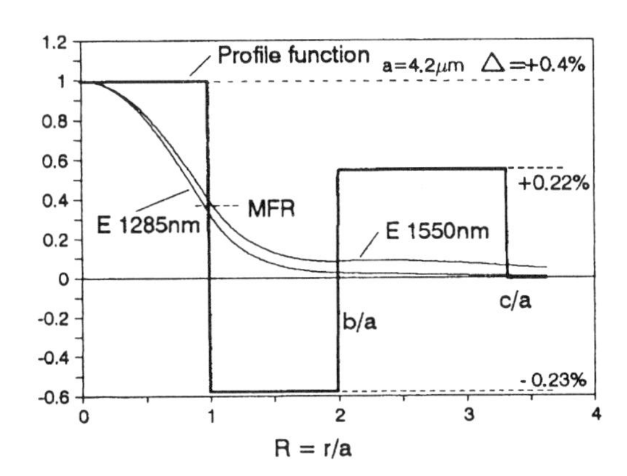

Figure 14.32
Dispersion-flattened fibre profile

Dispersion-Flattened Fibre Profiles

By treating the outer portions of a profile designed for dispersion-flattened response as the cladding region, the central difference method can be used for its analysis. The profile function given in Sharma and Banerjee's paper uses fluorine doping only. The author has adapted it for implementation in mixed germanosilicate and fluorine-doped material and the profile function and fields are shown in Figure 14.32. Such profiles are also known as W-profiles or triple-clad profiles. The delta-values and the radii were taken from the paper but the outer radius c/a was varied to give the dispersion response shown in Figure 14.33, with very low total dispersion at 1300 nm and 1550 nm and a variation of ± 6 ps/km/nm over the whole wavelength range. Notice that these low dispersions are also achieved with very reasonable MFRs, at both 1285 nm and 1550 nm, of about 4.2 μm making them no more difficult to joint than 1300 nm single-mode fibres. The effective cut-off wavelength for this profile is calculated to be 1265 nm which should enable both 1300 nm and 1550 nm signals to be propagated satisfactorily in the fundamental mode only.

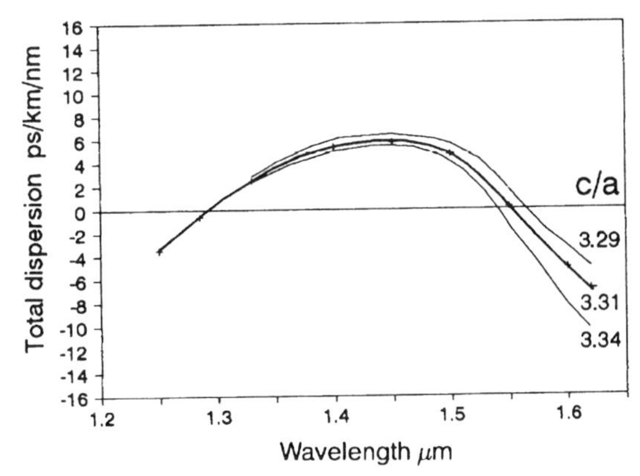

Figure 14.33
Calculated dispersion of dispersion-flattened fibre

The author has no experience in the manufacture of such fibres but suspects that they may be difficult to control in production since a variation of 1.5% ($\pm 0.1\,\mu$m in 13.9 μm) in c/a gives a variation of ± 2 ps/km/nm at 1550 nm as shown. This sensitivity is confirmed by Ainslie and Day [52]. Also the dispersions may be more sensitive to profile ripples than those previously analysed.

LOSS MECHANISMS

Glass Scattering and Absorption

The principal reasons for power loss in optical fibre transmission were given in the introduction to this chapter as UV absorption due to Ge–O bond vibration, IR absorption due to Si–O bond vibration and Rayleigh scattering of light from the glass lattice. In addition there are found to be wavelength dependent absorptions, $C(\lambda)$, principally from harmonics of several resonances of the Si–OH bond at about 2.7 μm wavelength and a wavelength independent loss, B, thought to be due to random small fibre dimensional changes. These losses may be expressed as follows

$$\alpha = \alpha_{\mathrm{UV}} + \alpha_{\mathrm{IR}} + \frac{A}{\lambda^4} + B + C(\lambda) \tag{14.83}$$

where A is the Rayleigh scattering coefficient. This depends on the degree of disturbance of the glass lattice by dopant atoms. In the case of doped silica glasses it may be shown that

$$A = 0.63 + R\sqrt{\Delta\%}\ \mathrm{dB/km} \tag{14.84}$$

R is an empirical factor which has been determined from measurements on fibre as 0.422 for germanium-doped silica and 0.35 for phosphorus-doped silica. The constant term 0.63 represents the Rayleigh scattering coefficient for pure undoped silica. Equation (14.84) can also be written as

$$A = 0.63 + \frac{10R}{\sqrt{2}\cdot n} \cdot NA\ \mathrm{dB/km} \tag{14.85}$$

which for a *step index* germanosilicate fibre becomes

$$A = 0.63 + 2.06 \cdot NA\ \mathrm{dB/km} \tag{14.86}$$

within 0.4%.

Note that n is the refractive index of pure silica which varies from 1.4528 to 1.4444 from 850 nm to 1550 nm (a variation of 0.78%). For a *near-parabolic* profile (see below) this becomes

$$A = 0.63 + 1.75 \cdot NA \tag{14.87}$$

The UV loss has been modelled by Shultz [53] for germanosilicate glass, as

$$\alpha_{\mathrm{UV}} = \frac{1.542m}{46.6m + 60} \exp\left(\frac{4.63}{\lambda}\right) \mathrm{dB/km} \tag{14.88}$$

where m is the mole fraction of germanium doping which corresponds to

$$dn = 0.001162 \times (\text{mole\% Ge}) \tag{14.89}$$

The value of dn to be used in both Equations (14.84) and (14.89) is that appropriate to the regions of the fibre core in which the transmitted modes spend the majority of their time. Thus for perfect step index fibres it corresponds to the constant dn of the core region. For imperfect single-mode fibres it would correspond to the ESI dn and for nearly parabolic graded index cores it corresponds to the dn at a normalized radius $r/a = 0.521$, which corresponds to 0.729 times the projected on-axis dn of the profile, see Equation (14.36).

The IR loss for silica glasses, has been modelled by Miya et al. [54], as

$$\alpha_{\text{IR}} = 7.81 \times 10^{11} \cdot \exp\left(\frac{-48.48}{\lambda}\right) \text{ dB/km} \tag{14.90}$$

In all the above equations the wavelength is in μm and the attenuations are defined by the ratio of the input and output optical powers

$$\alpha = 10 \log \frac{P_{in}}{P_{out}} \text{ dB} \tag{14.91}$$

The value of B is generally found to be 0.15 dB/km average, but it can vary about this value by a factor of two. The cause of this loss is unclear but it is thought to be due to effects at the cladding boundary, either random dimensional variations or related to locked-in thermal strains. There is some evidence that matching the thermal constants of the deposited cladding to the core, by the inclusion of germanium-doped layers at the boundary (index-compensated with additional fluorine) reduces the value of B significantly to about 0.06 dB/km. Also there is evidence that reducing the profile centre-dip by etching can reduce the value of B.

The components of attenuation for a germanosilicate graded index fibre with NA = 0.20 using the above modelling are shown in Figure 14.34. The total attenuation is also shown and includes the effect of hydroxyl contamination at a fairly low level of the order of 15 p.p.b. This excess attenuation due to hydroxyl contamination is modelled after Walker [55] and for an excess peak at 1.39 μm of 1 dB/km is shown in Figure 14.35. It will be seen that the excess attenuation for this level of hydroxyl contamination at 850 nm, 1300 nm and 1550 nm is very small. However it can become significant at levels of OH contamination giving excess attenuations greater than 5 dB/km at 1390 nm.

Hydroxyl Loss

Walker's modelling of the hydroxyl losses is based on Gaussian forms centred at wavelengths related to modes of vibration of the Si–OH bond. At about 1390 nm there are four components and at about 1250 nm, two components. The excess attenuation is given by

$$\alpha_{\text{OH}} = A_{\text{OH}} \sum_{1}^{n} A_n \, e^{-(\lambda-\lambda_n)^2/2\sigma_n^2} \tag{14.92}$$

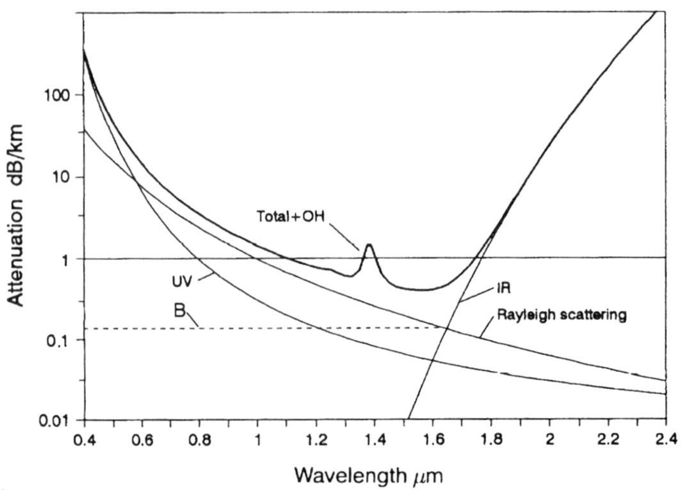

Figure 14.34
Attenuation components of graded index fibre

Walker determined the various parameters from measurements on 50 fibres as below

n	λ_n µm	σ_n	A_n
1	1.3473	0.0139	0.066
2	1.3789	0.0070	1.000
3	1.3900	0.0117	0.624
4	1.4016	0.0277	0.305
5	1.2426	0.0085	0.058
6	1.2562	0.0136	0.028
7	0.950	0.007	0.060
8	3.05	0.7	A_{POH}

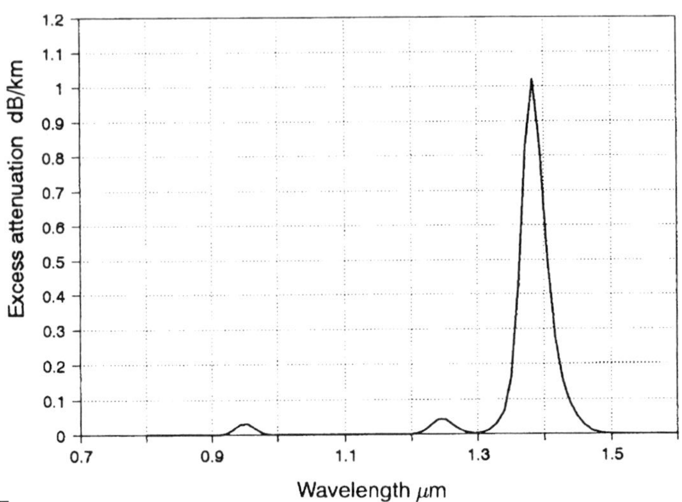

Figure 14.35
Excess attenuation due to hydroxyl contamination

The author found it necessary to add the terms of $n = 7$ when modelling multimode fibre down to 750 nm in order to account for the third overtone of the hydroxyl absorption at 950 nm. Walker also gives the terms of $n = 8$ to account for losses due to P–OH vibrations in fibres with significant phosphorus doping in the core. A_{OH} and A_{POH} in dB/km are adjusted to achieve the required overall excess peak heights at 1390 nm and 3050 nm (or at an appropriate wavelength in the band).

The width of the various Gaussian peaks is in practice modified by the width of the monochromator slit (i.e. the spectral width of the light source) used in the measurements. This can be allowed for by using the quadratic sum of the spectral width and the Gaussian width given, i.e.

$$\sigma = \sqrt{\sigma_n^2 + d\lambda^2} \tag{14.93}$$

where $d\lambda$ is the spectral width of the source, which is often of the order of 0.015 μm.

Using the modelling given above, this author has found it possible to match carefully measured fibre attenuations to within 0.03 dB/km for both multimode and single-mode designs. However macrobending and microbending effects due to spool winding have also to be allowed for. Depending on the spool size, fibre design and tension, these can easily amount to about 0.15 dB/km at 1550 nm.

It has been seen how, particularly in single-mode fibres, the field spreads considerably into the cladding region of the fibre. If the power flowing in the cladding (say about 7% of the total) is lost by excess attenuation compared with the core, this would amount to an extra loss of 0.3 dB. For this reason, since natural quartz is quite lossy, the cladding must consist of vapour deposited silica out to a sufficient radius from the core. This is usually about 7 μm for multimode fibre and about 20 μm for single-mode fibre.

Achievable Attenuations

In order to reduce fibre attenuation, the Rayleigh scattering loss must be minimized. From Equation (14.85) this means reducing the NA of the fibre. This limits the acceptance angle and increases the difficulty of launching power into the fibre. It also reduces the fibre guidance and increases bend losses. Typical specifications require the NA to be between 0.19 and 0.21 for 50 μm core diameters and between 0.26 and 0.29 for 62.5 μm core diameters. Thus for optimum bandwidth, *near-parabolic* profiles in germanosilicate fibres the following attenuations would be achievable

Fibre	NA	dn	A	λ	α_R	α_{UV}	B	α_{tot}
50/125	0.2	0.0138	0.931	850	1.78	0.48	0.1	2.37
				1300	0.33	0.07	0.1	0.50
62.5/125	0.275	0.0258	1.043	850	2.00	0.86	0.1	2.96
				1300	0.37	0.13	0.1	0.60

This assumes a fairly good value for B, and an excess hydroxyl peak at 1390 nm of less than 1 dB/km. The IR loss is negligible at these wavelengths. An improvement in B, or a slightly lower NA can shave about 0.05 dB/km from these minimum attenuations.

For the *single-mode* fibre profile of Figure 14.30. with B reduced to 0.05 dB/km the minimum predicted attenuations would be 0.38 dB/km at 1300 nm and 0.23 dB/km at 1550 nm. Since fluorine has a less effect on the Rayleigh scatter coefficient than germanium, and since it also eliminates the Ge–O UV absorption, a purely fluorine-doped depressed index fibre can achieve even lower attenuation values of about 0.25 and 0.15 dB/km at 1300 and 1550 nm.

Bend Losses in Fibres

When a fibre carrying a signal is bent uniformly with constant radius of curvature, the mode fields are distorted. As a consequence of this the local phase velocities on the outside of the bend increase to maintain the equiphase wavefronts. When this velocity exceeds the light velocity in the cladding, c/n (n is the cladding index), the field begins to propagate in the cladding and power is leaked away from the signal. Power coupling between adjacent modes also increases so that power diffuses out from all modes. These effects cause an excess loss known as macrobending loss which is proportional to the bend length and varies exponentially with bend radius.

If the bend length is short, of the order of 1 mm, the mode conversion effects at the beginning and end of the bend can can cause more loss than the bend itself. Bends of this kind, occuring randomly in curvature and length, are known as microbends. They can be caused for example by pressure of the fibre on a micro-rough surface such as a spool barrel, or by the pressure between layered turns of a fibre on a spool. These fibre axial distortions can be minimized by cushioning the bare fibre with plastic layers of carefully selected mechanical properties (e.g. the primary coatings required to protect the fibre surface against mechanical damage).

Macrobend Loss

Most published measurements of macrobend loss on commercially available matched cladding, non-dispersion-shifted, germanosilicate single-mode fibres fall within the shaded regions shown in Figure 14.36. At a given diameter the bend loss is less at 1300 nm than at 1550 nm, where the fibre is more weakly guiding. The macrobend loss variation with diameter falls into two regions which indicates that there are two loss mechanisms at work. Above about 40 mm diameter the slope of the curve is much less than in the region below 40 mm. It is difficult to make accurate measurements on the usual 10 or 20 m of fibre in this lower loss region so the actual slope may be understated in Figure 14.36. Measurement spread can be quite large in the lower loss region.

In the higher loss region this author has found that the macrobend loss can be modelled by the empirical formula

$$\alpha_M = \exp\left[8.5 - 519 \times D_{mm}\left\{\frac{1}{\lambda \times (\text{Mac\#})}\right\}^3\right] \text{dB/m} \qquad (14.94)$$

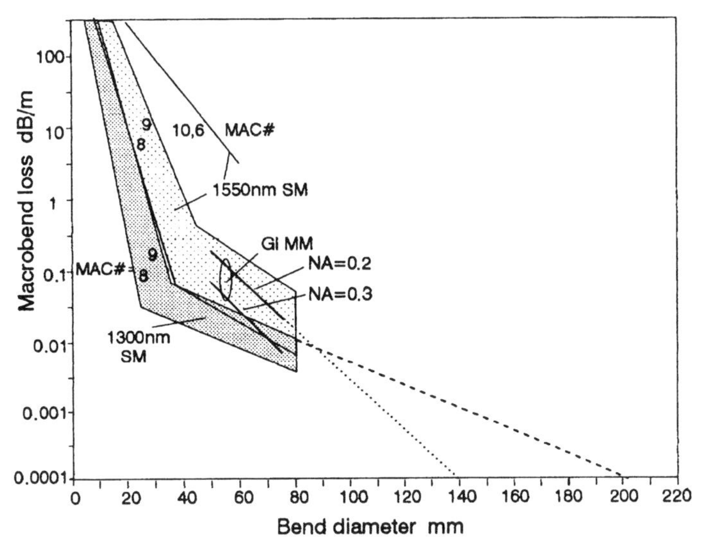

Figure 14.36
Macrobending loss versus bend diameter

where D is the bend diameter in mm, the wavelength is in μm and a parameter suggested by Su-Vu Chung [56] called the 'Mac number' is used

$$\text{Mac\#} = \frac{(2 \times \text{Mode field radius at 1300 nm})}{(\text{Eff. 3 m cut-off wavelength in } \mu\text{m})} = \frac{2 \times \text{MFR}}{\lambda_{ce}} \tag{14.95}$$

This reflects the fact that the further the signal wavelength is from the cut-off wavelength, the more weakly guiding is the fibre and the smaller the MFR due to higher index difference, the more strongly guiding the fibre is. In the figure the shaded areas in the high loss region are bounded by fibres with Mac numbers of 8 and 9 at both wavelengths. Also shown is the bend loss at 1550 nm of a more weakly guiding fibre with a Mac number of 10.6.

Extrapolating the lower loss region to large bend diameters as shown in the figure indicates that the bend diameter of the fibre must be greater than 200 mm, either in a cable or on a spool, to avoid penalties of greater than 0.1 dB/km. For very low loss fibres this diameter must be increased to greater than 260 mm to keep the penalty less than 0.01 dB/km. The author suggests that the loss mechanism predominating in the higher loss region is due to direct radiation from the fundamental mode, whereas in the lower loss region the mechanism is due to weak coupling between the fundamental mode and the very loss of LP_{11} mode. Some published measurements do not show this lower loss region. Depressed cladding fibres generally show lower bend losses than matched cladding fibres, but this is also in accord with a lower Mac number due to the MFR being smaller. Bends and interference ripples in the lower loss region due to power reflections at the cladding/substrate boundary, have also been reported in depressed cladding fibres.

Artiglia et al. [57] quote an approximate expression for the macrobend loss

$$\alpha_B = G\sqrt{\frac{w_\infty^3}{R}} \cdot \exp\left[\frac{-8R}{3\sqrt{2}\,k^2 n^2 w_\infty^3}\right] \text{ dB/m} \tag{14.96}$$

where G is a constant, R is the uniform bend radius, k is the wave number $2\pi/\lambda$ and

$$w_\infty = \sqrt{\frac{2}{kn_0\Delta\beta}} = \frac{1}{kn}\sqrt{\frac{2}{\sqrt{1+b\,dn/n}-1}} \tag{14.97}$$

where $\Delta\beta$ is the propagation constant difference between the fundamental mode and the radiation mode and b is the normalized propagation constant. The author has not had much success using this formula unless the exponent of w_∞ in the exponential term is reduced from 3 to 2.5 when a value of G of the order of 10 000 gives a reasonable match in the high loss region only. Note that R, λ and w must all be in μm.

Also shown in Figure 14.34 are the macrobend losses for typical multimode fibres. The macrobend loss of multimode fibres is not sensitive to wavelength but since the guidance of the modes reduces as the NA is reduced, it is quite sensitive to the NA. The author can find no information to indicate two loss regions corresponding to the single-mode losses. Extrapolation of the multimode lines indicates that a spooled (or in-cable) diameter of greater than 140 mm should not produce a significant attenuation penalty.

Microbend Loss

Microbend loss is caused by random distortions in the fibre axis causing field distortions which couple power from one mode into the next higher mode. Eventually a cut-off mode is reached and power is lost out of the fibre core. In the case of a single-mode fibre the first power transfer is into the cut-off LP_{11} mode. The random bends will have a spatial wavelength spectrum and an autocorrelation function which defines the most probable mean spatial wavelength. The stiffness of typical fibres limits their distortion at periodicities less than 1 mm. The coupling coefficients between modes rapidly decrease if the spatial wavelength is greater than the beat wavelength of modes in the fibre

$$\Lambda = \frac{2\pi}{\Delta\beta} = \frac{2\pi a}{\sqrt{2\Delta}} \tag{14.98}$$

In typical fibres this is about 1 mm for adjacent low order modes and because of the mechanical limitation of fibre stiffness, non-adjacent modes do not couple significantly. Thus spatial wavelengths of the curvature distortion near to 1 mm are the most serious in causing microbend losses.

Petermann [58] and Petermann and Kuhne [59] have given an analysis of microbending in single-mode, graded index and W-profile fibres. They give an approximate formula for microbending loss

$$\begin{aligned}\alpha_\mu &= \tfrac{1}{4}(kn_0 w_0)^2 \cdot \Phi(\Delta\beta) \\ &= \tfrac{1}{4}(kn_0 w_0)^2 \cdot \Phi\left(\frac{2}{kn_0 w_0^2}\right)\end{aligned} \tag{14.99}$$

This is strictly only true for a Gaussian shape of the fundamental field and w_0 is the mode field radius at $1/e$ of the axial field amplitude. However in most cases this holds well above $V = 2.0$ for both step index and truncated parabolic profiles.

$$w_0^2 = \frac{2\int_0^\infty E^2 r^3 \cdot dr}{\int_0^\infty E^2 r \cdot dr} \tag{14.100}$$

This uses the second moment of the fundamental field power to define the effective radial extent of the field and is evaluated at the operating wavelength. Petermann then chooses to describe the power spectrum of the fibre axis curvature as

$$\Phi\left(\frac{2}{kn_0 w_0^2}\right) = \Phi(\Omega) = \frac{A}{\Omega^{2p}} \tag{14.101}$$

where A is a constant and the exponent p controls the spectral decay of the curvature. $p = 0$ decribes a flat power spectrum with a very short correlation length, while $p = 1$ or 2 gives more realistic situations with larger correlation lengths. Equation (14.99) thus becomes

$$\alpha_\mu = \frac{A}{4}(kn_0 w_0)^2 \cdot (kn_0 w_0^2)^{2p} \tag{14.102}$$

As the field departs from the Gaussian shape, i.e. as the wavelength gets progressively longer (and further from the LP_{11} cut-off in single-mode fibres) Equation (14.102) becomes increasingly more approximate and Petermann proposed using a spotsize $w(p)$ in place of w_0, where

$$w_0 \leq w(p) \leq w_\infty \tag{14.103}$$

Near the LP_{11} cut-off the inequalities tend to equalities. As the wavelength increases the three spotsizes increase and diverge increasingly. w_∞ from Equation (14.97) is the maximum value of the spotsize. $w(p)$ increases from w_0 at $p = 0$ to w_∞ at $p = \infty$.

The value of p has been measured by various people for different surfaces. For the special case of the inner surface of a plastic tube used to contain the fibre loosely, a value of $p = 4$ was found. Artiglia et al. using a random assemblage of spheres of diameter 0.5–3 mm to compress the fibre, found $p = 2$. Other workers have found $p = 2$ to be applicable to the majority of micro-rough surfaces.

Petermann also found a fourth spotsize, for $p = -1$, the so-called 'strange spotsize'

$$\bar{w}^2 = \frac{2\int_0^\infty E^2 r \cdot dr}{\int_0^\infty (dE/dr)^2 r \cdot dr} \tag{14.104}$$

which governs the loss due to N randomly spaced joint offsets of r.m.s. value da in a length L of fibre

$$\alpha_j = (N/L) \cdot (da)^2/\bar{w}^2 \tag{14.105}$$

This 'strange spotsize' was later identified by Pask as

$$\bar{w}^2 = 2/w_{ff}^2 \tag{14.106}$$

where w_{ff} is the physically measurable r.m.s. far-field width. This spotsize is always slightly less than w_0. Marcuse [48] developed an empirical relation for w_0 for step index fibres, Equation (14.66), and Hussey and Martinez [49] extended this to $\bar{w}$, Equation (14.68).

These analyses enable the calculation of microbending loss for a randomly distorted fibre for a chosen value of p if $w(p)$ can be found. Yang et al. [60] suggested that taking the mean value

$$\overline{w(p)} = \tfrac{1}{2}(w_0 + w_\infty) \tag{14.107}$$

would be within 10% for the majority of fibre constructions but they worked with a value of $p = 3.2$.

Reduction of Microbending Loss

Petermann analysed the case of a fibre with a given degree of axial distortion. What is now required is to investigate how these distortions arise and what can be done to minimize them. This has been analysed by Gloge [61] and more recently by Grasso et al. [62].

A silica or glass fibre has to have its surface protected against abrasion and this is usually done by means of thin plastic coverings immediately after drawing (the primary coatings). The mechanical properties of these coverings are also important in modifying the sensitivity of the fibre axis to distortion by outside forces. The fibre and its coatings are modelled as a thin elastic beam pressed against a micro-rough surface as shown in Figure 14.37.

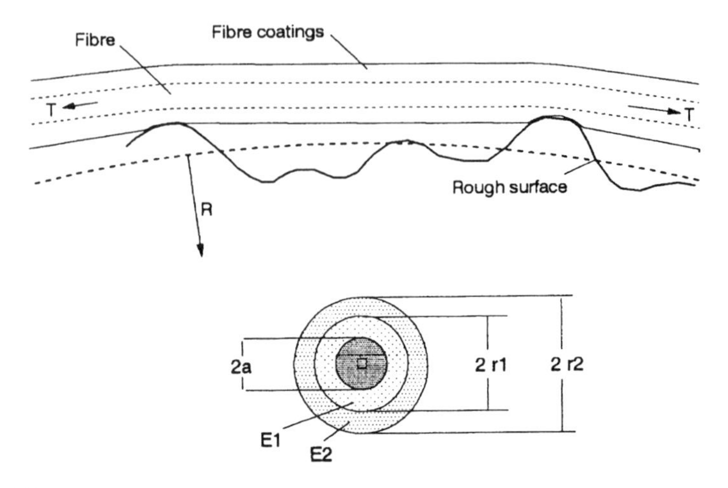

Figure 14.37
Coated fibre distorted by pressure against rough surface

The linear pressure, P, is calculated from the tension T in the fibre and the average radius of curvature R, of the surface

$$P = T/R \tag{14.108}$$

Both the peaks of the rough surface profile and the fibre coatings are subjected to a degree of compression which modifies the forces transmitted to and distorting the fibre. Assuming all the distortions are elastic, Grasso et al. calculated the curvature function in Equation (14.101) as

$$\Phi(\Omega) = \frac{\Omega^4 S(1 + \pi^2 D^4 \sigma^4/(4P^4))^{-1/4}}{(1 + (H/D)\Omega^4)^2} \tag{14.109}$$

where

n_0 = core refractive index
Ω = radian frequency of the curvature
P = linear pressure
S = power spectral density of surface roughness
H = fibre stiffness = EI
σ = r.m.s. of surface deformations = $\int_0^\infty S \cdot d\Omega$
E = Young's modulus of fibre
E_s = Surface compression Young's modulus
I = Fibre moment of inertia = $\pi a^4/4$
D = Surface lateral rigidity; $1/D = 1/D_c + 1/E_s$
D_c = Coating effective lateral rigidity. In the case of double coated fibre, if the outer coat is harder than the inner coat then approximately

$$D_c = E_1 + E_2\left(\frac{r_2 - r_1}{r_2}\right)^3$$

$E_{1,2}$ and $r_{1,2}$ are the moduli and radii of the inner/outer coatings as shown in Figure 14.37.

Ignoring the unity term in the parentheses in both the numerator and denominator of Equation (14.109) is allowable in practical cases. Then substituting for $\Omega = 1/(kn_0 w(p)^2)$ and inserting $\Phi(\Omega)$ in Equation (14.98) gives the microbending loss

$$\alpha_\mu = \frac{A}{4\sqrt{2\pi}} \times \left(\frac{S}{\sigma}\right) \times P \times \left(\frac{D}{H^2}\right) \times (kn_0 w_0)^2 (kn_0 w(p)^2)^4 \tag{14.110}$$

This consists of a constant term, a term describing the surface roughness, the linear pressure term, a fibre protection term and the spotsize terms of Equation (14.102) with $p = 2$ coming out naturally from the analysis.

The roughness power spectrum of a surface is obtained by measuring the surface irregularities and carrying out a fast Fourier transform on the results to obtain the amplitude of the harmonics versus the radian frequency. S is the relative amplitude of the surface roughness power spectrum at a given radian frequency (or spatial wavelength) and for microbending studies should be the value at a wavelength between 0.4 mm and 1.5 mm (15.7 rad to 4.2 rad) given by Equation (14.98). Frankly the author is confused by the definition of σ given in the paper (quoted exactly above apart from a symbol change) [62]. It corresponds to the area under the power spectrum curve and would therefore represent the effect of a flat spectrum.

Equation (14.110) shows that the microbending loss on a given surface is directly porportional to the pressure, P. So that for a given winding tension the microbend loss decreases as the mandrel radius increases. This author therefore speculates that the low loss region of macrobending in Figure 14.36 may be due to microbend loss on the mandrels used for these measurements.

The fourth term shows that to reduce the microbending susceptibility of the packaged fibre, the lateral rigidity of the inner coating should be low compared with the square of the fibre stiffness. This is achieved by using a fairly thick layer of soft material for the inner coat and a thin layer of hard material for the outer coat.

Finally, large spotsizes make the fibre more susceptible to microbending losses. Since the spotsizes increase with increasing wavelength, so will the microbending loss under given conditions.

Microbend Loss Measurements

To assess the effect of microbend loss on measurements of attenuation of primary coated fibres wound on spools, some measurements were carried out for the author by S. Spammer [63]. The spectral attenuation of a 2 km length of single-mode fibre with a single primary coat of soft acrylate was measured at three different tensions of winding on a 300 mm diameter barrel spool in a multilayer winding. These were compared with the spectral attenuation of the same fibre when loosely coiled on the floor. Microbending loss due to compression of the fibre at the multitude of crossing points in the winding was then derived from the differences in the measurements. The procedure was repeated at 100 g tension using a different fibre (with a lower Mac number) coated with a single coat of even softer silicone rubber.

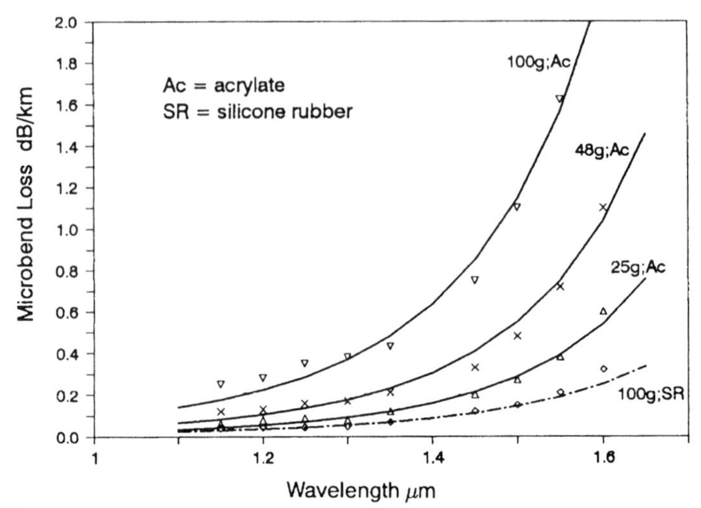

Figure 14.38
Effect of tension and primary coaating on microbend loss of multilayer winding

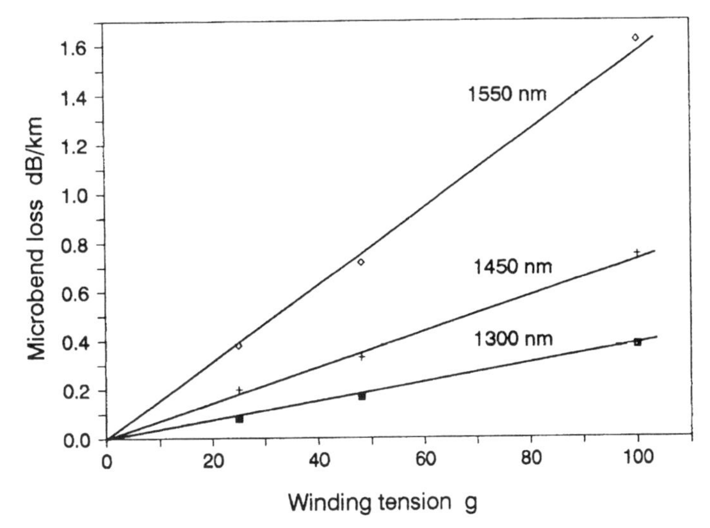

Figure 14.39
Microbend loss versus tension in multilayer winding

The results are shown by the symbols in Figure 14.38 and compare well with the lines calculated from Equation (14.110) using

$E = 72\,000$ MPa $\quad D_c = D/2 = 50$ MPa (acrylate), 12.5 MPa (silicone)
$I = 11.98 \times 10^{-6}$ $\quad$ (for two similar fibres pressed together)

Using the ESI parameters for the fibre, V and b were calculated, w_0 was calculated from the Marcuse formula and w_∞ was calculated from Equation (14.95). $w(p)$ was taken as the mean of w_0 and w_∞. Since it was not possible to measure the spectral roughness involved, the first two terms of Equation (14.110) were lumped together as a constant for the test arrangement and this was derived from the measurements as $A' = 3 \times 10^{-16}$, for the tension in grams, all diameters in mm, spotsizes and wavelength in μm and moduli (from the manufacturers' values) in MPa.

The linearity with winding tension is illustrated in Figure 14.39. Single layer windings were dificult to measure but appeared to have microbending losses, due to the roughness of the barrel, of about half that of the multilayer windings (consistent with $D_c = D$).

Thermal Buckling of Fibres

Another cause of loss in fibres arises as a result of fibre buckling, caused by reducing temperature as the fibre coatings contract more than the fibre. This causes microbending losses in the fibre. A fibre with a hard outer coat and a soft inner coat will buckle when the temperature drop is

$$dT_c = \frac{2[1/E_0A_0 + 1/E_2A_2]\sqrt{KE_0I_0}}{(a_2 - a_0)} \tag{14.111}$$

where E_n, A_n, I_n and a_n are the Young's moduli, areas, moments of inertia and thermal expansion coefficients of the layers, with $n = 0$ being the fibre and $n = 2$ the hard outer layer. K is the spring constant of the soft inner layer, given by Vangheluwe [64] as

$$K = \frac{4\pi E(1-\nu)(3-4\nu)}{(1+\nu)\left[(3-4\nu)^2 \ln\left(\frac{r_0}{r_i}\right) - \frac{(r_0/r_i)^2 - 1}{(r_0/r_i)^2 + 1}\right]} \tag{14.112}$$

where E is the Young's modulus and ν is the Poisson ratio of the inner soft coating of inner and outer radii r_i and r_0.

Suhir [65] gives an analysis of this mechanism which predicts the same buckling temperature but by more detailed analysis of the spring constant for dual coats, and by considering the fibre curvatures before the onset of buckling, predicts that microbending losses will start to increase rapidly at $dT_c/2$. While consideration of the effects of lateral forces on the fibre would favour the use of a thick soft inner coat so that distortions remain elastic, Suhir's analysis indicates that a thinner soft coat is desirable for resisting microbend losses under longitudinal thermal compression. An optimum thickness for the inner coat would appear to be between 20 and 30 μm.

Hydrogen Degradation

A number of workers, among them Barnes et al. [66] have investigated loss increments of optical fibres in an atmosphere containing hydrogen. These losses are due to hydrogen migrating to and dissolving in the silica of the fibre where it can further attach itself to unsatisfied oxygen bonds to produce an hydroxyl group and give Si–OH resonant losses. The hydrogen seems to originate from various plastic materials (particularly silicone rubbers) used in cable constructions or in the case of a cable containing dissimilar metallic elements, from electrolysis occuring from moisture ingress. The combination of steel and aluminium in the cable is very prone to hydrogen evolution by electrolysis. Barnes et al. give the following treatment.

Loss Increment Due to Hydrogen Degradation

The loss increment has three components α_1 due to interstitially dissolved hydrogen, α_2 due to hydroxyl formation and α_3 a wavelength dependent loss.

$$\alpha_1 = 0.014 \cdot p \cdot S(\lambda) \cdot \exp(1550/RT)\ \mathrm{dB/km} \tag{14.113}$$

where

p = partial pressure of hydrogen (atm)
T = absolute temperature (K)
R = gas constant (82.1 cm^3 · Atm/mol/K)

λ nm:	1290	1300	1310	1320	1330	1550
$S(\lambda)$:	1.8	1.3	1.0	0.8	0.7	2.7

Experimentally, within the 1300 nm window, α_2 and α_3 exhibit similar time and pressure dependence and appear to be related. They can therefore be combined as

$$\alpha_2 + \alpha_3 = 10\sqrt{p}\left[1 - \frac{1}{(1 + 2.33 \times 10^{-11} \cdot t \cdot \sqrt{p}\exp(-26\,100/RT)}\right] \text{dB/km} \quad (14.114)$$

where $t =$ time in seconds.

Using these formulae for a typical singlemode fibre, the total loss increments at 1300 nm after 25 years will be

Partial pressure of hydrogen (atm):		0.01	0.1	1.0	10.0
Attenuation increment	@ 0 °C:	0.00025	0.0025	0.025	0.252
(dB/km @1300 nm):	@ 20 °C:	0.00026	0.0026	0.026	0.256
	@ 40 °C:	0.00026	0.0026	0.026	0.260

α_1 is generally three times the vallue of $\alpha_2 + \alpha_3$. Because of this and $S(\lambda)$, the losses at 1550 nm are about double those at 1300 nm. The losses are most sensitive to the hydrogen partial pressure and less so to temperature and time (beyond a year or so).

Partial Pressure of Hydrogen in Cables

Most cables can be considered as two elements, the core and a cylindrical barrier surrounding the core. This barrier tends to inhibit the out-diffusion of hydrogen generated by the core. The partial pressure of hydrogen accumulating in the cable is given by

$$p = \left(\frac{H}{Q}\right)[1 - (1 - p_0 Q/H)\exp(-Q \cdot t/v)] \rightarrow \frac{H}{Q} \quad \text{at large } t \quad (14.115)$$

where

$p_0 =$ initial partial pressure of hydrogen (atm)
$H =$ hydrogen evolution rate (cm^3/cm/s)
$Q =$ hydrogen permeation rate through barrier
$= 10^{-7}$ for APL sheath; 2.6×10^{-12} for 0.5 mm welded steel;
5×10^{-5} for a 3 mm Pe sheath.
$v =$ free volume per unit length of cable (cm^3/cm)

The hydrogen evolution rate must be determined experimentally for all the materials used in the cable. For a typical non-metallic 12-fibre cable with a 3 mm thick polythene sheath this author has calculated that the expected loss increment at 1300 nm due to hydrogen accumulation would be less than about 0.003 dB/km over 25 years even at 80 °C, $p \leq 0.1$ atm.

MANUFACTURE OF OPTICAL FIBRE

Most processes for producing optical fibres start with a large diameter preform containing a doped core region, of higher refractive index, and a cladding region. Fibre

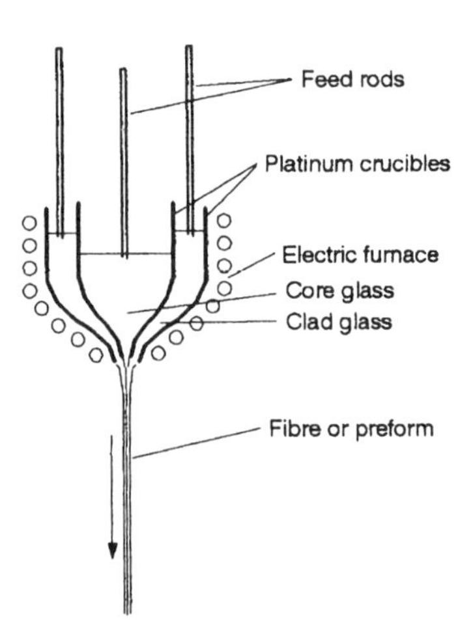

Figure 14.40
Double crucible method of fibre or preform manufacture

is then drawn from this preform in a separate operation on a drawing tower. The fibre is coated with plastic protective primary coats before reeling on to spools.

Double Crucible Method

Early work on glass fibres for optical transmission used two concentric platinum crucibles containing the molten core and cladding glass, Figure 14.40. The feed rods were made from recrystallized materials to be as free from contaminants as possible. By adjusting the exit orifices either fibres or preforms could be drawn. This method was able to achieve attenuations of about 6–10 dB/km at 850 nm. Mixing of the glasses by diffusion of ions at the molten core/cladding interface produced enough rounding of the index profile to achieve useful bandwidths of the order of 30 MHz · km (see Figure 14.10).

Vapour Deposition Methods

It was not until silica glasses were deposited from the vapour phase of highly purified silicon tetrachloride (with doping from the vapour of germanium tetrachloride) that really low attenuation fibres could be produced. This was first done by Corning Glass by the outside vapour deposition method (OVD), and shortly after by an inside deposition method (MCVD or modified chemical vapour deposition) developed by STL. The Japanese developed the vapour axial deposition method (VAD), which was also an outside deposition method, a little later. The outside deposition methods are illustrated in Figure 14.41.

In the OVD method, a 1.5 m long rotating ceramic rod support tube is heated by a traversing oxyhydrogen flame. Into the burner gases are introduced the vapour of silicon tetrachloride and a controlled flow of the dopant vapour. This deposits layers of doped silica in the form of fine sintered particles. When sufficient layers have been

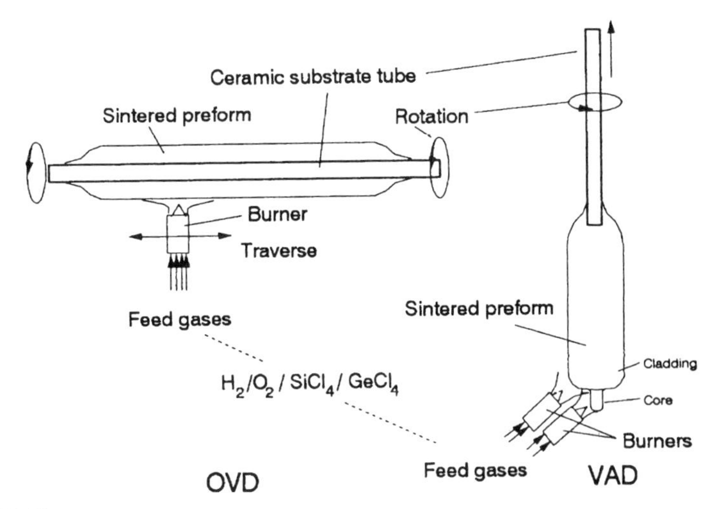

Figure 14.41
Outside vapour deposition methods of preform manufacture

deposited, the sintered preform is removed from the ceramic support, dried in chlorine gas, vitrified and collapsed radially into a solid preform rod. Control of the doping for each layer builds up the required refractive index profile. If sufficient thickness of optical cladding is deposited, the preform may then be sleeved with a commercial purity silica tube.

The VAD method is similar except that the sintered preform is built up axially on the end of a rotating ceramic rod which is automatically raised to keep the deposition point in line with the fixed burners. This enables much larger preforms to be deposited, but the control of the refractive index profile is more difficult, requiring gradation of dopant vapour between several burners. Again the sintered preform is dried and vitrified after removal from the support rod.

The MCVD method [67], Figure 14.42, separates the heating and deposition gas flows, which enables a lower quality of hydrogen to be used for the oxyhydrogen burner.

The substrate tube, or start-tube, of commercial purity silica becomes an integral part of the final preform,.and hence of the fibre, and the deposition proceeds from outside-in with the deposited optical cladding and then building up the dopant concentration for the inner layers to achieve the desired refractive index profile. The silica soot particles produced in the hot interior of the tube are deposited on the cooler walls of tube ahead of the advancing flame region by a process of thermophoresis (the migration of silica particles in the gas stream from hotter to cooler regions).

As the flame advances, the deposited soot is first sintered and then vitrified. As this happens in the presence of a relatively high chlorine concentration of 3 to 10% (from the oxidation of the tetrachlorides), drying of the doped silica deposit is facilitated.

The chemical reactions occuring are

$$SiCl_4 + O_2 = SiO_2 + 2Cl_2 \tag{14.116}$$

$$GeCl_4 + O_2 = GeO_2 + 2Cl_2 \tag{14.117}$$

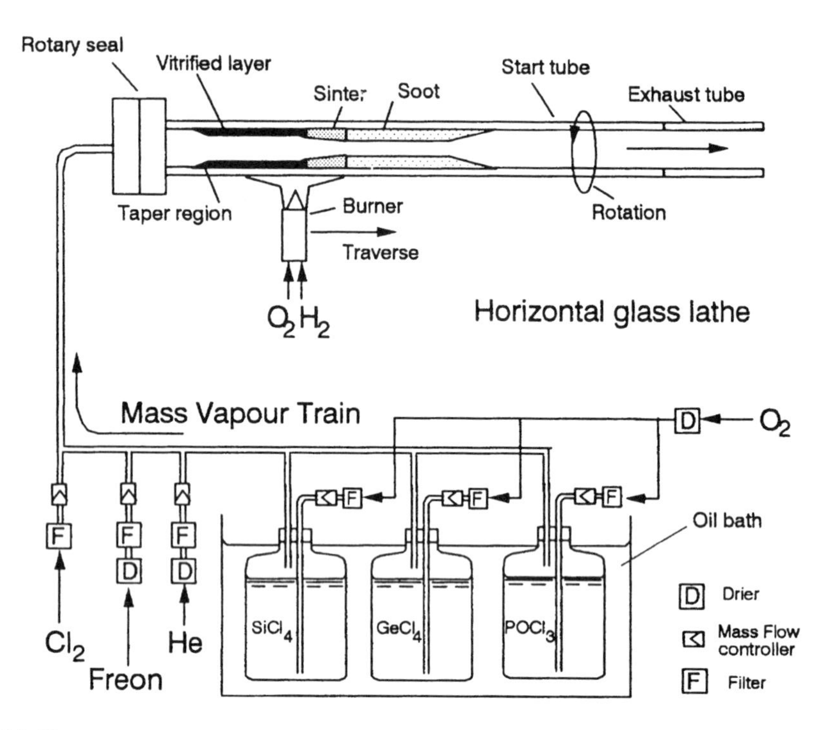

Figure 14.42
MCVD line and mass vapour train

These are equilibrium reactions which proceed from left to right as the temperature increases. At 1750 K the deposition of silica in the first reaction is almost 100% complete. In isolation, the second reaction would give almost complete deposition of germania at about 2000 K. However this is about the melting point of the fused quartz start-tube. Also when occuring at typical dopant concentrations in the presence of the first reaction, the high concentration of liberated chlorine tends to force the second reaction to the left. Consequently a maximum deposition efficiency of germania (of about 20%) occurs at about 1830 K where the reaction becomes equilibrium limited. An excess of oxygen helps to promote the deposition of both silica and germania. The temperature of the hot zone, monitored by an optical pyrometer, is controlled to be about 1900 K on the outer surface of the the start-tube.

In order to facilitate the vitrification of the deposited germanosilicate layers a small amount of phosphorus is also incorporated in the deposit from the reaction of phosphorus oxychloride and oxygen

$$2POCl_3 + 2O_2 = P_2O_5 + 3Cl_2 \tag{14.118}$$

All the dopants reduce the melting point and viscosity of the deposited material. During the deposition of the optical-quality cladding layers, the incorporation of phosphorus increases the refractive index and for matched- or depressed-cladding indices this must be compensated for the incorporation of fluorine dopant (usually from 'Freon' – CCl_2F_2).

The vapours of $SiCl_4$, $GeCl_4$ and $POCl_3$ are conveyed to the reaction zone from their containers by bubbling a stream of dry oxygen through them. The boiling points of silica tetrachloride and germania tetrachloride are 58 °C and 84 °C and of phosphorus oxychloride 108 °C. To improve the efficiency and stability of vapour pick-up, the bubbler vessels are kept in a temperature-controlled oil-bath at about 40 °C. All these chemicals react fairly violently with water.

The control of hydroxyl (–OH) contamination is of great importance. All the reactants must be very dry and free from hydrocarbons, additionally the silicon chloride must be free from silane ($SiHCl_3$), which has a b.p. close to that of the chloride and is difficult to remove by distillation. It may be removed with UV-irradiated chlorine. Hydroxyl contamination of the start tube and the possible migration of –OH into the core is controlled using a sufficiently thick deposited cladding. The rotary seal at the headstock needs to be carefully designed with dry nitrogen barriers to prevent ingress of moisture.

The pipelines carrying the reactants must be of polymeric material (frequently PTFE) to avoid pick-up of metallic ions and should be trace heated to avoid condensation of the vapours. Additionally they may need to be coaxially protected with dry nitrogen barriers to prevent ingress of moisture. The OH level incorporated into the deposited silica is controlled by the reaction

$$2H_2O + 2Cl_2 = 4HCl + O_2 \tag{14.119}$$

The HCl is volatile and is swept away in the gas stream thus promoting the reaction from left to right. The presence of 10% chlorine is sufficient to reduce the OH incorporation into the glass by a thousandfold, resulting in an excess attenuation of about 0.04 dB/km/p.p.m. H_2O in the gas stream (or 0.04 dB/km/p.p.b. OH in the glass). Thus the gas stream water content should be limited to less than 20 p.p.m.

Near the headstock of the lathe, at the start of a deposition pass, the vitrified layer is tapered, (both in thickness and dopant concentration) owing to the finite length of the hot zone. Adding helium to the reactant gases shortens this taper significantly thus improving the preform yield of uniform fibre. The use of helium in the ratio of $He/O_2 = 2$ also increases the deposition rate of silica to about 1 g/min. The helium appears to act as a heat transfer agent.

During sintering and vitrification of the deposited core layer, some of the germania dopant can be lost from the inner surface by reconversion to the tetrachloride. This reduces the refractive index at the interface with the next deposited layer and thus causes ripples in the refractive index profile.

Fused silica start-tubes typically are 1.2 m long with an internal diameter of 19 mm and an external diameter of 25 mm. This gives an average cross-sectional area of 207 mm^2 with a typical spread of 10 mm^2. When drawn to 0.125 mm diameter fibre this gives the possibility of achieving length yields of greater than 16 km per preform.

Good uniformity, circularity and straightness of the start-tube is essential. Before setting-up on the glass lathe, the tubes are cleaned with solvent to remove any grease contamination and may also be acid etched, and then washed and dried. The tube is aligned in the headstock of the lathe and is then flared at the tailstock end and fused to a larger diameter exit tube. This in turn connects to exhaust piping leading to a scrubber before exit to the atmosphere.

The first pass on the lathe is generally an etch of the interior wall using a Freon/oxygen gas mixture at a high temperature which also flame polishes the outer surface. These procedures ensure removal of any debris or contaminants which could lead to bubbles or impurities. The flame polish also helps to promote fibre strength by elimination of surface flaws in the start-tube. Some silica is lost from the outside of the tube during flame polishing; this is known as 'burn-off'. The Freon etch also tends to reduce the wavelength independent attenuation constant, B.

Typical rotational speeds are about 30–40 r.p.m. and traverse speeds of 2–3.5 mm/s are used. For multimode graded index fibre about eight cladding passes are typical and 35 to 60 doped core passes with carefully controlled increases of germania doping, depending on growth rates. For single-mode fibre many more cladding passes of the order of 15 to 30 are used, generally at the highest achievable growth rate. This is because of the fundamental field spread into the cladding in single-mode operation. The last few cladding layers may have germania doping (compensated by fluorine doping) for thermal strain matching with the core layers. The number of core passes at a constant germania doping level varies from two to eight depending on whether preform sleeving is to be used (see later).

Finally the tubular preform must be collapsed into a solid rod preform before removal from the lathe. The flame temperature is increased to the melting point of the fused silica start-tube and the collapse is generally carried out in several passes. In the first collapse path, chlorine is fed into the tube to promote drying and the internal pressure is maintained at about 0.1 mbar to counteract the flame pressure. On the second pass, Freon is introduced at the same pressure to etch the last deposited core layer. This has been shown to reduce the wavelength independent attenuation constant, B. On the third pass, the tube is completely collapsed near the tailstock end and the collapse traverse is reversed and slowed to about 0.2 mm/s to produce complete collapse up to the headstock end. Uniform collapse is dependent on the initial tube geometry and the pressure difference between the inside of the tube and the flame pressure. Burn-off of the germania dopant from the innermost layer during collapse causes a central dip in the refractive index profile, partly compensated for by the etch pass.

Typical times to complete a preform are 1 h setting up, 4 h for deposition and 3 h for collapse, a total of 8 h. This corresponds to an estimated fibre output yield of 2 km/h. This process produces a collapsed preform of about 17 mm diameter with a deposited core/clad region of 7 mm diameter and a single-mode core of 1.16 mm diameter.

Figure 14.43 shows the geometry involved in MCVD deposition and collapse. For a constant deposition rate G g/min the thickness of the deposited layer is given by

$$t = \frac{100G}{T\rho\pi D_i} \text{ mm} \tag{14.120}$$

where T is the traverse speed in mm/s, ρ is the density in g/cm^3 (2.1) and D_i is the internal diameter (mm) on which the deposit is laid. The external diameter of the deposited layers after collapse is given by

$$d_n = \sqrt{4Dt + d_{n-1}^2} \tag{14.121}$$

where D is the diameter of the deposited layer, t is the layer thickness and d_n is the diameter after collapse of the nth layer out from the axis.

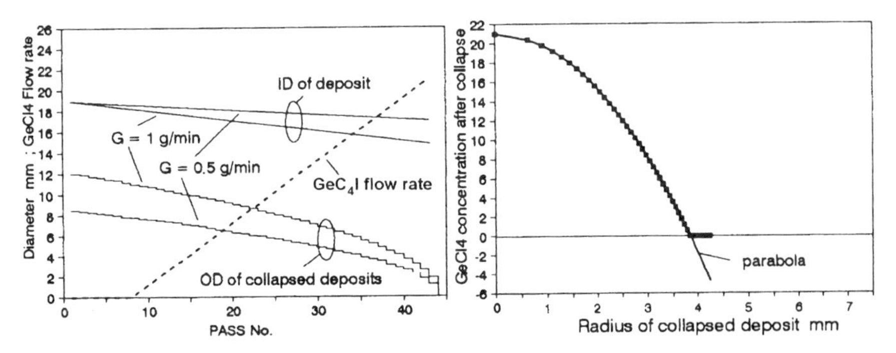

Figure 14.43
Geometry of deposition, collapse and doping for MCVD preforms

If the flow rate of the dopant is linear with the pass number, then when plotted against the radius of the deposit after collapse, it will lie exactly on a parabola due to the square root in Equation (14.119). If the deposition efficiency of the dopant is constant this results in a truncated parabolic refractive index profile (exponent $\alpha = 2$). For other power profiles the flow rate of dopant should be made more or less than linear with pass number. The slope of the dopant flow rate versus pass number determines the axial refractive index difference.

Obviously a high growth rate of deposited silica is desirable to shorten fabrication times. High growth rate is promoted by several factors.

- High reaction rate from increased temperature (100% above 1750 K for silica; equilibrium limited for germania above 1830 K)
- Increased reactant flow (excess oxygen and higher bubbler temperatures)
- Increased helium/oxygen ratio
- Lower total flow rate of reactants (to avoid sweeping the soot out of tube)
- Increased efficiency of thermophoretic deposition (cool the outside of the tube ahead of the flame with water or cold gas flow)

There is a limit to increasing the growth rate which is basically concerned with efficient and bubble-free vitrification of the deposited soot. This can be countered by faster traverse and an increased number of deposited layers. Helium also helps to reduce bubble formation.

With so many variables and the necessity of maintaining a constant deposition rate, it is necessary to maintain a high degree of control. All the flow rates are controlled by mass flow controllers under computer control. Once a set of parameters has been determined and fine-tuned by experiment, they are stored in the computer as a 'recipe' for future use. The computer can also be used to change deposition conditions at the headstock end in order to reduce the taper in deposit thickness and refractive index.

The bigger the preform the more the fibre that can be drawn from it, thus reducing non-productive set-up times. Increasing the diameter of the start-tube brings difficulties however. Firstly, higher ouside temperatures are required to ensure that the axial temperature is high enough for reactions and secondly, the thermophoretic force becomes insufficient to capture the soot efficiently on the tube walls. The deposition efficiency is given by

$$\varepsilon \approx 0.8(1 - T_w/T_r) \quad (114.122)$$

where T_r is the reaction temperature and T_w is the wall temperature, both in K. Thus cooling the tube wall ahead of the flame with water or a cold gas stream can increase the efficiency of capture.

A better method for making larger preforms, particularly for single-mode fibre, is to deposit a larger core in the initial start tube, and after collapse to the solid preform to sleeve it with another fused silica tube. This is done on a separate lathe with the sleeve being collapsed using an oxyhydrogen burner while the space between is evacuated to prevent the formation of bubbles. This takes about 1.5 h. In this way a 23 mm diameter preform capable of yielding over 37 km of fibre can be produced with a fabrication yield increased to 3.9 km/h.

A modification of the MCVD method uses a radio-frequency induced plasma in the excess oxygen at atmospheric pressure. This creates a 'fire-ball' in the centre of the tube and enables larger diameter tubes (>40 mm) to be used. The fire-ball, at about 10^4 °C, is responsible for the reaction temperature and the very efficient thermophoretic deposition, while sintering and vitrification is carried out by an external oxyhydrogen burner. Growth rates of 5 g/min can be achieved using this method.

After the preform is produced, it is inspected for dimensional comformity of the outside and core diameter, for straightness and for any core or cladding bubbles. Its refractive index profile is measured at several positions along the length which enables the spotsize and cut-off wavelength of single-mode fibres to be predicted and the profile exponent alpha to be determined for graded index preforms.

Fibre Drawing

Figure 14.44 is a sketch of a fibre drawing tower. To assist in handling, a silica rod stub is fused to the preform tailstock end. This is held in a chuck at the top of the tower which is attached to a controlled feed mechanism. The preform is heated in a furnace a few centimetres from the headstock end, where any core tapers exist. This releases a 'drop' which pulls a fibre from the preform under gravity. This drop is detached and the fibre is led through a diameter gauge and the two coating cups holding the first and second primary coating materials, which are cured by heat or by UV light focused on to the coating by elliptical mirrors. A coating diameter gauge and an eccentricity gauge monitor the coating. A belt capstan is then used to draw the fibre from the preform at up to 120 m/min and at a low controlled tension (25 g). This requires the tower to be about 10 m high to enable the fibre to cool before coating. A cold clean nitrogen blast below the furnace can assist in this cooling

The fibre speed is controlled by feedback from the diameter gauge to maintain a constant fibre diameter, and the preform feed speed is also automatically adjusted to keep the tip of the preform in the furnace hot zone. The temperature of the hot zone is about 1950 °C. Higher temperatures (and hence lower tensions) can adversely affect the spectral loss but may be necessary for high-speed drawing. Reducing the tension from

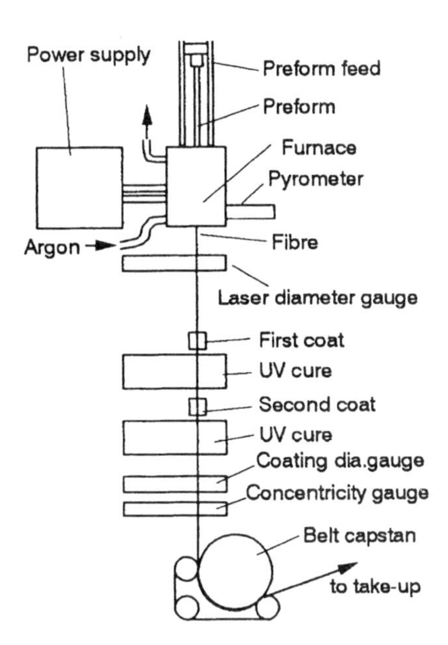

Figure 14.44
Fibre drawing tower

25 g to 17 g increases the 1300 nm attenuation by 0.05 dB/km in a single-mode fibre. The furnace is usually a graphite element resistance type or a zirconia induction type. If it is a graphite furnace the element is run in a clean inert argon flow to prevent oxidation (nitrogen has the possibility of forming cyanides with the carbon).

In order to preserve the fibre strength, great care must be taken to avoid any dust particles damaging the fibre before coating. For this reason the fibre drawing is usually carried out in a 'surgically clean' atmosphere in a 'clean room'. Clean air is provided under slight excess pressure through sub-μm filters to the clean room and constant monitoring of the dust count of the atmosphere is required. The graphite element can also produce dust particles. This can be mediated by 'burning in' the new element at a moderate temperature for 24 h, or by fusing silica soot into its surface, or by operating it behind a refractory baffle. The counter-flow of argon also helps in this respect. A zirconia element is considered more dust-free.

Proofstrain Testing

The primary coated fibre is subjected to a proofstrain test to monitor its strength. This consists of winding it between two capstans running at the same rotational speed but differing in diameter by 1%. Alternatively the fibre can be run over a pulley exerting 8 N tension on the fibre. In either case the strain must be applied for 1 s to 1 m of fibre. This corresponds to 0.7 GN/m^2 or to 100 kpsi. This test ensures the integrity of the fibre by breaking the fibre, thus eliminating any weak points below these levels. Obviously fibre weak spots must be at a minimum to make this test economically viable. Following the proofstrain test a comprehensive set of optical tests are applied to the fibre.

FIBRE STRENGTH

Griffith's Crack Theory

Even a pristine, new, undamaged surface of a brittle material has a random scattering of surface defects. These intrinsic flaws appear at a certain statistical distribution related to the surface area (i.e. a constant number of flaws of a given depth per m^2). As the surface area of the test sample is decreased, e.g. by reducing the diameter of a fixed length test sample, the tensile strength increases. Thus small diameter fibres have a greater strength than the same material in the form of a rod of larger diameter.

Consider a section of fibre with a single flaw of depth c, subject to a constant stress σ, as shown in Figure 14.45.

The highest value of strain in the fibre occurs at the tip of the flaw where the curvature is greatest. The stress intensity at the tip is

$$K = \sigma Y \sqrt{c} \tag{14.123}$$

where Y is a geometric factor equal to 1.57 for a semi-elliptical flaw shape and 1.24 for a semicircular flaw shape. Under the influence of this stress the atomic bonds of the material start to break and the flaw depth increases at a velocity given by

$$\frac{dc}{dt} = (AK)^n \tag{14.124}$$

where A is a constant and n is the static fatigue parameter. When the flaw depth reaches a critical value, catastrophic failure of the fibre occurs.

For a silica fibre $K = 0.825\,\text{MPa} \cdot \text{m}^{1/2}$ and with a Young's modulus $E = 72\,000\,\text{MPa}$, under a constant applied stress giving a strain of 1%, the critical semi-elliptical flaw depth is

$$c = \{K/(Y\sigma)\}^2 = \{0.825/(1.57 \times 72000 \times 0.01)\}^2 = 0.53 \times 10^{-6}\,\text{m}$$

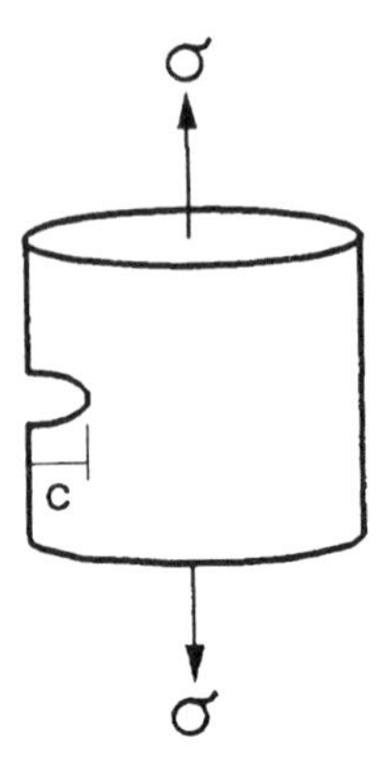

Figure 14.45
Fibre surface flaw

Substituting Equation (14.121) into (14.122) and integrating to obtain the time for a flaw depth c_i to grow to the critical value c_c gives

$$\int_{c_i}^{c_c} c^{-n/2} dc = (A\sigma Y)^n \int_0^t dt$$

$$I = \frac{1}{(1-n/2)}\{c_c^{(1-n/2)} - c_i^{(1-n/2)}\} = (A\sigma Y)^n \cdot t \tag{14.125}$$

whence

$$\log\sigma = -\frac{1}{n}\log t + \frac{1}{n}\left\{\frac{I}{(AY)^n}\right\}$$
$$= -\frac{1}{n}\log t + \log\sigma_0 \tag{14.126}$$

Thus a log–log plot of σ against t will give a straight line of slope $-1/n$. Time is measured in seconds and stress in N/mm^2. σ_0 is the stress at a time of 1 s.

The static fatigue parameter n, is a function of the fibre surface quality and also of the fibre environment. In the presence of water, the growth rate of flaws under stress is increased due to a phenomenon called stress corrosion. The silicon-oxygen bond is attacked as follows

$$\text{Si–O–Si} + H_2O = \text{Si–OH} + \text{Si–OH} \tag{14.127}$$

The effect is to decrease the value of n in wet or moist environments. Measurements have determined that for plastic coated silica fibres

$n = 30$ at 20 °C and 40% relative humidity
$n = 22$ at 20 °C and 99% relative humidity (wet)
$n = 15$ at 100 °C in water

The effect of this on fibre static fatigue lifetime is shown in Figure 14.46 where the static stress in the 1 m fibre samples is plotted against the mean time to failure.

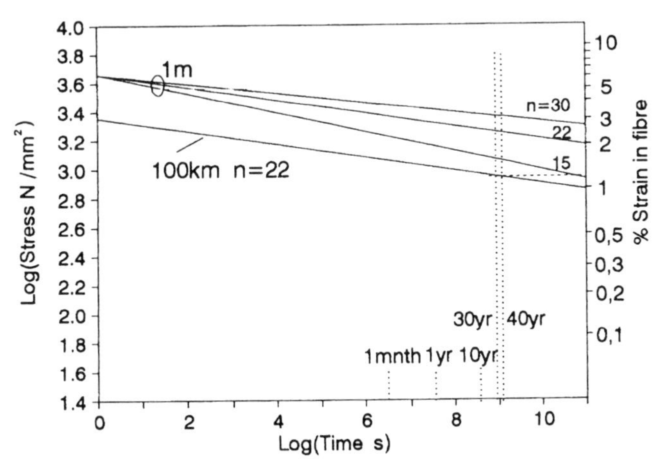

Figure 14.46
Fibre static fatigue lifetimes

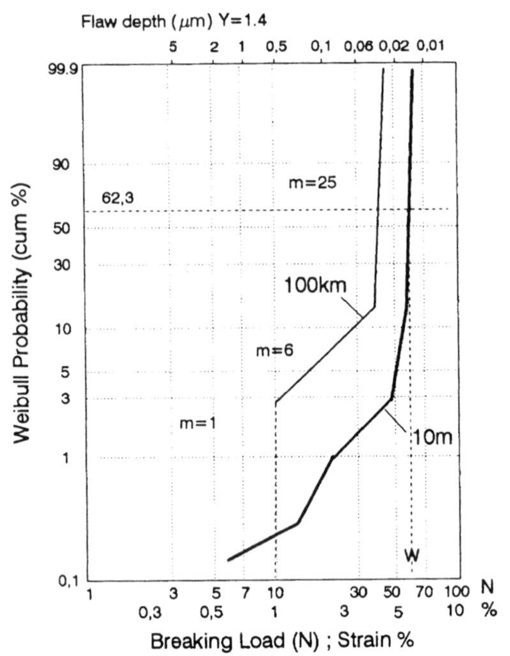

Figure 14.47
Weibull plot of 10 m fibre breaking loads

The failure probability in this graph is 50%. The value of σ_0 is obtained by extrapolating measurements back to $t = 1$ s and the value of n was found to be 30 from the slope of the line for 20 °C and 40% RH. The spread of the measurements is about 25%. The lines for $n = 22$ and 15 were calculated from Equation (14.126). In order to estimate the effects on longer lengths of fibre, measurements of breaking loads on a large number (typically 500) of 10 m fibre samples were plotted on a Weibull chart as shown in Figure 14.47. The Weibull distribution is given mathematically by

$$F = 1 - \exp\left(-L\left(\frac{\sigma}{W}\right)^m\right) \tag{14.128}$$

where F is the cumulative failure rate at stress σ. W is the characteristic strength at $F = 62.3\%$. The plot can be approximated by straight-line segments of different slope m. Each segment is probably due to an independent distribution of flaw depths from different sources. The highest slope segment at the higher stresses is that due to the intrinsic flaws in the material. At a given failure rate the stress can be scaled for different lengths by

$$\frac{\sigma_2}{\sigma_1} = \sqrt[m]{\frac{L_2}{L_1}} \tag{14.129}$$

as shown in the diagram for 100 km of fibre. Since the fibre is poofstrained at 1%, all flaws of depth greater than 0.5 μm will be eliminated. This occurs at a failure level of about 3%. A mean value of 1.4 was used for Y in this diagram.

The same length scaling can be used for the static fatigue diagram, Figure 14.46, as shown for 100 km of wet fibre at 20 °C ($n = 22$). A spread of 25% in the measurements has also been applied, to estimate the highest probability of failure. Thus for 100 km of fibre under a constant strain of 1.2% the lifetime due to *intrinsic* flaws can be estimated as being between 30 and 40 years. Such constant strain conditions are usually associated with constant structural features of the fibre cable, for example the curvature strain caused by stranding the fibre at a constant lay length about a fixed diameter. (see Chapter 3, Equation (3.3)). For a typical distribution of *extrinsic* flaws the Weibull plot shows that in 100 km of fibre there will be about a 3% residue of flaws up to 0.5 μm depth after proofstraining to 1%. The flaw growth under a constant load can be calculated from Equation (14.125)

$$\frac{c_f}{c_i} = \left[\frac{(A\sigma Y)^n t \cdot x}{c_i^x} + 1\right]^{1/x} \tag{14.130}$$

where $x = (1 - n/2)$ and c_f is the depth to which an initial flaw depth of c_i grows in t s. Applying this to the 0.5 μm flaw shows that under 0.46% constant elongation it will probably break after 40 years, as shown in Figure 14.48. Thus for safety, 100 km lengths of this quality of fibre should be limited to less than 0.46% strain to obtain 40-year lifetimes under wet conditions at 20 °C.

Fibre Cleaving

To prepare a fibre end for testing or jointing, the fibre is cleaved. This is done by scoring the surface with a ceramic or diamond edge and subjecting the fibre to longitudinal stress, when the artificial 'flaw' propagates through the fibre perpendicularly to the axis of the applied stress. The surface produced by this cleave is a mirror finish until the propagation velocity exceeds the velocity of sound in the material, when

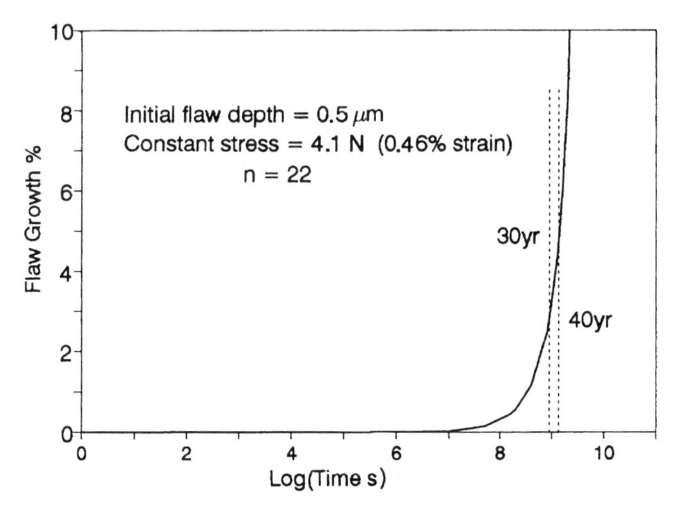

Figure 14.48
Fibre flaw growth under constant strain

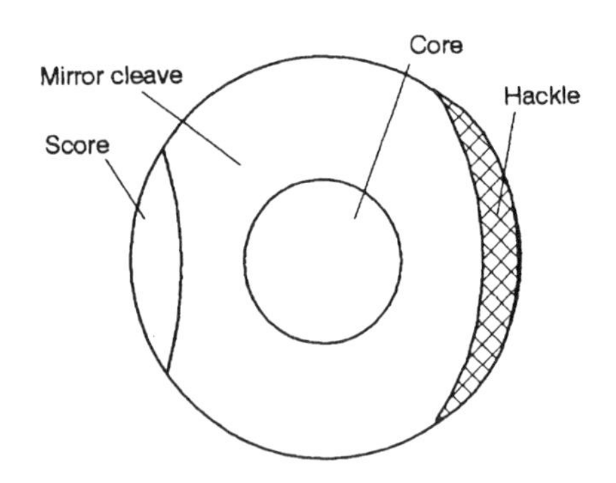

Figure 14.49
Cleaved fibre end face

it becomes rough or 'hackled'. For silica the velocity of sound is about 5900 m/s and the flaw propagation velocity is

$$v = (A\sigma Y\sqrt{c})^n \times 10^{-6} \text{ m/s} \tag{14.131}$$

with $A = 9.7 \times 10^{-4}$, and c in μm. For $n = 30$, σ must be limited to about 2.2/ 0.01227 = 179 N/mm^2 to limit the hackle region. For the cleave to take a reasonable time, like 1.3 s, the score should be 17 μm deep. This results in the fibre having a mirror finish over the greater part of its surface, as in Figure 14.49.

The stress can conveniently applied by bending the fibre to achieve the equivalent amount of strain at the score. For a 0.125 mm fibre, bending to a 50 mm diameter will produce a strain of 0.25% on the outer surface equivalent to 2.2 N longitudinal force. However, this bending of the fibre during cleaving can result in 'lips' and 'chips' on the outside of the fibre and it is therefore preferable to apply a purely linear stress to the fibre after scoring. The major portion of the time for cleaving is taken by the early score growth to twice its initial depth as shown in Figure 14.50.

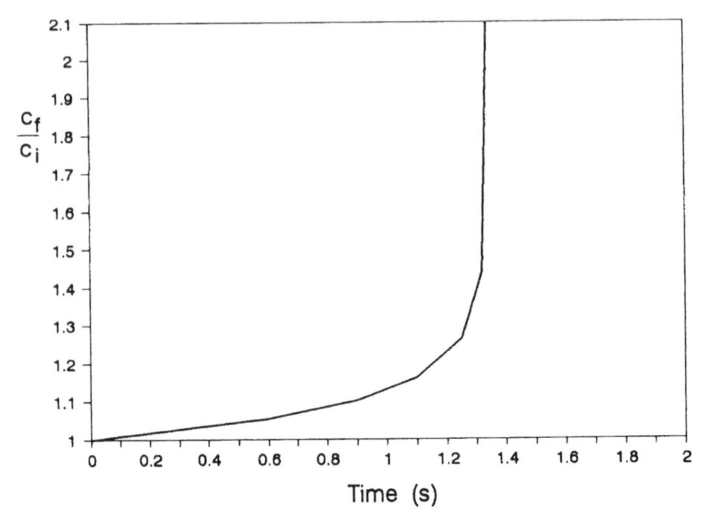

Figure 14.50
Flaw growth in cleaving fibre

OPTICAL CABLING AND INSTALLATION

Secondary Protection of Fibres

As manufactured, the fibre has a diameter of 125 μm and is provided with primary coatings up to a diameter of 250 μm. It therefore needs secondary protection before cabling. The three principle ways in which this is done are illustrated in Figure 14.51.

Firstly the fibre can be covered with a tight extrusion of a tough plastic such as nylon to a diameter of about 1 mm. This is known as tight buffering (Figure 14.51(a)). Although the Young's modulus of nylon is only about 1600 compared with 72 000 N/mm^2 for the fibre, the area of the plastic is 50 times that of the fibre. Therefore the *EA* product of the plastic is 1160 compared with 883 for the fibre. Consequently as the extruded plastic cools onto the fibre it contracts and produces longitudinal compression in the fibre. This can amount to 0.25% compression depending on the cooling conditions. This is advantageous since it allows the buffered fibre to withstand 0.7% strain without straining the fibre more than the allowable 0.46%.

The second method of protection is to enclose the fibre in a loose tube of about 3 mm o.d. and 2 mm i.d. By feeding in a small excess length of fibre it takes up a wavy formation which can add about 0.5% to the length of fibre. Again this helps to buffer the fibre against tensile strains in the tube. This is called loose-tube protection (Figure 14.51(b)). Up to six or eight fibres can be accomodated in the same tube. The fibres are usually given an oscillatory lay before entering the tube extrusion. In order to prevent fibre migration, the tube may be filled with a thixotropic gel. This also helps in protecting the fibres against water ingress.

The ultimate in compact fibre protection is the ribbon formation invented by Western Electric. Twelve fibres with their primary protection are laid close and parallel and are laminated between plastic tapes. These ribbons can be stacked together to

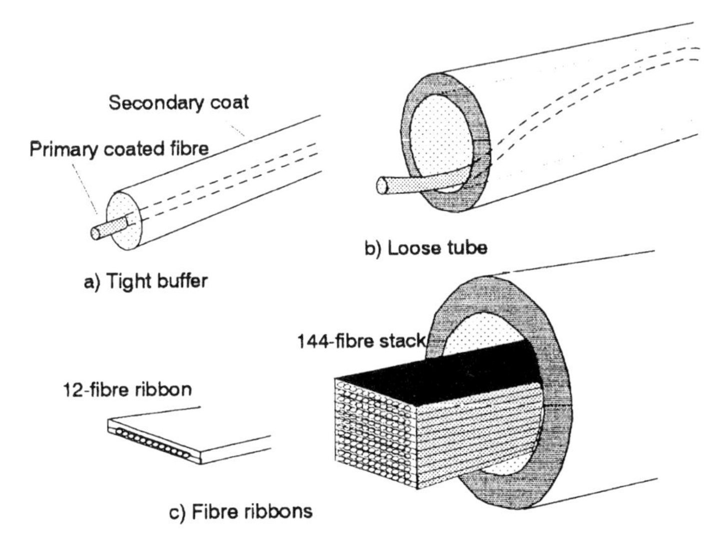

Figure 14.51
Secondary protection of fibre

accomodate 144 fibres in an area of about 25 mm^2 (5 mm × 5 mm) and enclosed with a long helical lay in a tough outer plastic tube of about 12 mm outside diameter. This may then be double steel wire armoured and oversheathed to produce the final cable.

Fibre Cabling

To maintain one of the important advantages of fibre, namely its immunity to electromagnetic fields and lightning strikes, the fibre cable should be non-metallic. To provide tensile strength for installation, strength members such as glass-reinforced plastic (GRP) rods and/or polyaramid strands (e.g. Kevlar) are included during cabling. Typical constructions are shown in Figure 14.52, where the tight-buffered or loose-tubed fibres are stranded around a GRP rod and provided with a layer, or interstitials, of polyaramid strands. Both these cables would be of about 12 mm outside diameter. The tight-buffered fibre cable may be provided with a gel filling to protect against water ingress. Alternatively, a tape coated with a water-swellable powder can be lapped over the fiber layer. If water enters the cable through damage, the powder absorbs it and swells to produce a block preventing further water penetration.

Unlike stranding of ductile copper wire cables, the stranding of fibres requires closely controlled low tensions without any snatching. This is usually ensured by the use of servo-controlled pay-offs and take-ups. Also to avoid torsional strain in the fibres, the pay-off reels must be accurately detorsioned in accordance with Equation (3.4).

To minimize the curvature strain in the fibre, the lay length of the stranded fibres should be related to the pitch circle diameter of the layer in accordance with Equations (3.2) and (3.3). For a 12-fibre tight-buffered cable, the pitch circle diameter will be about 4 mm. A lay length of 50 mm will produce a fibre curvature radius of 34.4 mm and a constant strain on the outside of the fibre of 0.19%. This must be taken into account when assessing the fibre static fatigue lifetime.

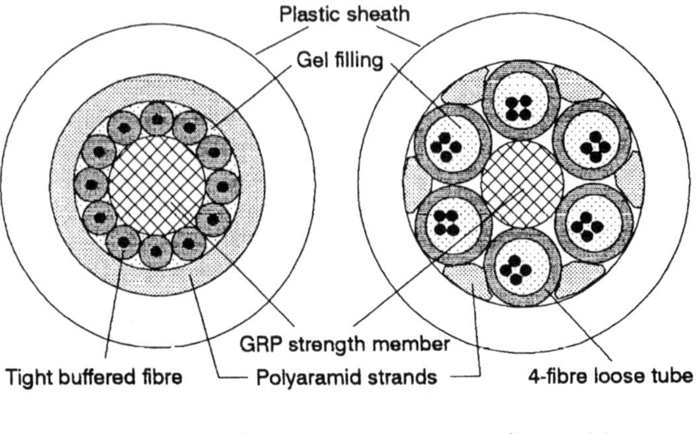

Figure 14.52
Non-metallic fibre cable constructions

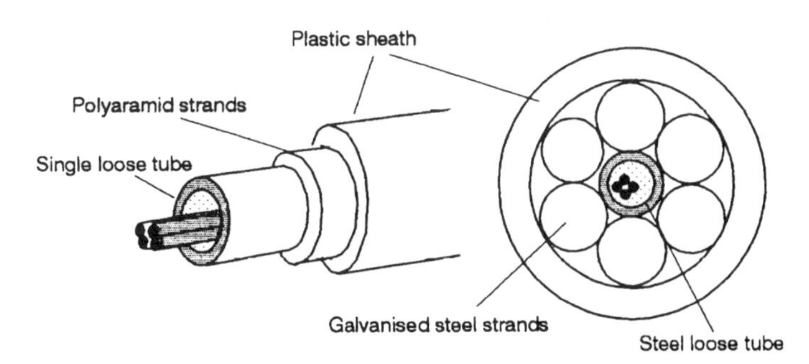

Figure 14.53
Optical fibre aerial cables

Aerial Cables

Owing to the light weight of optical fibre cable, a popular method of installation is a self-supporting aerial cable. Since polyaramid strands have about twice the tensile strength (mass-for-mass) of steel strands, a non-metallic structure such as shown in Figure 14.53 is capable of span lengths of the order of 250 m. To improve the grip of the sheath on the polyaramid strands, an adhesive is often applied before sheathing. For longer spans, such as on aerial power transmission structures, this cable can be helically wrapped around the ground-wire after installation. Plastic sheaths subjected to high electric field strengths tend to deteriorate and special grades of plastic are needed in such installations.

For long self-supporting applications, the metallic construction also shown in Figure 14.53 is often used. Span lengths of the order of 1000 m are feasible. Although such metallic cables likely to be struck by lightning, the large mass of steel protects the cable from serious damage from the majority of strikes (see Chapter 12). A further development of this concept for power transmission routes, is to incorporate an aluminium or plastic loose-tube construction into the centre of the aluminium strands of the ground-wire itself. This is known as the ground-wire embedded (GWE) construction.

Fibre Identification

To identify the fibres in a cable, the tight-buffer is coloured by the inclusion of colour masterbatch during extrusion. For loose-tube construction the primary-coated fibres are coloured with surface-applied inks before tubing, in a separate operation. The tubes themselves can also be coloured.

Attenuation Monitoring

During the various operations of secondary protection and cabling, the attenuation of the fibres is carefully monitored to avoid building in extra losses due to micro- or macro-bending. This is usually carried out with an optical time-domain reflectometer (OTDR) which will be described later. There is sometimes a reduction of loss during cabling which is due to the removal of spool macrobending losses when the fibre is unwound. Generally the changes in attenuation of the fibres can be limited to acceptably low values for each type or wavelength of fibre, by the careful control of manufacturing processes.

Cable Testing

In addition to comprehensive optical testing of the fibres after cabling, various mechanical tests are carried out on a type-test basis. These include crush, impact and bend testing, looking for mechanical damage of the cable or fibres and the variation of attenuation as the cable is subjected to three or four temperature cycles between −5 °C and 40 °C (or other agreed limits). In addition a 50 m or 100 m length of cable is subjected to tensile tests while monitoring mechanical extension of the cable. The fibre extension during this test is monitored by phase changes in the modulation of an optical signal sent down the fibres.

Installation of Optical Fibre Cables

All the installation methods and precautions discussed in Chapter 11 are applicable to optical cables. However the mass of optical cables, particularly the non-metallic constructions, is of the order of 100 g/m i.e. about 20 times less than the average ducted copper conductor cables. Hence longer lengths can generally be handled at lower tensions. For example a 1 km length of 100 g/m fibre cable would experience a tension rising to about 600 N at the pulling end of a straight-pull duct route. The cable construction should therefore be such as to limit the extension of the fibre to less than about 0.5% under these conditions. As noted in Figure 11.2 the tension relaxes after pulling, to about half this value midway along the cable. It is essential to follow the manufacturer's recommendations on the strength of the cable and the minimum bending radius allowed. Additionally there must be no snatching of the cable during installation; the brittle nature of the fibres, as opposed to the ductility of copper conductors, is very unforgiving of snatches. It is usual to draw fibre cables into ducts using a mechanical fuse link at the pulling end and to use a tension-controlled winch.

Self-supporting aerial optical fibre cables are usually circular. Hence they do not need to be twisted to reduce aerodynamic lift and galloping, as do figure-of-8 sheathed aerial cables. Directly buried optical cables require protection from crushing and penetration damage in the same way as copper cables. As this generally involves the use of steel wires or tapes, it of course increases the mass and removes the lightning immunity of otherwise non-metallic constructions.

Another method of installing optical connections within or between buidings is the 'blown fibre' technique. In this method a stranded bunch of primary-coated fibres is conveyed along a previously installed empty plastic duct, by a stream of air injected into one end of the duct at velocity of about 5 m/s. The end of the fibre bundle needs to be protected by a soft buffer to avoid the fibres digging into the duct wall at bends and then snagging. The fibre bundle floats in the air stream and is not therefore subjected to any tension.

Fibre Jointing

The most permanent method of jointing silica fibres is by fusing them together using an electric arc. This method is illustrated in Figure 14.54. The secondary protection is first removed with a purpose-designed stripping tool. Then the primary coatings are removed with a solvent and cotton wool. The fibre ends are carefully cleaved and mounted in V-grooves on the fusion jointing equipment. The fibres are aligned in the lateral directions while viewing them under about 250 times magnification and then the ends are brought close together but not touching. A pre-fusion arc of controlled intensity and duration is used to round the edges of the fibres and to produce slightly convex end faces. This helps to prevent the trapping of air and formation of bubbles. The fibre ends are then brought together and a higher intensity of arc of longer duration is used to fuse the fibre ends together. The fused fibres are removed from the jointing equipment and the joint and neighbouring fibre is protected. The bare fibre may have a primary coat applied using a solvent-based polymer and then the joint can be protected with heat shrinkable polymer tubes perhaps containing a stiffening element such as a steel rod.

Joint losses of from 0.1 dB to 0.3 dB can be achieved in such joints. Manually controlled fusion jointing equipment can be used if the operator is using it constantly, but the skill level required is very high. More expensive equipment is automated under microprocessor control and with several stored programmes, to give highly reproducible results.

The cost of fusion jointing may not be warranted in some applications and several proprietary mechanical jointing systems are available which give excellent low attenuation joints. These have proven stable and reliable over lifetimes exceeding five

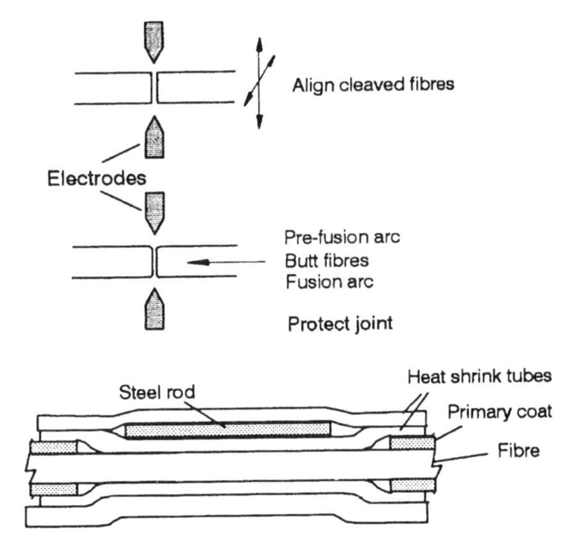

Figure 14.54
Fusion jointing of optical fibres and joint protection

years. These mechanical joints generally use a ceramic capillary tube to hold and align the fibres and use an index matching fluid or gel to avoid Fresnel reflections, so that joint losses of 0.1 dB to 0.3 dB are readily achieved.

The accuracy of alignment of singlemode fibre cores is more critical than for multimode fibres. For singlemode fibres an axial misalignment of the cores d causes a transmission loss of

$$\alpha = 4.34\left(\frac{d}{w_0}\right) \text{dB} \tag{14.132}$$

A mismatch of the spotsize between jointed fibres causes a *undirectional* transmission loss, in the direction of larger to smaller spotsize, of

$$\alpha = 10 \log\left(\frac{2w_1w_2}{w_1^2 + w_2^2}\right)^2 \text{dB} \tag{14.133}$$

Joint Closures

Unlike copper cables is not possible to make neat and tidy joints in optical fibre cables. It is always necessary to have considerable slack in the fibres, if only to reach the jointing equipment. Also it may be necessary to remake the joint to achieve the desired low loss. The slack fibre must be accomodated in the joint closure without excessive bend loss and without causing the fibre mechanical stress. This is usually achieved with a lightweight frame called a 'fibre organizer' around which a few turns of the fibre are arranged. The minimum bend diameter is usually about 120 mm. From Figure 14.36 it will be seen that this would give a loss of the order of 0.01 dB/m or about 0.02 dB for six turns and the fibre strain would be less than 0.1%. The joint closure itself can be any of the types discussed in Chapter 11 but because of the small diameter of the cables, the dome joint of Figure 11.15 is particularly suitable and popular.

OPTICAL MEASUREMENTS

Refractive Index Profiles

If a line at 45° to its axis is viewed through a cylindrical rod of uniform refrative index it will appear to be tilted as shown in Figure 14.55. The tilt is a function of the refractive index difference. If the rod is replaced with a preform in a liquid matching the cladding index, the deflection of the line will be proportional to the core refractive index. By measuring the deflection across the core radius the refractive index difference of the core can be calculated and plotted against the radius as shown in the figure. A well-known proprietary instrument for performing the equivalent of this technique with spatial filtering is the York refractive index profiler P102.

On a fibre however, the dimensions are too small for this technique to be applied. The near field at a given radius on a fibre end face is proportional to the refractive index at that radius. If the intensity of light at that spot is measured by using a sufficiently small

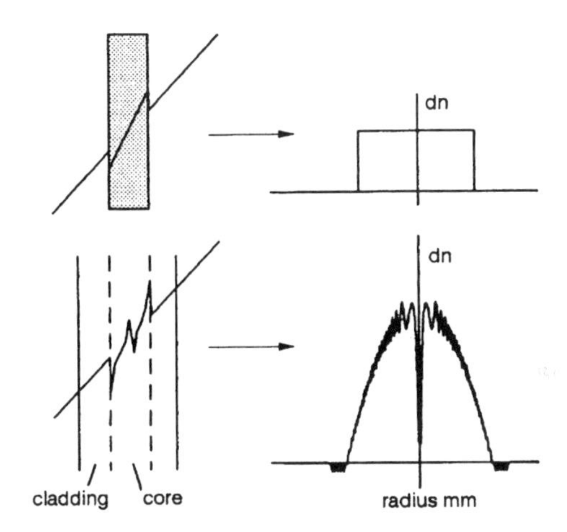

Figure 14.55
Preform refractive index profiling

focused spot, the refractive index difference can be plotted against against radius. To get sufficient resolution and sensitivity is however difficult. An alternative method put into a practical form by White [70] is to use the refracted near-field technique sketched in Figure 14.56. A fine spot from a laser is focused on to the fibre end face through a high NA lens system. The fibre is stripped of its coatings and is in an optical cell containing a liquid of slightly higher refractive index than the cladding. The light refracted out of the fibre cladding is collected by a large diameter lens system. Light which is due to leaky modes exiting the fibre, is blocked by an opaque disc. The total light power passing the disc is proportional to the refractive index of the fibre face at the illuminating spot. By scanning the spot across the end face of the fibre a plot of refractive index difference against radius can be drawn. By calibrating the equipment, the actual index values can be obtained. A commercial embodiment of the technique is produced by SIRA. In both methods of measurement it is important that the wavelength of light used is near to the operating wavelength of the fibre.

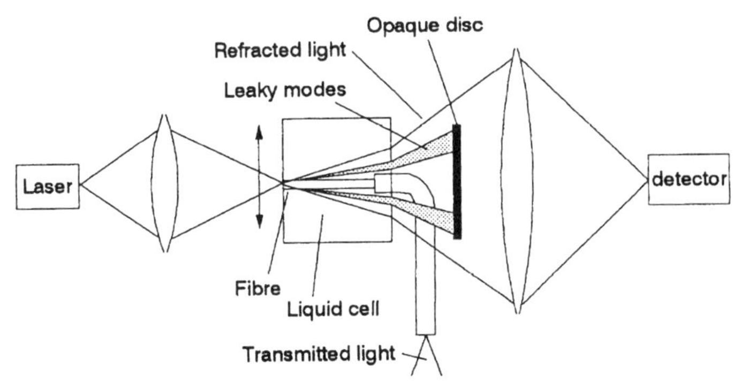

Figure 14.56
Refracted near-field technique for fibre refractive index profiling

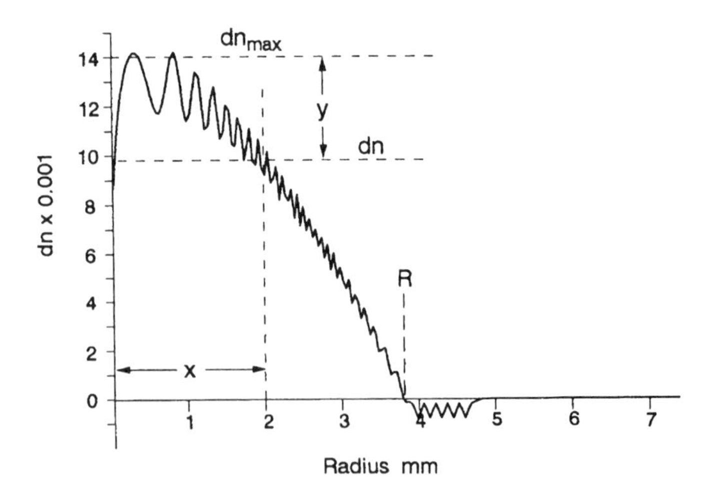

Figure 14.57
Determination of profile exponent

To determine the profile exponent, α, from the refractive index plot, the maximum value of the power law on the axis, $dn_{\max}$, must be estimated. Then from Figure 14.57, since

$$y = x^{\alpha}$$

hence

$$\alpha = \frac{\log(1 - dn/dn_{\max})}{\log(x/R)} \tag{14.134}$$

where R is the truncation radius of the profile.

Insertion Loss Measurements

If the output of an optical source, P_i, is measured with an optical power meter to calibrate it, and then a fibre of known length L km is inserted between the two instruments and the output power P_0 is measured, the attenuation of the fibre can be calculated from

$$\alpha = \frac{10}{L} \log \frac{P_i}{P_0} \text{ dB/km} \tag{14.135}$$

As a measurement of attenuation this method can only be approximate for three reasons. Firstly the power launched into the fibre may differ from the power launched directly into the power meter, secondly no attempt is made to achieve equilibrium power distribution between the modes of the fibre and thirdly some power may be carried for a considerable distance in leaky modes before it is lost from the fibre. However the longer the fibre, the less important these errors become. Consequently this method of measurement is quite suitable for long installed lengths of fibre.

Attenuation Measurements

In order to typify the attenuation of a fibre in such a way that it can be used to calculate the actual fibre loss in any length, certain precautions are necessary to eliminate the errors present in insertion loss measuments.

Since the variations in connecting the receive end of the fibre to the detector are generally small because of the larger area of the detector, the output of the long length of fibre is compared with the power output of a short length of the same fibre by cutting back to a few metres. In this way the launch conditions can be left undisturbed.

Secondly, equilibrium power distribution is achieved at the launch end in one of several ways. The simplest is to make the 'short' length several kilometres long to enable a natural mode distribution to occur. This wastes measurement power and is bulky. An alternative way is to speed up the establishment of equilibrium by using a short length of fibre which is subjected to physical distortion such as by winding into a small diameter coil (e.g. 25 mm diameter for a metre length of fibre) or by subjecting a short length to severe microbending. A more controlled alternative is to launch the light through three fusion jointed 1 m lengths of fibre, the first being graded index, the second a small core diameter step index (e.g. single-mode fibre) and the third a graded index fibre. The effect of this is to partially strip power from the higher order modes. All these are known as 'mode scramblers'.

A similar effect can be achieved by a controlled launch into the fibre by restricting the area and NA of the launch onto the fibre end so that the lower order modes are selectively excited. The usual conditions are 80% of the core area and 70% of the numerical aperture. These methods are illustrated in Figure 14.58.

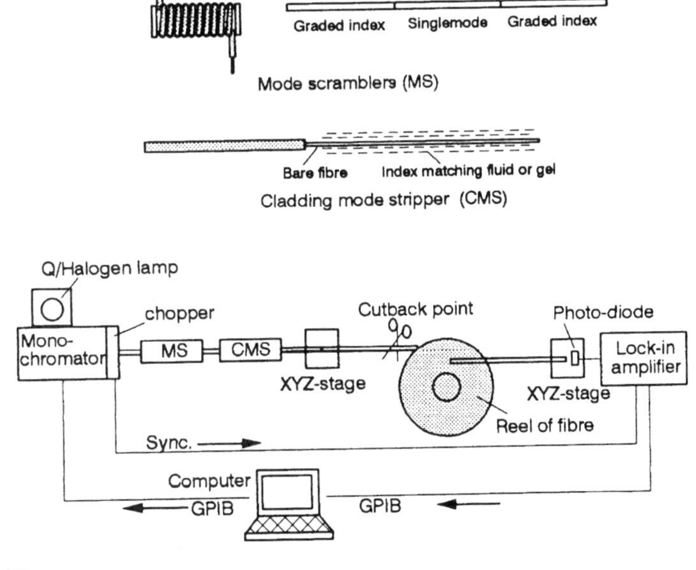

Figure 14.58
Attenuation measurement by cut-back method

In order to avoid P_i being artificially large owing to some power travelling the short distance in cladding modes (which would make the attenuation calculated from Equation (14.133) too large), these must be stripped from the launch fibre. This is done by removing the primary and secondary coatings from a few centimetres of fibre and immersing it in index matching fluid to encourage the rapid loss of any leaky modes from the fibre surface. This is known as 'cladding mode stripping'. For single-mode fibre, the mode scrambling is unnecessary but cladding mode stripping is required.

If the measurements are only required at a single wavelength or at a few wavelengths, either solid-state lasers or a quartz–halogen lamp with interference filters can be used as the source. For measurements across a spectral range, the source is usually a quartz–halogen lamp with an adjustable monochromator in the form of a reflective diffraction grating. This enables a wavelength range of not quite 2:1 to be achieved. The wavelength range is covered twice, once for the long length measurement and then again for the cut-back measurement. The ratios of the long and short measurements, and Equation (14.135), are then computed. It is convenient if the measurements are stored and subsequent calculation carried out in a computer with a plotter output. The computer can also be used for controlling the monochromator wavelength.

The detector uses a large receptor-area photodiode of appropriate wavelength range whose output is amplified. Avalanche photodiodes are generally used for their intrinsic gain and to cover the whole wavelength range with good sensitivity. To avoid noise from ambient lighting, the light from the monochromator is 'chopped' with a rotating blade chopper and the amplifier is a lock-in type which is synchronized to the chopper frequency.

For alignment of fibres with each other, or with lensed-launch systems, or with the detector, the fibres are held in V-grooves which can be manipulated with μm movements of XYZ stages. With good experimental technique accuracies of 0.01 dB/km can be achieved on 5 km lengths of fibre. The more automated the measurement and fibre alignments the greater the accuracy of the measurement is likely to be. Several proprietary arrangements are available such as the PhotonKinetics Models FOA-2000 and FOA-2200.

Bandwidth Measurement

The direct measurement of bandwidth on multimode fibres can be carried out by a swept frequency modulation of an LED source. The arrangement shown in Figure 14.59(a) replaces the halogen lamp/monochromator in Figure 14.58(c). The mode scrambler and mode stripper must also be used. Again the variation of the output of the source is cancelled out by the cut-back technique. Modulation frequencies of up to 1 GHz from the variable h.f. oscillator are possible which allows high bandwidth fibres to be measured if the length is of the order of 5 km. Note that the use of the mode scrambler ensures equilibrium mode power distribution and the measurements should be corrected by the square root of length. The l.f. oscillator acts as the light-chopper.

An alternative technique, shown in Figure 14.59(b), is to use a very short, almost Gaussian, pulse from a solid-state laser focused on to the fibre end by a microscope objective to give a limited aperture launch. The output pulse is captured on a digital storage oscilloscope and compared with the launched pulse by the cut-back technique.

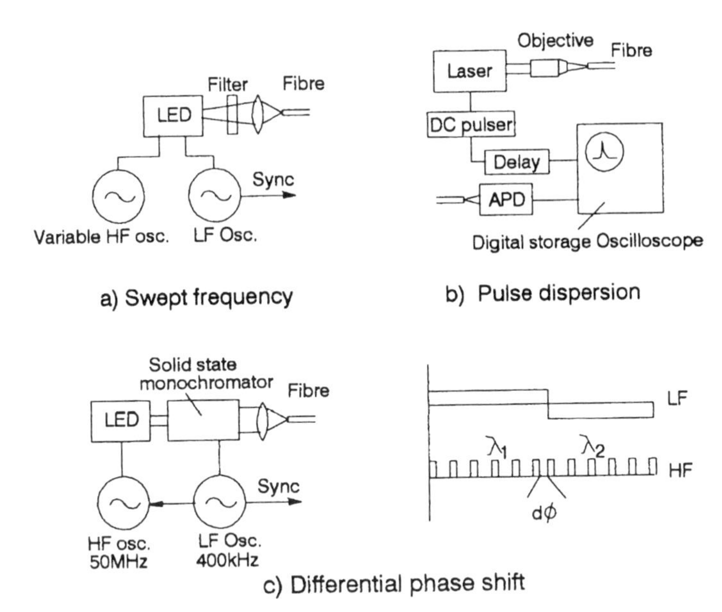

Figure 14.59
Bandwidth, dispersion and chromatic dispersion measurement methods

A computer is used to apply a fast Fourier transform to both pulses. The logarithmic ratio of the transforms gives the fibre frequency response in dB, with the bandwidth given by the −3 dB point.

Chromatic Dispersion of Single-Mode Fibre

One way of measuring the dispersion of single-mode fibre is to measure the relative time delay of a pulse modulated source as the wavelength is varied. A plot of $dt/d\lambda$ versus λ is obtained and the numerical differentiation of this curve gives the total chromatic dispersion of the fibre. However the most accurate and reproducible method of measuring the total chromatic dispersion of singlemode fibres is shown in Figure 14.59(c). An LED with a spectral range of about 100 nm is used as the source and is modulated at a high frequency such as 50 MHz. A solid-state monochromator is used to define the wavelength of measurement. The signal from a low-frequency oscillator, at about 400 kHz, shifts the monochromator wavelength by a small amount on alternate half-cycles of the low frequency. The received h.f. modulation is thus phase shifted by the fibre dispersion in alternate half-cycles. This phase shift, $d\phi$, is measured relative to the h.f. modulation reference and for a known wavelength shift, $d\lambda$, gives the dispersion directly at each centre wavelength used. If the high-frequency modulation is f,

$$\frac{d^2t}{d\lambda^2} = \frac{1}{2\pi f} \cdot \frac{d\phi}{d\lambda} \tag{14.136}$$

The dispersion is plotted over the wavelength range and the wavelength of zero dispersion can be found by interpolation. The chromatic dispersion is scaled linearly with fibre

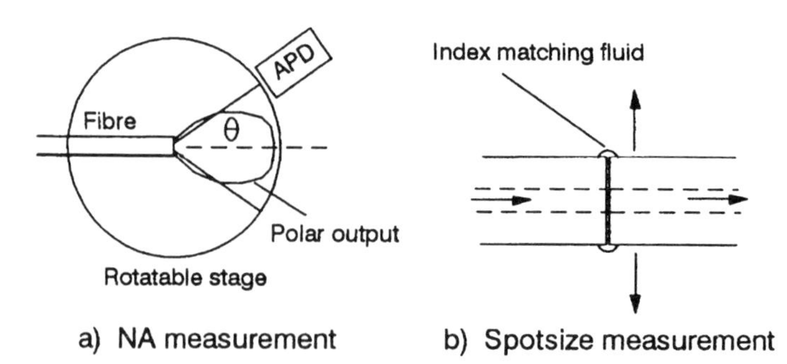

Figure 14.60
Numerical aperture and spotsize measurement

length. Commercial equipment embodying this patented double-modulation technique is available from EG & G and PhotonKinetics. Reproducibiliy of 0.05 nm in zero dispersion wavelength and 0.2% in dispersion slope can be achieved by this technique.

Numerical Aperture

To measure the numerical aperture of multimode fibres a 3 m length of fibre is illuminated to give a full-fill across all modes. Cladding modes are stripped and the fibre end face is situated at the centre of a rotatable stage carrying the photodetector. The output at the on-axis position is measured and the photodetector is rotated around the fibre until the output falls to some arbitrary value such as 5% (defined by specification). The half-angle θ in Figure 14.60(a) between the stage positions on each side of the fibre is recorded. The numerical aperture is the sine of this half-angle.

Mode Field Radius of Single-Mode Fibre

The mode field radius, or spotsize, of a single-mode fibre is measured by the offset technique shown in Figure 14.56(b). A 3 m length of fibre is cleaved at its mid-point. The opposing faces of the cleave are mounted in an XYZ stage and aligned under 200× magnification. A drop of index matching fluid (e.g. methyl salicylate) is applied to the faces and they are brought to within 5 μm of each other. The maximum power transfer when the fibre cores are aligned is recorded. The power transferred when the fibres are offset on either side of this position by known steps of 1 μm each is also recorded. To overcome noise a Gaussian curve is fitted to these recorded values and the offset giving $1/e^2$ drop in power is the mode field radius. Both lengths of fibre should have leaky-mode stripping applied.

If the MFR is measured over a range of wavelengths either side of the cut-off wavelength, it will be observed that it reduces as the wavelength reduces until the cut-off wavelength is reached, when it starts to increase owing to the excitation of the LP_{11} mode. An estimate of the cut-off wavelength can be obtained from the intersection of

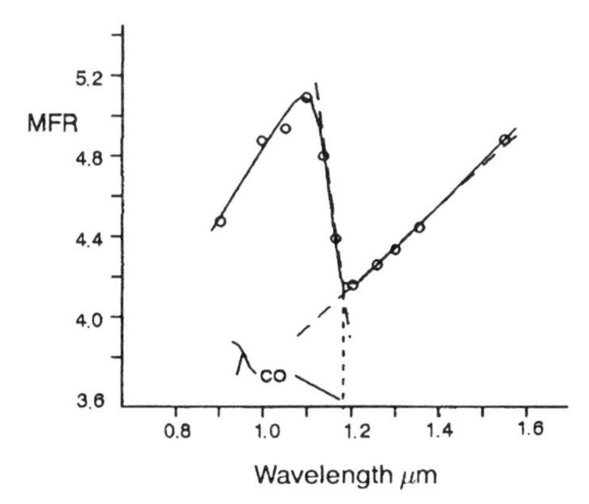

Figure 14.61
Cut-off wavelength determination from spotsize measurement

the extrapolations to the two portions of the curve (Figure 14.61). The MFR in this offset joint method is given by

$$w_0 = \frac{d}{2} \ln\left(\frac{P}{P_{\max}}\right) \tag{14.137}$$

This equation also gives the loss to be expected in a fibre joint if the cores are offset by a radial distance d.

Alternative ways of determining the MFR by measurements in the far-field of the fibre can be more convenient. One of these is the variable aperture far-field (VAFF) method. The power radiated through an axial aperture of radius R at a distance z from the fibre is

$$P_R = P_{\max}\left[1 - \exp\left(\frac{-2R^2}{w_z^2}\right)\right] \tag{14.38}$$

where $P_{\max}$ is the total radiated power (i.e. with R very large) and w_z is the $1/e^2$ intensity point in the tranverse direction in the far field. By using the Kirchoff–Huygens diffraction integral it can be shown that

$$w_z^2 = w_0^2\left[1 + \left(\frac{\lambda z}{\pi w_0}\right)^2\right]$$

For $z \gg w_0 \gg \lambda$ this becomes

$$w_z = \frac{\lambda z}{\pi w_0} \tag{14.139}$$

Substituting in Equation (14.138) gives

$$P_R = P_{\max}\left[1 - \exp\left\{-2\left(\frac{\pi w_0 R}{\lambda z}\right)^2\right\}\right] \tag{14.140}$$

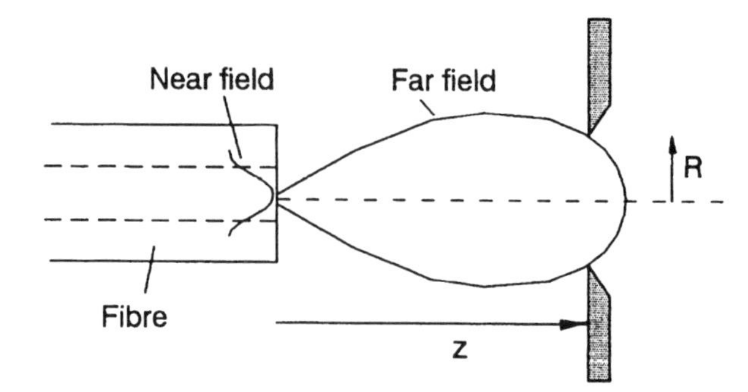

Figure 14.62
Variable aperture far-field method for mode field radius measurement

whence

$$\ln\left[1 - \frac{P_R}{P_{\max}}\right] = -mR^2 \tag{14.141}$$

where

$$m = 2\left(\frac{\pi w_0}{\lambda z}\right)^2 \quad \text{or} \quad w_0 = \frac{\lambda z}{\pi}\sqrt{\frac{m}{2}} \tag{14.142}$$

By plotting $\ln[1 - P_R/P_{\max}]$ against R^2 a straight line of slope $-m$ is obtained, and hence w_0 can be calculated from Equation (14.142).

The test arrangement is sketched in Figure 14.62. The fibre is illuminated with monochromatic light and cladding mode stripped. It is viewed on axis at a fixed distance z, through a series of precision apertures mounted in a rotatable disc, and the power received is recorded. m is derived from the measurements and w_0 is calculated.

Effective Cut-Off Wavelength of Single-Mode Fibre

The theoretical cut-off wavelength of a single-mode fibre is the wavelength at which there is no trace of the LP_{11} mode. As this is very difficult to measure (for example, should it be less than one-hundredth or one-thousandth of the LP_{01} mode), a standard test method is defined. The spectral attenuation of a 3 m length of fibre (arranged in a 300 mm diameter loop to stabilize bend losses) is compared with the attenuation of the same fibre when a 30 mm diameter mode stripping loop is added. The difference in the curves will look like the sketch in Figure 14.63. The effective cut-off wavelength is defined as the wavelength where the difference in loss is 0.1 dB. Below the effective cut-off, the curve would be expected to rise by 6 dB (since there are 4 LP_{11} modes compared with the 2 LP_{01} modes), however the 300 mm diameter bend imposes bend losses on the LP_{11} modes. This is equivalent to a rise in the baseline at lower wavelengths.

From a number of investigations this author has found that the effective cut-off measured in this way is 11.7% lower than the calculated theoretical cut-off of the fibre. Also if the cut-off wavelength is measured on longer lengths of fibre, using the 30 mm

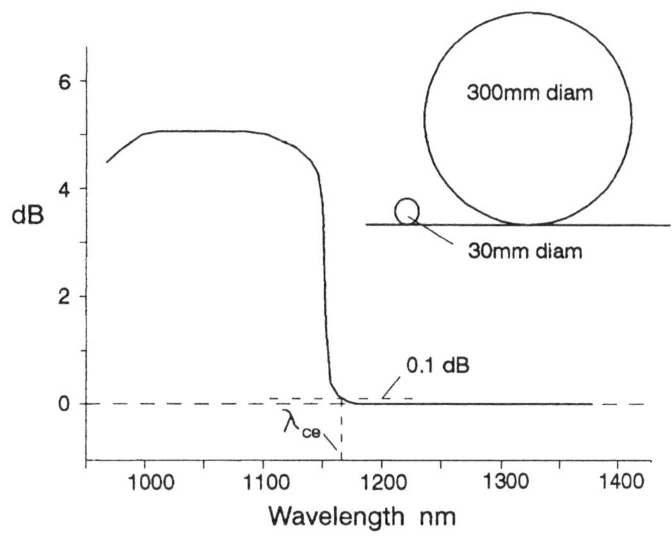

Figure 14.63
Measurement of effective cut-off wavelength

diameter mode stripping loop, the effective cut-off wavelength decreases exponentially by about 100 nm as the length increases beyond 1000 m (Figure 14.64). So that attempting to measure cut-off on a cabled length of fibre always gives lower values than on 3 m of the same fibre.

Optical Time-Domain Reflectometer

A powerful measurement technique for looking at fibre and joint losses is time-domain reflectometry. A short laser pulse is launched into the fibre through a beam splitter. Light from the pulse is backscattered toward the launch point and is diverted by the

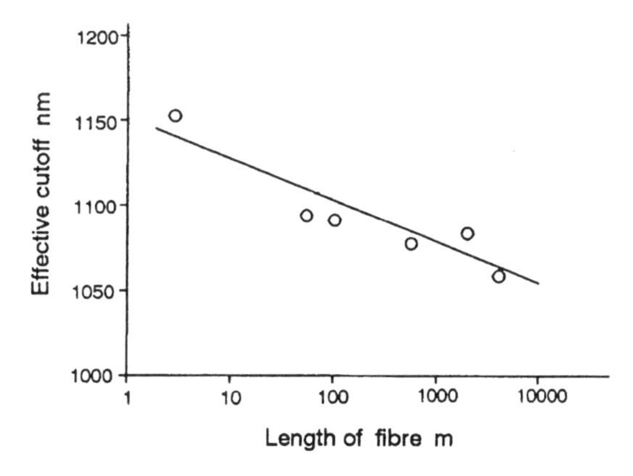

Figure 14.64
Effective cut-off wavelength versus fibre length

beam splitter to an APD. The signal from the APD is amplified and averaged for each point on the fibre. The averaged signal is displayed on a timebase which represents the distance down the fibre. Since the backscattered signal is proportional to the there-and-back attenuation, the display shows a uniform slope proportional to the fibre attenuation constant. In Figure 14.65 the OTDR arrangement is shown and the display from a number of jointed fibres is sketched. Each part of the display has a slope corresponding to the fibre attenuation constant and each joint shows a small discrete loss. At the end of the fibre, the Fresnel reflection of 4% of the received pulse usually swamps the display and clearly indicates the fibre end.

Since the loss is due to twice the fibre attenuation the loss scale is usually calibrated directly in one-way decibel loss. A highly accurate timebase in conjunction with an accurate knowledge of the group refractive index of the fibre at the wavelength of the laser enables very accurate distance measurement down the fibre. Various adjustable markers are generally provided which give digital read-outs of the curve at their set positions.

For a single uniform fibre, the relative power returned to the fibre launch point is given by

$$10\log\frac{P}{P_s} = 10\log R_b - 2\alpha L \text{ dB} \tag{14.143}$$

where P is the received power, P_s is the launched pulse power and the the last term is the double fibre length attenuation. R_b is the backscatter coefficient of the fibre. For a single-mode fibre

$$R_b = \frac{3vTA}{4(knw_0)^2} \tag{14.144}$$

and for a multimode fibre

$$R_b = \frac{1}{4}vTA\cdot\frac{dn}{n} \tag{14.145}$$

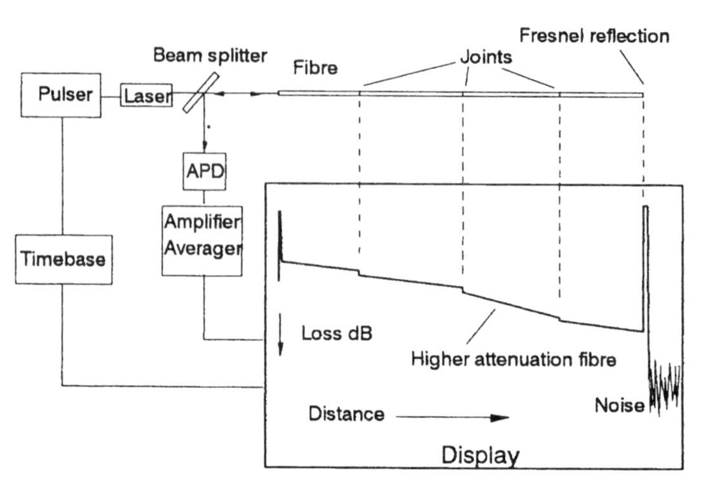

Figure 14.65
Optical time-domain reflectometer

where v is the group velocity (c/n) and T is the time duration of the pulse. A is the Rayleigh scattering coefficient (Equation (14.84) *et seq*), $k = 2\pi/\lambda$, and w_0 is the mode field radius of the single-mode fibre. $10 \log R_b$ is of the order of -50 dB, hence the necessity of a high-power launch pulse and signal averaging to reduce noise on the display.

If R_b is uniform throughout the fibre the display has a uniform slope accurately equal to the cut-back attenuation coefficient measured on the fibre. The actual value of R_b merely moves the display uniformly up or down. If however R_b changes with fibre length, as might happen if a fibre is drawn from a preform in which there is a taper of diameter or dn, there will be an extra slope change in addition to that of the attenuation coefficient. The sign of this slope error depends on the sign of the taper and will therefore reverse when the fibre is tested from the opposite end. A uniform increase of core diameter with length from the test end increases the slope of the display, and vice versa. A uniform increase of dn with length from the test end reduces the slope of the display and vice versa. In the presence of such tapers, the mean of two displays taken from both ends of the fibre accurately corresponds to the attenuation coefficient measured by the cut-back technique

Interpretation of OTDR Joint Losses

Because the OTDR trace is actually a backscatter trace, losses shown at joints are not always what they seem. In Figure 14.66 are shown some representative cases. At (a) is a perfect joint between identical fibres. There is no transmission loss and no backscatter change so the OTDR trace shows nothing. In (b) is shown a joint with the cores offset

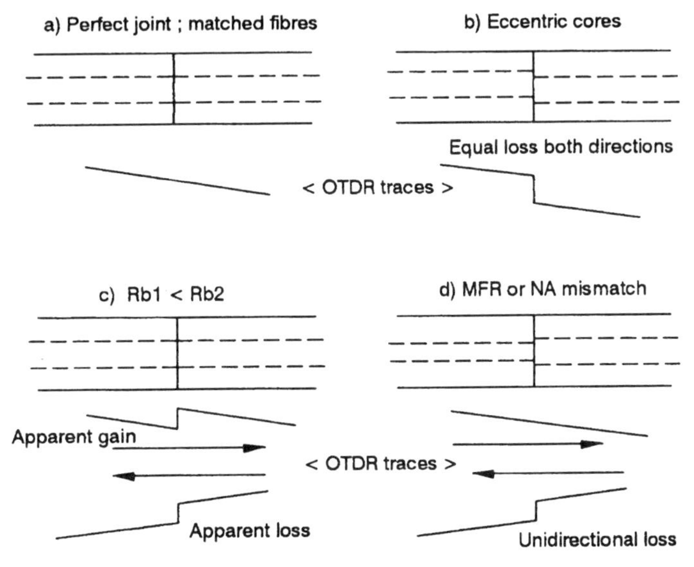

Figure 14.66
OTDR traces of fibre joints

which has a the same transmission loss in both directions. The OTDR trace is also symmetrical, with the sent pulse and the backscatter both having the same loss across the joint. Because the OTDR loss scale is halved, the indicated loss is the true one-way loss. At (c) is shown the effect of jointing two fibres of different backscatter coefficients. Tested in the direction of lower to higher backscatter, the OTDR trace shows an apparent gain. In the opposite direction there is an equal apparent loss. The algebraic mean of the two directions gives the correct result, i.e. there is no transmission loss due to the joint. At (d) a joint between fibres with a mismatched core diameter, MFR or NA will show no transmission loss going from smaller to larger, but in the opposite direction will show a transmission loss. The OTDR trace similarly will show a unidirectional loss but because only the sent pulse, or the backscatter, is attenuated the measured loss will be half the true one-way transmission loss (due to the halving of the loss scale).

SOME OPTICAL FIBRE DEVICES

Fibre Couplers

One great advantage of optical fibres is that there is virtually no coupling between parallel fibres, i.e. no crosstalk. However it is sometimes required to couple one fibre to another. This can be done by polishing flats on to a radiused fibre, penetrating well into the optical cladding. If these flats are aligned and cemented together, the field in the cladding of one fibre will penetrate into the cladding and core of the other. With care

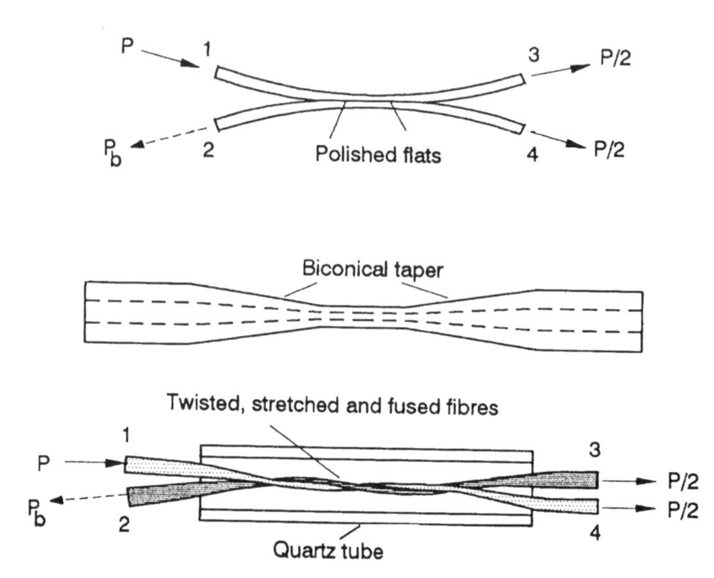

Figure 14.67
Fibre couplers

the coupling can be great enough to split the power sent into one input port, equally between the two output ports (Figure 14.67).

An alternative method is to make use of biconically tapered fibres. As the core diameter of a fibre diminishes, the power in the higher order modes leaks out into the cladding. Conversely an increasing core diameter is able to accept power into new high order modes. In the case of single-mode fibres, the decreasing core diameter forces more power into the cladding. By twisting two or more fibres together for two to six twists, heating them to fusion temperatures while stretching them by 20 mm to 60 mm, the claddings will fuse and biconical tapers will be given to all the fibres, promoting coupling between them. Equal coupling between as many as 100 fibres has been demonstrated using this method [68].

The coupler can be protected from damage by a quartz tube. The power loss between an input port and any of n output ports, is $10 \log n$ dB with a small additional loss of about 0.3 dB. The power out of an unused input port is only backscattered power at about 60 dB below the input signal. Such coupling devices can be used as star-couplers in fibre distribution networks.

Optical Fibre Laser Amplifiers

A optical fibre with a core doped with appropriate rare-earth elements can form the basis of a laser amplifier. The dopant atoms are raised to an excited state by pumping with low wavelength power from a diode laser and are stimulated to release this power at a longer wavelength by a low level signal of the same longer wavelength passing through the core, thus amplifying it. By suitable choice of dopants and codopants, the pump laser can operate at about 850 nm and the lasing amplification can occur at 1300 nm. The arrangement is sketched in Figure 14.68.

The doped fibre is end-pumped through a fibre coupler. With a suitably high level of doping of about 0.3 mole percent of a 1:4 mixture of erbium and ytterbium only about 250 mm of the lasing fibre is required to achieve gain of the order of 20 dB. Attaining a sufficiently high level of doping is not possible from the gas phase and a solution doping technique was developed by Townsend et al. [69]. After depositing an optical cladding in an MCVD preform, the flame temperature is lowered so that some layers of deposit in the form of lightly sintered silica soot are formed. The preform is removed from the lathe, soaked in an aqueous solution of the rare-earth halides for an hour and then rinsed with acetone to remove water. Final drying is carried out at 600 °C in a chlorine gas flow after replacing in the lathe, for about 40 min. Preform collapse and

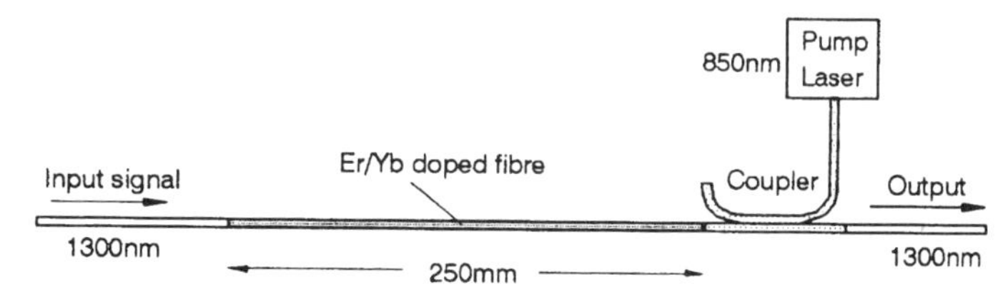

Figure 14.68
Fibre laser amplifier

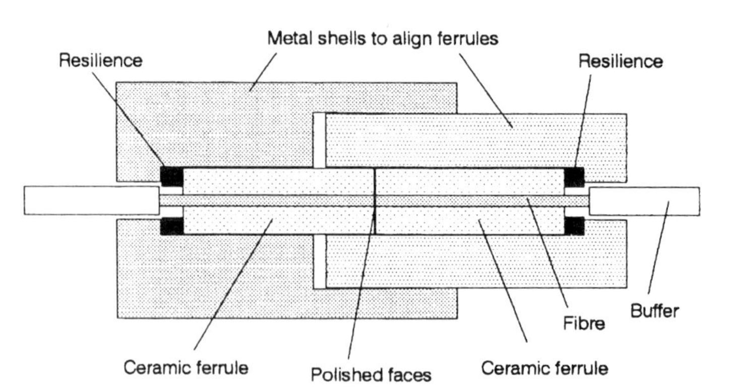

Figure 14.69
Elements of optical fibre connectors

fibre drawing are normal. This produces a fibre with a broad absorption peak around 850 nm. Losses in the 1300 nm band are about 60 dB/km, but only 250 mm of fibre is used for the laser. Such fibre laser amplifiers are simpler than regenerative repeaters which require optical to electrical conversions, but being analogue amplifiers, they do amplify any signal noise equally.

Fibre Connectors

There are probably a dozen or more fibre optic connectors available commercially. Most of them are adaptations of well-known electrical connector shells that carry a concentric ceramic ferrule with a precision concentric bore. The bare fibre is inserted into the ferrule and retained with a suitable resin or hot-melt adhesive. The fibre end is cleaved and then polished to a flush, flat mirror finish. The essential elements of connectors are shown in Figure 14.69. The metal shells align and butt the ceramic ferrules under gentle pressure. For single-mode fibres alignment accuracies of better than 0.1 μm and less than 0.5° angular misalignment are required. Note that connector losses, often better than 0.5 dB, must be measured by a transmission technique, not by an OTDR.

References

1. Loeb, *Review of Modern Physics*, 8, 267, 1936
2. Harrison J C, The metal foil–polythene cable sheath and its use in the BPO, *Proc. Inst. P.O. Elec. Engineers*, 1968
3. McNamee A, *Unpublished BICC report*, 1965
4. Foch A, Electrotechnique general, *cours de l'Ecole Nationale superieure des Telecommunications*, 1945
5. Prache P M, Resistance et reactance interne en courant alternatif des conducteurs de section circulaire, *Cables et Transmission*, janv. 1953
6. Levasseur A, Calcul rapide de l'effet Kelvin par une nouvelle formule valable en toutes circumstances, *Revue generale de l'Electricite*, dec 1929
7. Schelkunoff S A, The electromagnetic theory of coaxial transmission lines and cylindrical shields, *Bell System Technical Journal*, Oct 1934
8. Carsten H R F, The capacitance and thermal conductance of screened multi-pair systems in high-frequency cables, *Proc. IEE*, July 1944
9. Robinson D R, *Unpublished BICC (CRED) report*, June 1966
10. Rosen A, *Proc. IEE*, 1971
11. Shebes M R, Radiotecknika, Moscow 1949
12. Spencer H J C, Optimum design of local twin telephone cables with aluminium conductors, *Proc. IEE*, **116**, 4, April 1969
13. Madsen, P H, Dimensioning of cables and circuits for telephone equipment, *Teleteknik*, June 1960
14. Kaden H, Ueber die Betriebs und Kopplungs Kapazititaten zwischen den Leitungssytem einer Vierers, *Archiv fur Elektrotechnik*, 9, 1935
15. Baranov N, Influence des pas de torsion sur l'inductance mutuelle entre deux paires spirallees d'une voie de transmission telephonique, *Revue generale de l'Electricite*, juin 1943 (and experimental verification of formulae, *loc. cit.* mar. 1944)
16. Pearson N T, Design of lay-schemes, *unpublished BICC report*, 1956
17. Dwight H B, *Elec. Engineering* 1936
18. Boyce C F, *Proc. SAIEE*, 1952
19. Croze R, Simon L and Caire J P, *Transmission telephonique, theorie des lignes, Ecole superieure des telecommunications*, edition Eyrolle, 1973
20. *Textbook of line communication Vol. 1*, The Royal Signals, 1947
21. Dummer G W A and Blackband W T, *Wires and RF Cables*, Pitman, 1961 (beware of mistakes in this book)

22. CCITT, *Directives concerning the protection of telecommunication lines against harmful effects of electricity lines*, 1963
23. Rosen A, *Protection of railway lineside cables from induced voltages*, Soc. Railway Engineers (reprinted in ATE Journal 1958)
24. A Rosen *Internal BICC Report*, 1959
25. Meyers A L, *Letter, Elect. Review, London*, **120**, 791, 1942
26. Mildner R C, Private report RAE Farnborough, 1944
27. Hinchliffe J D S, Pearson N T and Goldberg J, Extension to braid theory, *unpublished BICC Report*, 1956
28. Knowles E D and Olsen L E, Cable shielding effectiveness testing, *IEEE Trans. on Electromagnetic Compatibility*, Feb. 1974
29. CCITT, *Recommenations for the protection of underground cables against corrosion*, 1960
30. Golde R H (ed) *Lightning*: Vol 1 *Physics of Lightning*, Vol 2 *Lightning Porotection*, Academic Press, London, 1977
31. CCITT, *Recommendation for lightning protection of cables*, 1974
32. Sunde, *Lightning effects in Transmission Systems*, D van Nostrand, 1956
33. Coleman, *Internal report*, Electrical Research Association, 1957
34. *Proceedings 26th IWCS*, 1977
35. *Proceedings 30th IWCS*, 1981
36. Kao K C and Hockman G A, Dielectric-fibre surface waveguides for optical frequencies *Proc. IEE*, **113**, 7
37. Olshansky R and Keck D B, *Appl. Opt.*, **15**, 483, 1976
38. Fleming J W, Material dispersion in lightguide glasses, *Elect. Lett.*, 20 April 1978
39. Adams M J, Payne D N, Sladen F M E and Hartog A H, *Elect. Lett.*, Sept. 1978
40. Sladen F M E, Payne D N and Adams M J, *Elect. Lett.*, June 1979
41. Gloge D and Marcatili E A J, Multimode theory of grade-core fibres, *Bell System Tech. J.*, **52**, 1563, Nov. 1973
42. Irving D H and Karbowiak A E, *Elect. Lett.*, Jan. 1979
43. Gloge D, Weakly guiding fibres, *App. Opt.*. **10**, Oct. 1971
44. White K I and Nelson B P, *Elect. Lett.*, May 1979
45. Rudolph H D and Neumann E G, Approximation for the eigenvalues. SI fibre, *Nachrichtentech Zeitung*, **29**, 1976
46. Hussey C D and Pask C, Theory of the profile-moments description of single-mode fibres, *Proc. IEE*, **129**, Pt. H No. 3, June 1982
47. Martinez F and Hussey C D, Enhanced ESI for prediction of waveguide dispersion in single-mode optical fibres, *Elect. Lett.*, **20**, 24, Nov. 1984
48. Marcuse D, Loss analysis of single mode fibre splices, *Bell System Tech. J.*, **56**, 1977
49. Hussey C D and Martinez F, Approximate analytical forms for the propagation characteristics of single-mode optical fibres, *Elect. Lett.*, **21**, 23, Nov. 1985
50. Sammut R A and Pask C, Simplified numerical analysis of Optical fibres and Planar waveguides, *Elect. Lett.*, Dec. 1980
51. Sharma A and Banerjee S, Chromatic dispersion in singlemode fibres with arbitrary index profiles, *J. Lightwave Tech.*, Dec. 1989.
52. Ainslie B J and Day C R, Review of sm fibres with modified dispersion characteristics, *J. Lightwave Tech.*, Aug. 1986
53. Shultz P C, *6th Int. Cong. Glass*, 1977
54. Miya V et al., *Trans. IECE Japan*, July 1980
55. Walker S S, Rapid modelling and estimation of spectral loss in optical fibres, *J. Lightwave Tech.*, Aug. 1986
56. Su-Vu Chung, *IWCS Proc.*, 1988
57. Artiglia M et al., Bending loss characterisation in single-mode fibres, *ECOC*, 1987

58. Petermann K, *Optical and Quantum Electronics*, 9, 1977
59. Petermann K and Kuhne R, *J. Lightwave. Tech.*, Jan. 1986
60. Yang R, Wu C and Yip G L, *Optics Letters*, Jun. 1987
61. Gloge D, *Bell System Tech. J.*, Feb. 1975
62. Grasso G et al., *Proc. IWCS*, 1988
63. Spammer S, *Basis of Master's thesis*, 1988
64. Vangheluwe D C L, *App. Optics*, July 1984
65. Suhir E, *J. Lightwave Technology*, July and Aug. 1988
66. Barnes S R, Pitt N J and Hornung S, *ECOC*, 1985
67. Nagel S R, MacChesney J B and Walker K L, An overview of the modified chemical vapour depostion (MCVD) process and performance, *IEEE J. Quant. Elecronics*, QE-18, 4, April 1982
68. Katsuyuki Imoto et al., *J. Lightwave Tech.*, May 1987
69. Townsend J E, Poole S B and Payne D N, *Elec. Lett.*, Jan. 1987
70. White K I, *Optical and Quantum Elec.*, II, Mar. 1982

Index